Modern Aspects of Power System Frequency Stability and Control

Modern Aspects of Power System Frequency Stability and Control

Andrew Dixon, BSc, MSc, PhD, CEng, MIET, MIEEE

ELSEVIER

ACADEMIC PRESS
An imprint of Elsevier

Academic Press is an imprint of Elsevier
125 London Wall, London EC2Y 5AS, United Kingdom
525 B Street, Suite 1650, San Diego, CA 92101, United States
50 Hampshire Street, 5th Floor, Cambridge, MA 02139, United States
The Boulevard, Langford Lane, Kidlington, Oxford OX5 1GB, United Kingdom

British Library Cataloguing-in-Publication Data
A catalogue record for this book is available from the British Library

Library of Congress Cataloging-in-Publication Data
A catalog record for this book is available from the Library of Congress

ISBN: 978-0-12-816139-5

For Information on all Academic Press publications
visit our website at https://www.elsevier.com/books-and-journals

Publisher: Joe Hayton
Acquisition Editor: Lisa Reading
Editorial Project Manager: Thomas Van Der Ploeg
Production Project Manager: R. Vijay Bharath
Cover Designer: Christian J. Bilbow

Typeset by MPS Limited, Chennai, India

Contents

Biography xxiii
Foreword xxv
Preface xxvii
Glossary xxxi
List of symbols used li

1. The need for frequency control 1

1.1 An introduction to frequency control 1
1.1.1 The subject of the book 1
1.1.2 Frequency stability and control as a system requirement 1
1.1.3 Combining the old with the new 1
1.1.4 Nomenclature: what is meant by "the frequency" in this book? 2
1.1.5 Satisfactory normal operation of the power system 3
1.1.6 The operating condition of synchronous machines 3
1.1.7 Deviations of the system frequency from normal ("nominal") 3
1.1.8 A further comment about system impedances 4
1.1.9 The continuous requirement to regulate the system frequency 4
1.1.10 Methods of regulating the system frequency 4
1.1.11 Control of frequency in the modern era 5
1.1.12 The wider system potentially at risk 5

1.2 An introduction to the "system requirements" 5
1.2.1 A definition of the "system requirements" 5
1.2.2 The variation of the "system requirements" with the "daily load cycle" 5
1.2.3 The degree of integrity of a power system and the "system requirements" 6

1.3 The system requirements for the "intact" ("prefault") system 6
1.3.1 A definition of the "intact system" 6
1.3.2 The approximate balance of power 6
1.3.3 The daily load cycle 7
1.3.4 Modern changes to the shape of the daily load profile 8

1.4 System requirements following the loss of a generating set 8
1.4.1 Introduction 8

1.4.2 The primary causes of the loss of generation from the system 9
1.4.3 Rate of change of frequency relays 10
1.4.4 Loss of generation: following the initial fall in frequency 10
1.4.5 Loss of generation: automatic governor action 10
1.4.6 The selection or otherwise of governing action 12
1.4.7 Loss of generation: the effect of frequency-dependent loads 12
1.4.8 Partial loading of generating plant 13
1.4.9 Loss of generation and generation rescheduling 15

1.5 System requirements following the loss of a demand block 16

1.5.1 Demand (load) trips 16
1.5.2 Loss size 16
1.5.3 The causes of "demand (load) trips" 16
1.5.4 The task of the control engineer 17
1.5.5 Actions available to the control engineer during a "high-frequency event" 17
1.5.6 Demand (load) shedding 17

1.6 The system requirements following the loss of lines, transformers, busbars, and other transmission equipment 18

1.6.1 The loss in bulk of generation and demand 18
1.6.2 The largest losses 18
1.6.3 The loss of equipment other than generators or loads 18
1.6.4 The loss of lines 18
1.6.5 The loss of transformers 19
1.6.6 The loss of busbars 19
1.6.7 Summary 19

1.7 The monitoring of the system frequency in real time 19

1.7.1 Introduction 19
1.7.2 Daily demand variations 19
1.7.3 Unplanned changes 20

1.8 Modern challenges in frequency control 20

1.8.1 Introduction 20
1.8.2 Hybrid systems 20
1.8.3 Newer types of generation 21

1.9 Modeling of the power system for analyzing frequency behavior 21

1.9.1 Introduction 21
1.9.2 Aims 21
1.9.3 The "swing equation" 22
1.9.4 Low inertia systems 22

1.10 The next chapter: what can provide frequency control? 22

References 22

2. What can provide frequency control? 23

2.1 Traditional providers of frequency control 23

2.1.1 Introduction 23

2.1.2 Governor response: synchronous generators 23
2.1.3 The use of the jettisoning of blocks of demand (load) for response 24
2.2 Frequency response: a clarification of terminology 24
2.3 Continuous response 25
2.3.1 A definition of "continuous response" 25
2.3.2 An example of "continuous response" 25
2.3.3 Measurement of the system frequency 25
2.3.4 The effect of continuous response on the "system frequency" 25
2.4 "Step-change" response 26
2.5 New providers of frequency control 27
2.5.1 High-voltage direct current links 27
2.5.2 "Wind turbines" (wind farms) 27
2.5.3 "Solar photovoltaic" (solar farms) 28
2.5.4 High-voltage-connected batteries 29
2.5.5 Demand-side management 29
2.5.6 Smart grids 31
2.6 The issue of system inertia 32
2.6.1 The factors necessary for achieving effective "frequency control" 32
2.6.2 A short discussion on the need for "inertia" in power systems 33
2.7 The next chapter: per unit systems for frequency analysis 36
References 36

3. Per unit systems for frequency analysis 39

3.1 Per unit systems: an introduction 39
3.1.1 What is a per unit system? 39
3.1.2 Dimensions of "per unit" quantities 39
3.1.3 Why do we need per unit quantities? 39
3.2 Per unit systems: individual machines 40
3.2.1 The individual machine 40
3.2.2 An example of defining a "per unit" quantity for frequency studies 40
3.3 Per unit systems and the power system at large 41
3.3.1 Active power outputs 41
3.3.2 System mega-voltampere rating 41
3.3.3 Inertia ("moment of inertia") J_{ind} 41
3.3.4 The incorrectness of adding together the inertias 42
3.3.5 Adding together the kinetic energies of the individual machines 42
3.3.6 Calculating an "average power factor" 43
3.3.7 Defining an "*H*-constant" for the system as a whole 43
3.4 What is the practical use of the above analysis? 44
3.4.1 The answer to the question 44

3.4.2 Looking further ahead in the book 44
3.4.3 In the next section 44
3.5 The fundamental relationship between "frequency stability" and "system inertia" 44
3.5.1 An intuitive definition 44
3.5.2 Can we run an "AC" system without any inertia? 44
3.5.3 The "system frequency" in "inertia-less systems" 45
3.6 In the next chapter: the initial analysis of the frequency stability problem and a derivation of the "swing equation" 45
3.6.1 The fundamental system power balance 45
3.6.2 Analysis of the frequency stability problem "from first principles" 45
Reference 45

4. Initial analysis of the frequency control problem and a derivation of the swing equation 47

4.1 Introduction 47
4.1.1 The ultimate aim of this chapter: deriving the "swing equation" 47
4.1.2 In the next sections 47
4.2 The clarification of some terminology 47
4.2.1 The "intact power system" 47
4.2.2 Longer term (planning) timescales 48
4.2.3 Shorter term (operational) timescales 48
4.2.4 In the next section 48
4.3 Elements of the fundamental balance in the intact power system 48
4.3.1 The power balance across the wider system 48
4.3.2 The individual machine power balance 49
4.4 The imbalance following a system loss 49
4.4.1 Internal power balance 49
4.4.2 Rotational kinetic energy 50
4.4.3 Residual rotational power 50
4.4.4 A step by step derivation of the "swing equation" 50
4.4.5 A common assumption, but is it really needed? 51
4.4.6 In the next section 51
4.4.7 The "per unit" form of the "swing equation" 51
4.5 The next chapter: available techniques for calculating frequency response requirements 52
4.5.1 Approaching the calculation of the frequency response requirements 52
4.5.2 Some modeling details 52
4.5.3 Available solution techniques 53
Reference 53

5. Techniques available for calculating frequency response requirements 55

5.1 Introduction to techniques available for calculating frequency response requirements 55
5.1.1 A description of the objective of this chapter 55
5.1.2 Requirements during "normal operation" 55
5.1.3 Evolutionary time periods "postevent" 55
5.1.4 Putting together a usable form of the "swing equation" 55
5.1.5 The available techniques for solving the "swing equation" 56
5.2 The power system frequency and frequency requirements during "normal operation" 56
5.2.1 A description of the "quiet" system 56
5.2.2 "Major frequency events" on the system 56
5.2.3 "Minor frequency events" on the system 57
5.2.4 The need for control during "quiet periods" 57
5.3 The time periods of frequency evolution following a system loss 58
5.3.1 Immediately postfault 58
5.3.2 The time period from 1 second up to about a minute postfault 62
5.3.3 The response time period of more than about a minute postfault 63
5.4 Approaching the solution of the "swing equation" 63
5.4.1 The general aims of solving the "swing equation" 63
5.4.2 The "evolution" of the power system frequency: the basics 64
5.4.3 The frequency response requirements: a definition 65
5.4.4 How do we go about calculating the system frequency response requirements? 65
5.4.5 The modeling of a "multimachine" system 70
5.4.6 The variation of demand with frequency 74
5.4.7 Forms of the variation of responsive power with frequency 74
5.4.8 The general mathematical form of the "swing equation" to be solved 78
5.4.9 Approximating the frequency term $2 \times H \times \omega \times d\omega/dt$ on the right-hand side of Eq. (5.14) 79
5.4.10 Approximating the "general responsive profile" term 80
5.4.11 "A bit of tidying up" 80
5.5 Available techniques for solving the "swing equation" 81
5.5.1 Analytical methods 81
5.5.2 Numerical methods 84
5.5.3 The "control diagram" method 85

5.6 The next chapter: analytical methods for solving the "swing equation" 85
5.6.1 The "direct solution of the differential equation" and "Laplace transform" methods 85
5.6.2 Worked examples to be provided in MATLAB 86
References 86

6. Analytical methods for solving the "swing equation" 87

6.1 Introduction to the analytical solution methods 87
6.1.1 Direct solution of the differential equation ("Method 1") 87
6.1.2 Solution by Laplace transforms ("Method 2") 87
6.1.3 The common objective in the case of the analytical methods 87
6.1.4 Recalling the equation to be solved—the "swing equation" 88
6.2 Method 1: The direct solution of the differential equation 88
6.2.1 Solving the mathematical problem: what does Eq. (6.1) mean? 88
6.2.2 Some remarks on statements made in the preceding sections 89
6.2.3 The differential equation [Eq. (6.1)] with the additional simplification noted earlier 90
6.2.4 Solving the differential equation 91
6.2.5 A check on the accuracy of the solution for $f(t)$ given by Eq. (6.18) 93
6.2.6 Noting the form of the solution given by Eq. (6.18) 94
6.2.7 Relating the constants A_1, B_1, C_1, and D_1 of Eq. (6.21) back to H, k_1, a_0, and a_1 of Eq. (6.2) 94
6.2.8 An alternative method for relating A_1, B_1, C_1, and D_1 back to H, k_1, a_0, and a_1 96
6.2.9 The parameter H 98
6.2.10 The parameter k_1 98
6.2.11 The parameters a_0 and a_1 98
6.2.12 More about the variation of demand with frequency 99
6.3 MATLAB example simulation 1: using the "direct solution of the differential equation" method to solve the "swing equation" 100
6.3.1 Objective 100
6.3.2 Introduction to the first illustrative example 100
6.3.3 Steps to be taken in the illustrative example 101
6.3.4 A statement of the problem 101
6.3.5 The general solution to the problem 102
6.3.6 The input data to the MATLAB code 102
6.3.7 The output data from the MATLAB code 104
6.3.8 The suggested MATLAB code 104
6.3.9 Instructions for the user 104
6.3.10 In the next section 105

6.4 Method 2: solution of the "swing equation" by Laplace transforms 105
6.4.1 Introduction 105
6.4.2 Processing the individual terms 105
6.4.3 The need to make a further approximation in the "Laplace transform" method 107
6.4.4 Taking Laplace transforms of the individual terms of [Eq. (6.56)] 109
6.4.5 Definition of the Laplace transform 109
6.4.6 The Laplace transform of constant terms 110
6.4.7 The Laplace transform of terms linear in f 110
6.4.8 The Laplace transform of the unit step function $u(t - t_{trip})$ 110
6.4.9 The Laplace transform of terms linear in t 110
6.4.10 The Laplace transform of terms proportional to t^2 111
6.4.11 The Laplace transform of terms of higher order in t 111
6.4.12 The Laplace transform of terms proportional to the first derivative of f with respect to t 111
6.4.13 Putting it all together 111
6.4.14 Dealing with the $f \cdot df/dt$ term 112
6.4.15 Trying to use the *convolution theorem* for Laplace transforms on the $f \cdot df/dt$ term 112
6.4.16 Trying to use "integration by parts" on the $f \cdot df/dt$ term 113
6.4.17 The need to make an additional approximation 114
6.4.18 Approximating the $f(t) \cdot df/dt$ term 114
6.4.19 Approximating the *GRP* term 115
6.4.20 The $u(t - t_{trip})$ term 115
6.4.21 The final stages of the calculation to find $f(t)$ for $t \geq 0$ 116
6.4.22 Laplace transforming the equation Eq. (6.93) 117
6.4.23 Taking inverse Laplace transforms 117
6.4.24 Putting the solution together 118
6.4.25 The form of the solution given in Eq. (6.105) 119
6.4.26 Some comments on the solution 120
6.4.27 A comparison of the Method 2 "Laplace transform" solution with the solution from Method 1 ("direct solution of the differential equation") 120
6.5 MATLAB example simulation 2: using the "Laplace transform" method to solve the "swing equation" 123
6.5.1 Objective 123
6.5.2 Introduction to the second illustrative example 123
6.5.3 Steps to be taken in the second illustrative example 124
6.5.4 A statement of the problem 124
6.5.5 The general solution to the challenges 125
6.5.6 The main input data to the MATLAB code 127
6.5.7 The output data from the MATLAB code 127
6.5.8 The suggested MATLAB code 128
6.5.9 Instructions for the user 131
6.5.10 In the next section 132

6.6 Advantages and disadvantages of the "direct solution" and "Laplace transform" methods 132
6.6.1 Direct solution of the differential equation 132
6.6.2 Laplace transform solution 132
6.7 The next chapter: numerical methods for solving the "swing equation" 133
6.7.1 Introduction 133
6.7.2 Summary 133
References 133

7. Numerical methods for solving the "swing equation" 135

7.1 An introduction to "numerical methods" for solving the "swing equation" 135
7.1.1 A resume of what we covered in the previous chapter 135
7.1.2 What we will cover in this chapter 135
7.1.3 What kind of mathematical problem do we have? 136
7.1.4 The definition of the "Taylor series expansion" 137
7.1.5 The use of the "forward difference" and the "backward difference" in numerical methods 138
7.1.6 The definition of "forward difference methods" 138
7.1.7 The definition of "backward difference methods" 138
7.1.8 The definition of the "forward difference operator" 138
7.1.9 The definition of the "backward difference operator" 139
7.1.10 Defining the time step for the numerical solution of the "swing equation" 139
7.2 Some important practical considerations 139
7.2.1 "Single-step" and "multistep" integration methods 139
7.2.2 The effect of the size of the time step on the accuracy and stability of the numerical solution 140
7.2.3 The effect of the length of the simulation time on the numerical solution 141
7.2.4 The "truncation errors" associated with numerical integration schemes 141
7.2.5 "Truncation error" formulas for numerical integration methods 143
7.3 The numerical techniques available for solving first-order ordinary differential equations 143
7.4 "Euler's method"—the simplest method 144
7.4.1 Introduction to "Euler's" method 144
7.4.2 The "truncation error" associated with "Euler's" method 144
7.4.3 The "numerical stability" of "Euler's" method 144
7.4.4 Formulating "Euler's" method for solving the "swing equation" 145
7.5 MATLAB example simulation 3: using "Euler's" method to solve the "swing equation" 147
7.5.1 Introduction 147

7.5.2 A statement of the problem 147
7.5.3 The general solution of the problem 148
7.5.4 The input data to MATLAB 148
7.5.5 The output data from MATLAB 148
7.5.6 The suggested MATLAB code 148
7.5.7 Instructions for the user 149
7.5.8 In the next section 149
7.6 The "modified Euler's" method 150
7.6.1 An introduction to the "modified Euler's" method 150
7.6.2 Formulation of the "modified Euler's" method 150
7.6.3 The "truncation error" associated with the "modified Euler's" method 151
7.6.4 Formulating the "modified Euler's" method for solving the "swing equation" 152
7.7 MATLAB example simulation 4: using the "modified Euler's" method to solve the "swing equation" 154
7.7.1 Introduction 154
7.7.2 A statement of the problem 154
7.7.3 The general solution to the problem 155
7.7.4 The input data 156
7.7.5 The output data 157
7.7.6 The MATLAB code 157
7.7.7 Instructions for the user 158
7.7.8 In the next section 158
7.8 The "Runge–Kutta" type methods 158
7.8.1 An introduction to the "Runge–Kutta" type methods 158
7.8.2 The first-order "Runge–Kutta" method (="Euler's" method) 159
7.8.3 Higher order "Runge–Kutta" methods 159
7.8.4 "Truncation errors" associated with the various "Runge–Kutta" methods 159
7.8.5 The "stability" of the various "Runge–Kutta" methods 161
7.8.6 Formulating the *classic explicit fourth-order* "Runge–Kutta" *method* for solving the "swing equation" 161
7.9 MATLAB example simulation 5: using the classic explicit fourth-order "Runge–Kutta" method to solve the "swing equation" 164
7.9.1 Introduction 164
7.9.2 A statement of the problem 164
7.9.3 The general solution of the problem 164
7.9.4 The input data 165
7.9.5 The output data 165
7.9.6 The MATLAB code 165
7.9.7 Instructions for the user 167
7.9.8 In the next sections 167

7.10 The "Milne–Hamming multistep predictor–corrector" method 167
7.10.1 A general introduction to "multistep predictor–corrector" methods 167
7.10.2 A brief background to the "Milne–Hamming" method 168
7.10.3 The definition of the "Milne–Hamming" method 169
7.10.4 The "truncation errors" associated with the "Milne–Hamming" method 169
7.10.5 A comparison of the "stability" of the "Milne–Hamming" method with that of other methods 170
7.10.6 Formulating the "Milne–Hamming method" for solving the "swing equation" 170
7.11 MATLAB example simulation 6: using the "Milne–Hamming" method to solve the "swing equation" 172
7.11.1 Introduction 172
7.11.2 A statement of the problem 173
7.11.3 The general solution to the problem 174
7.11.4 The input data to the *MATLAB* code 175
7.11.5 The output data from the *MATLAB* code 175
7.11.6 The *MATLAB* codes 175
7.11.7 Instructions for the user 178
7.11.8 In the next section 178
7.12 The "Adams–Bashforth–Moulton" multistep predictor–corrector method 179
7.12.1 A brief background to the "Adams–Bashforth–Moulton" method 179
7.12.2 The definition of the *fourth-order* "Adams–Bashforth–Moulton" method 179
7.12.3 The "truncation errors" associated with the fourth-order "Adams–Bashforth–Moulton" method 180
7.12.4 The "stability" of the "Adams–Bashforth–Moulton" method 180
7.12.5 Formulating the "Adams–Bashforth–Moulton" method for solving the "swing equation" 181
7.13 MATLAB example simulation 7: using the "Adams–Bashforth–Moulton" method to solve the "swing equation" 183
7.13.1 Introduction 183
7.13.2 A statement of the problem 183
7.13.3 The general solution of the problem 184
7.13.4 The input data 184
7.13.5 The output data 184
7.13.6 The MATLAB codes 185
7.13.7 Instructions for the user 188

7.14 **Advantages and disadvantages of the numerical methods** 189
7.14.1 Advantages 189
7.14.2 Disadvantages 189
7.15 **The next chapter: the "control diagram" method** 189
References 189

8. The "control diagram" method 191

8.1 **A short introduction to the "control diagram" method** 191
8.1.1 A third approach to solving the "Swing Equation" 191
8.1.2 The use of the "control diagram" to represent a differential equation 191
8.1.3 Development of the "control diagram" representation 192
8.2 **Recalling the "swing equation"** 192
8.2.1 The "standard form" of the "swing equation" 192
8.2.2 A simplifying substitution 192
8.2.3 In the next subsection 193
8.3 **The representation of the "swing equation" in "state-space" format** 193
8.3.1 An introduction to "state-space" format 193
8.3.2 Application of the "state-space" format to the "swing equation" 193
8.3.3 An interpretation of some of the terms in Eqs. (8.3a)–(8.3e) inclusive 194
8.3.4 A note on the application of the "state-space" formulation to general "stability and control" problems 196
8.4 **Conversion from the "state-space" format to the "transfer function" format** 196
8.4.1 An introduction to the "transfer function" format 196
8.4.2 Converting the "state-space" format of the "swing equation" given in Eq. (8.3) to "transfer function" format 196
8.4.3 The "gain" and "time constant" associated with the system described by Eq. (8.3) 197
8.5 **Conversion from the "transfer function" format to the "control diagram" format** 197
8.5.1 An introduction to the "control diagram" format 197
8.5.2 Converting the "transfer function" representation of the "standard swing equation" of Eq. (8.5) to "control diagram" format 198
8.5.3 An alternative "control diagram" format for the "standard swing equation" [Eqs. (8.1a)–(8.1d)] 198
8.5.4 A comparison of Diagram 8.2 with Diagram 8.1 198
8.6 **Iterating the values in the "control diagram" representation of the "swing equation" to find a solution for the "system frequency" as a function of time** 199
8.6.1 The overall objective of the process 199
8.6.2 The "initial condition" in the "control diagram" method 199

8.6.3 How the "control diagram" reacts to the status of the "input" 200
8.6.4 How to perform a "single iteration" using the "control diagram" method 200
8.6.5 How to perform "multiple iterations" of the "control diagram" method 201

8.7 MATLAB example simulation 8: an example of the "control diagram" method: modeling a generator "ramped response" 201
8.7.1 Introduction 201
8.7.2 A statement of the problem 202
8.7.3 The general solution of the problem 202
8.7.4 The input data 202
8.7.5 The output data 204
8.7.6 The suggested *MATLAB* code 204
8.7.7 Instructions for the user 206

8.8 Comparison of the "control diagram" approach with some of the other methods available 206
8.8.1 Introduction 206
8.8.2 Similarity of the "control diagram" approach to the "analytical methods" 206
8.8.3 Similarity of the "control diagram" approach to the "numerical methods" 206
8.8.4 Uniqueness of the "control diagram" approach 206

8.9 The next chapter: some important practical applications 206
8.9.1 Introduction 206
8.9.2 *[I]* The calculation of the "rate of change of frequency" following a loss of *generation* or *demand* 207
8.9.3 *[II]* The calculation of the "system response requirements" following a loss of *generation* (a "low frequency excursion") 207
8.9.4 *[III]* The calculation of the "system response requirements" following a *loss of demand* (a "high frequency excursion") 208
8.9.5 *[IV]* The calculation of the "system response requirements" during "normal operation" ("relatively small variations" in the "system frequency") 208

References 209

9. Some important practical applications 211

9.1 An introduction to some important practical applications 211
9.1.1 Different uses of the "swing equation" 211
9.1.2 The rate of change of frequency calculation 211

9.2 The system "rate of change of frequency" assessment 211
9.2.1 *[I]* The calculation of the "rate of change of frequency" following a loss of *generation or demand* 211
9.2.2 A procedure for calculating the "rate of change of frequency" 212

9.3 MATLAB example simulation #9: the "rate of change of frequency" calculation 216
9.3.1 Introduction 216
9.3.2 The statement of the problem 217
9.3.3 The general solution to the problem 217
9.3.4 The input data to the MATLAB code 218
9.3.5 The output datum from the MATLAB code 218
9.3.6 The suggested *MATLAB* code 218
9.3.7 Notes 219
9.3.8 General instructions for the user 219
9.4 The "system response requirements" for a "low frequency excursion" 219
9.4.1 *[II]* The calculation of the "system response requirements" following a loss of generation (a "low frequency excursion") 219
9.4.2 A procedure for calculating the "system response requirements" following a *loss of generation* (a "low frequency excursion") 219
9.5 MATLAB example simulation #10: calculating the "response requirements" for a "low frequency excursion" 222
9.5.1 Introduction 222
9.5.2 The statement of the problem 222
9.5.3 The general solution of the "low frequency excursion" problem 223
9.5.4 The input data 228
9.5.5 The output data 228
9.5.6 The MATLAB code 228
9.5.7 Instructions for the user 231
9.6 *[III]* The "system response requirements" for a "high frequency excursion" 232
9.6.1 Calculation of the "system response requirements" following a *loss of demand* (a "high frequency excursion") 232
9.6.2 A procedure for calculating the "system response requirements" following a *loss of demand* (a "high frequency excursion") 232
9.7 *MATLAB example 11*: calculating the "response requirements" for a "high frequency excursion" 233
9.7.1 Introduction 233
9.7.2 A statement of the high-frequency problem 233
9.7.3 The general solution of the "high frequency excursion" problem 234
9.7.4 The input data 236
9.7.5 The output data 236
9.7.6 The suggested MATLAB code 236
9.7.7 Instructions for the user 238

9.8 The "system response requirements" during "normal operation" 239
9.8.1 *[IV]* Calculation of the "system response requirements" during "normal operation" ("relatively small variations" in the "system frequency") 239
9.8.2 A procedure for calculating the "response requirements" during "normal operation" 239
9.9 *MATLAB example 12*: calculating the "response requirements" during "normal operation" 240
9.9.1 Introduction 240
9.9.2 A statement of the problem 240
9.9.3 The general solution of the problem 240
9.9.4 The input data 241
9.9.5 The main output data 242
9.9.6 The suggested *MATLAB* code 242
9.9.7 General instructions for the user 243
9.10 The next chapter: the challenges of operating systems with high penetrations of renewables (low-inertia systems) 243

10. Challenges of operating systems with high penetrations of renewables (low-inertia systems) 245

10.1 An introduction to low-inertia systems 245
10.1.1 The original concept of the alternating current transmission system 245
10.1.2 The developing challenge of the system inertia 245
10.1.3 The frequency stability of "alternating current" machines in power systems 246
10.1.4 The "erosion" of inertia from power systems 246
10.1.5 The closure of "fossil-fuel" power stations 246
10.1.6 Synthetic inertia 246
10.1.7 Fast frequency response 247
10.1.8 The "virtual synchronous machine" 247
10.1.9 Energy storage devices 247
10.1.10 In the next section 247
10.2 Low-inertia systems: the modeling of high-voltage direct current links in system frequency studies 248
10.2.1 An introduction to "high-voltage direct current links" 248
10.2.2 The effect of "high-voltage direct current links" on the "system inertia" 248
10.2.3 The provision of "frequency response" by "high-voltage direct current links" themselves 249
10.2.4 In the next section 249
10.3 *MATLAB example simulation #13*: the effect of "high-voltage direct current links" on the system frequency 249
10.3.1 Introduction 249
10.3.2 A statement of the problem 250

10.3.3 The general solution of the problem 250
10.3.4 The input data 251
10.3.5 The output data 252
10.3.6 The suggested *MATLAB* code 252
10.3.7 Instructions for the user 253
10.4 Low-inertia systems: the modeling of "wind farms" in system frequency studies 253
10.4.1 An introduction to "wind farms" 253
10.4.2 The effect of "wind farms" on the "system inertia" 257
10.4.3 The provision of "frequency response" by "wind farms" themselves 258
10.4.4 In the next section 258
10.5 MATLAB example simulation #14: the effect of "wind farms" on the "system frequency" 258
10.5.1 Introduction 258
10.5.2 A statement of the problem 258
10.5.3 The general solution of the problem 259
10.5.4 The complete input data 260
10.5.5 The output data 261
10.5.6 The suggested *MATLAB* code 261
10.5.7 Instructions for the user 262
10.6 Low-inertia systems: the modeling of "solar photovoltaic farms" in frequency studies 262
10.6.1 A definition of a "solar photovoltaic farm" 262
10.6.2 Secondary applications of "solar photovoltaic farms" 262
10.6.3 The status of a "solar photovoltaic farm" in the intact system 262
10.6.4 The effect of tripping a "solar photovoltaic farm" 263
10.6.5 The provision of "frequency response" by "solar photovoltaic farms" themselves 263
10.6.6 In the next section 263
10.7 MATLAB example #15: the effect of "solar photovoltaic farms" on the "system frequency" 263
10.7.1 Introduction 263
10.7.2 A statement of the problem 263
10.7.3 The general solution of the problem 264
10.7.4 The input data 266
10.7.5 The output data 267
10.7.6 The suggested MATLAB code 267
10.7.7 Instructions for the user 268
10.8 The next chapter: "smart grids" 268
References 268

11. "Smart grids" and the "system frequency" 271

11.1 A brief introduction to "smart grids" 271
11.1.1 A note on the source material currently available on "smart grids" 271

11.1.2 What is a "smart grid"? 271
11.1.3 What can a "smart grid" do? 272
11.2 The possible structure of "smart grids" 274
11.2.1 Introduction 274
11.2.2 Proposed architectures for "smart grids" 275
11.2.3 In the next section 275
11.3 The optimization of "smart grids" 275
11.3.1 Introduction 275
11.3.2 Potentially useful optimization techniques 276
11.4 The potential effect of "smart grids" on the "system frequency" 277
11.4.1 Introduction 277
11.4.2 Combined heat and power schemes 278
11.4.3 Solar photovoltaic cell technology 279
11.4.4 Wind turbines in distribution systems (domestic wind turbines) 279
11.4.5 Biomass plants 280
11.4.6 Controllable domestic loads 281
11.4.7 Controllable commercial loads 281
11.4.8 Industrial loads 281
11.4.9 Control architectures of "smart grids" 282
11.4.10 Multiagent systems 282
11.4.11 Microgrids 283
11.4.12 In the next section 283
11.5 MATLAB example simulation 16: "smart grids" and the "system frequency" 283
11.5.1 Introduction 283
11.5.2 A statement of the problem 284
11.5.3 The general solution of the problem 284
11.5.4 The input data 285
11.5.5 The output data 286
11.5.6 The suggested *MATLAB* code 286
11.5.7 Instructions for the user 287
11.6 The next chapter: conclusion 287
References 287

12. Conclusions 289

12.1 A summary of the first 11 chapters of this book 289
12.1.1 A summary of Chapter 1, The need for frequency control 289
12.1.2 A summary of Chapter 2, What can provide frequency control? 290
12.1.3 A summary of Chapter 3, Per unit systems for frequency analysis 290
12.1.4 A summary of Chapter 4, Initial analysis of the frequency control problem and a derivation of the swing equation 291

12.1.5 A summary of Chapter 5, Techniques available for calculating frequency response requirements 291
12.1.6 A summary of Chapter 6, Analytical solutions 292
12.1.7 A summary of Chapter 7, Numerical methods for solving the "swing equation" 293
12.1.8 A summary of Chapter 8, The "control diagram" method 294
12.1.9 A summary of Chapter 9, Some important practical applications 294
12.1.10 A summary of Chapter 10, The challenges of operating systems with high penetrations of renewables ("low inertia systems") 295
12.1.11 A summary of Chapter 11, "Smart grids" and the "system frequency" 296

12.2 What is the best method for solving the "swing equation"? 296
12.2.1 A summary of the methods covered in this book 296
12.2.2 The "direct solution of the differential equation" method 297
12.2.3 The "Laplace transform" method 297
12.2.4 The "Euler's" method 297
12.2.5 The "modified Euler's" method 297
12.2.6 The "Runge–Kutta" methods 298
12.2.7 The "Milne–Hamming" methods 298
12.2.8 The "Adams–Bashforth–Moulton" methods 299
12.2.9 The "control diagram" method 299
12.2.10 The verdict 299

12.3 Does a shortfall of inertia on the system matter? 300
12.3.1 The original design structure of a purely "AC" system 300
12.3.2 The mixed "AC" and "DC" system: the decline of inertia 300
12.3.3 Mitigation measures to counter a shortage of inertia on the system 300

12.4 Where do we go from here? 300
12.4.1 Basic trends 300
12.4.2 Manipulation of the system demand 301
12.4.3 Adding inertia to the system 301
12.4.4 The "Virtual Synchronous Machine" 301
12.4.5 "AC" or "DC"? 301
12.4.6 Solar technology 301

Appendix A: The "exact solution" of the "swing equation" 303
Appendix B: The "exact solution" of the "Swing Equation" when the power set point is not the nominal power output 305
Appendix C: List of MATLAB simulation exercises 307
Index 309

Biography

Dr. Andrew Dixon obtained his PhD degree in applied mathematics from the University of St. Andrews, Scotland, in 1988 and his master of science degree in electrical power engineering from the University of Bath, England, in 2005.

He joined the newly formed National Grid Company at the National Grid Research and Development Centre in Surrey, England in April 1990, working for 2 years on writing a computer program enabling the optimization of grid system voltage levels.

In 1992 he moved to the (then) National Grid Headquarters in London, England to become a transmission network planning engineer at the system development branch there. Shortly afterward, the National Grid Headquarters relocated to Coventry in the Midlands region of England. At the new National Grid Headquarters at Coventry, Dr. Dixon continued to work on voltage optimization and spent approximately 1 year participating in the writing of the Electricity Seven Year Statement and in the development of the Great Britain Grid Code.

In 1998 he switched roles again by moving to the National Control Centre in Berkshire to work as a planning engineer for 13 week ahead studies. He was also involved in analysis program development for system operations and real-time analysis program support. Also, while at the National Control Centre in Berkshire he had the experience of leading a team of planning engineers.

In 2002 Dr. Dixon moved back to the Midlands of England to take up the role of asset management engineer, developing analysis tools and performing calculations to estimate the thermal capabilities of transformers. In his next role, he joined a specialist technical group and performed analytical work in the fields of frequency stability and subsynchronous resonance. The frequency stability work centered around on designing a computer program to calculate frequency response requirements and the subsynchronous resonance work related to developing a tool to calculate possible subsynchronous resonance levels on the British Grid System.

His penultimate posting at National Grid was once again to work on the UK Grid Code, this time relating to the development of key clauses relating to the control of subsynchronous resonance energy levels on the British Grid system. His final role at National Grid was concerned mainly with providing contributions to a new document published industry wide that sought to identify in advance technical issues so that remedial measures could be put into place in advance.

Dr. Dixon emigrated to the United States in 2015–16. He became self-employed and, among other things, he wrote this book.

Foreword

It has become a tradition when writing the foreword to a book of this kind to start off by saying something like "this is a new era for electricity transmission," or "changes in electricity transmission are faster now than at any time since the supergrid was invented," or something similar. In one sense the foreword to this book is no different from the rest. We are seeing changes to how electricity is generated, transmitted, and distributed that are gathering pace at an astonishing rate and with hitherto unknown diversity.

However, in another sense, the foregoing remarks are a considerable understatement in terms. The reason for this is that now we are facing changes that are utterly unlike those that went before. Our comfortable, fossil fuel–fired corpus of generation in use day in, day out is now a thing of the past. Instead, we are looking at a variety of "green" power sources, some of which quite literally change with the wind. In addition, new ways of controlling household demands have appeared, potentially creating situations in which daily demand profiles have become unlike those upon which we have relied on in the past.

Yet, perhaps above all, we must never lose sight of the fundamental properties upon which our electricity system relies to operate safely and efficiently. One of these fundamental properties is the system inertia. Without inertia, no alternating current system can operate stably. Inertia is also the most important factor in the recovery or otherwise of an AC electrical power system following the loss of generation or demand.

The modern electrical power system is made up, not only of AC elements, but of DC elements too. Thus many connections are now made, not with AC overhead lines or cables, but by HVDC links. We could say that we have a kind of hybrid AC and DC system. We no longer have a purely AC system. One of the consequences of this situation is that the traditional providers of system inertia that we have relied on in the past are being replaced by providers that do not possess inertia in the conventional meaning of the word. Hence, if we are to maintain the stability of the system, then we must provide alternative sources of inertia.

In this book, Dr. Dixon, with whom I worked personally at the British Electricity Control Centre near Reading for several years and who has over 26 years' of experience in the British Electricity industry in various roles, takes us on a detailed journey through the modeling and applications of frequency control and stability that culminates in the pertinent question, "Where do we go from here?"

Rachel Morfill
Electricity Transmission, National Grid, Berkshire, United Kingdom
2018

Preface

In this book the author has endeavored to draw upon his experience in the fields of applied mathematics, power system design, and power system operations that include a career spanning more than 26 years in the UK electricity industry. The author hopes that the book will be of interest and use to the engineers working in the field of power system frequency stability and control, to researchers in academia and to students alike.

In power system theoretical studies, frequency control and stability are, in some ways, regarded as well-understood subjects. In the field of power system operations, however, they, of course, take on a magnified importance. To ensure the integrity of modern power systems, particularly in view of recent technological developments that are changing the frequency behavior of power systems and the way they must be controlled, the subjects of frequency control and stability must be reexamined extremely carefully.

The book is divided broadly into five sections, containing material as follows:

The first section of the book, comprising Chapter 1, *The need for frequency control, and* Chapter 2, *What can provide frequency control? forms an introduction to the rationale behind frequency control*:

> Chapter 1, The need for frequency control, introduces frequency control and addresses the question of why we need frequency control. The concept of system requirements is also introduced, together with the monitoring of frequency, modern challenges in frequency control and asks how we should model the power system for frequency control purposes.
>
> Chapter 2, What can provide frequency control? asks the question "what can provide frequency control?" with a special emphasis on the challenges posed by modern low-inertia systems. Different kinds of frequency control profile are discussed and finishes up with a description of the issue of system inertia.

The second section of the book, comprising Chapter 3, Per unit systems for frequency analysis, *and* Chapter 4, Initial analysis of the frequency control problem and a derivation of the swing equation, *enters the field of the modeling of the power system for the purposes of analyzing the frequency.*

> Chapter 3, Per unit systems for frequency analysis, examines the familiar per unit systems for individual synchronous machines, and how these

definitions can be extended to cover whole power systems that contain many machines for the purposes of frequency analysis. The fundamental relationship between frequency stability and inertia is also discussed. An initial analysis of the frequency control problem is included in this chapter and a basic derivation of the familiar "swing equation" is given.
Chapter 4, Initial analysis of the frequency control problem and a derivation of the swing equation, contains a discussion of the elements that make up the fundamental power balance in the intact system and the consequences of the imbalance that occurs when a generating set or a block of load is lost from the system.

The third section of the book, which is the largest, consists of Chapters 5–8. *This section of the book is concerned with the different techniques that are available for solving the "swing equation" and subsequently obtaining the frequency response requirements for a system. Each method is covered in considerable detail.*

Chapter 5, Techniques available for calculating frequency response requirements, considers the power system frequency during normal operation, and it looks at what happens when a power imbalance occurs on the system and the available techniques for solving the resulting mathematical problem.
Chapter 6, Analytical solutions, describes two analytical methods that can be used to solve the "swing equation," the "direct solution of the differential equation" method and the method by Laplace transforms.
Chapter 7, Numerical methods for solving the "swing equation," examines a range of numerical solution methods, starting with the simplest and finishing up with the most complex and sophisticated, the simplest method considered being Euler's method and the most complex the Adams–Bashforth–Moulton method.
Chapter 8, The "control diagram" method, considers an approach using control diagrams. The "swing equation" is represented by a block diagram and iterations performed to find the optimal solution for a given case.

The fourth section of the book, which consists only of Chapter 9, Some important practical applications, *but this is a very important chapter because it discusses probably the four main calculations that the control engineer needs to do concerning the system frequency.*

Chapter 9, Some important practical applications, covers four of the most important applications of the solution of the "swing equation" which are of use to the control engineer in real-time frequency management. They are (1) the calculation of the rate of change of frequency following the loss from the system of an amount of generation or a block of load; (2) the calculation of the system frequency requirements for a "low-frequency excursion" (following the loss of a generation set); (3) the calculation of the system frequency requirements for a "high-frequency excursion" (following the loss of a block of load); and (4) the calculation of the system frequency requirements during "normal operation."

The fifth section of the book is devoted to current issues and developments.

Chapter 10, The challenges of operating systems with high penetrations of renewables (low-inertia systems), contains a discussion of the challenges posed to control engineers by the high penetration of inertia-less generation in modern power systems. This chapter considers in turn the influence on the power system frequency of three relatively recent additions to power systems: (1) high-voltage direct current links, (2) wind farms, and (3) solar farms.

Chapter 11, "Smart grids" and the "system frequency," looks at a very topical area at the current time: smart grids. How will, for example, the computerized optimization of power system loads affect the system frequency and our ability to control it in a satisfactory way, such as avoiding instabilities?

Chapter 12, Conclusions, concludes the book by first reviewing the different methods for solving the "swing equation" and comparing them for effectiveness and suitability in the operational planning and control environment. Which is the best method? Second, we ask the perhaps fundamental question at present, does a shortfall of inertia on the system matter? Does it need compensating for, and, if so, what do we have at our disposal to achieve this? Finally, where do we go from here?

MATLAB example simulations

The book contains a total of 16 MATLAB example simulations to assist the reader in understanding the techniques and issues described in the book. Please try them.

Acknowledgments

The author wishes to thank Graham Stein, Richard Ierna, Yun Li, Nikola Gargov, and Ziming Song for providing some helpful feedback in the drafting stage and Rachel Morfill for writing the Foreword.

Andrew Dixon
Middlebury, CT, United States
November 2018

Glossary

Chapter 1: The need for frequency control

AC alternating current

afternoon peak a peaking of the daily load profile in the afternoon due to high concentrations of solar power demand connected to the system

afternoon plateau in a traditional daily load profile, a leveling out of the daily demand occurring in the afternoon

automatic demand-tripping scheme a scheme in which demand (load) is tripped when the system frequency falls to a preset level

automatic governor action the response of an automatic speed governor to changes in the system frequency

automatic voltage regulator a controller incorporated in a synchronous generator that regulates the system voltage

cascade tripping the phenomenon in which generators are tripped one after another in response to a single initial fault

daily demand (load) cycle the variation of the system demand (load) during the day

DC direct current

deloaded of a generator, operating at less than full output

demand forecasting error the correction to the estimated demand that must be made to account for imperfect knowledge of the system demand in advance

demand (load) trip the unplanned loss of a block of demand (load) from the system

droop characteristic the negative slope of a nonisochronous automatic generator governor system

evening peak the highest point of the system demand (load) traditionally observed at about 5–6 p.m. on a working day

frequency error the difference between the actual frequency and the nominal frequency, or its integral with respect to time

frequency-dependent load a load (demand) whose value depends on the value of the system frequency

frequency response erosion the phenomenon whereby less than the whole responsive capability of a generator is available of the system frequency being different from the nominal frequency

frequency-responsive mode of a generator, a selectable mode in which frequency response is activated

frequency unresponsive mode of a generator, a selectable mode in which frequency response is *not* activated

generator trip the unplanned disconnection of a generating set from the system

***H*-constant** the inertia constant of a machine that has rotating parts, such as a synchronous generator

high-frequency event a fault condition in which the system frequency rises suddenly above the upper limit of its normal operating range due to the loss of demand (load) from the system

inertial response the natural response of a rotating machine to a sudden mismatch between generation and load

intact system the system in the absence of fault outages (operations); the system in the absence of any outages at all (long-term planning)

intelligent controls control systems with the capacity to make human-like decisions

isochronous governor a governor control system that is programmed to bring the system frequency back to a selected set-point frequency if a mismatch between generation and demand (load) occurs

load following the continuous tracking of generation to try to meet the load (demand) exactly

loss an unplanned disconnection of equipment, usually of generation or demand

loss of mains of a distribution system-connected generator, loss of connection to the main supply system

low-frequency event a fault condition in which the frequency falls suddenly below the lower limit of its normal operating range due to the loss of generation from the system

low-inertia system an alternating current system that has little inertia or insufficient inertia to operate in a stable fashion

manual action a control action that is not automatic

morning pickup traditionally the phenomenon in the early working day when the demand (load) of the system rises sharply

morning plateau traditionally the phenomenon in midmorning when the demand (load) levels out

national grid the high-voltage transmission system

negative feedback a feedback signal that subtracts from the original feedforward signal

nominal frequency the standard designed operating frequency of a power system

nonisochronous governor a governor control system that operates on a slope profile

nonresponsive generation a generator that does not have any controls activated to respond to changes in the system power balance—it only responds due to inertial response

nonsynchronous source a source of power generated without rotating machines

overfrequency trip the tripping for its own protection of a generator when the system frequency has become too high

overnight minimum traditionally the time during the night when the system demand (load) is a minimum

photovoltaic (PV) absorbing the energy from light and converting it into electrical current

pole slipping the phenomenon in a synchronous machine whereby the traction between the magnetic fields of rotor and stator is lost, resulting in the spinning of the rotor at super-synchronous speeds in an uncontrolled fashion

positive feedback a feedback signal that adds to the original feedforward signal

power system frequency the frequency at which power transfers are designed to occur; the nominal frequency of the system

prefault system the system in the absence of any faults

protection maloperation the operation of protection when it is not intended to operate

reserve power held on deloaded machines for use for various purposes such as correcting demand forecasting errors

response the power held on deloaded generators for the purposes of opposing frequency changes due to the sudden loss of generation or demand from the system

responsive generation a generator that holds reserve and has controls activated to respond to changes in the power balance

RoCoF rate of change of frequency

slope the gradient of the response characteristic of a nonisochronous governor

smart grids a system of intelligent controls and equipment that optimizes the usage of power by consumers

speed rotational angular velocity

speed governor a control system of a generator that reacts to changes in the power balance on the system by regulating the rotational angular velocity of the machine

swing equation the equation that defines the dynamic behavior of the rotor of a generating machine when there is a mismatch between generation and demand on the system

system demand the system active load plus ohmic losses; alternatively, the same as active load when losses are disregarded

system operator (SO) the entity responsible for overseeing the day-to-day operation of a power system

system requirements the daily response needs of a system

trigger level the preset value of frequency or some other variable at which a response is activated

underfrequency trip the tripping for its own protection of a generator when the system frequency has become too low

Chapter 2: What can provide frequency control?

active power output the usable power output of a generating machine, usually measured in MW

artificial intelligence a method of computer programming that seeks to reproduce the decision-making processes that humans use to solve problems

automatic voltage regulator a control system on synchronous generators that seeks to stabilize the system voltage

back-to-back convertor a convertor made up of two AC–DC convertors connected in series with a very short connector between them

commutation failure the breakdown of the electronic firing sequence within an HVDC convertor

continuous response the kind of frequency response in which the changes in responsive output are gradual

daily demand (load) cycle the variation of the system demand (load) during the day

demand (load) tripping the method by which blocks of demand are tripped intentionally to provide frequency control

demand-side management (DSM) the method by which consumer energy demand is incentivized financially and by means of educational initiatives

DFIG doubly fed induction generator—a design of wind turbine generator that employs a back-to-back convertor

EU European Union
EV electric vehicle
droop characteristic the negative slope response characteristic of some synchronous generators
droop profile the same as "droop characteristic"
energy-efficient resources energy resources that have low manual energy inputs and/or losses such as wind and solar energy
expert system (ES) a method of artificial intelligence that makes decisions using a database of expert knowledge
fast frequency response frequency response with very short response times produced by means of electronics
frequency control the regulation of the system frequency by synchronous and nonsynchronous sources
frequency response the change in the active power output of a source in response to changes in the system frequency
frequency-responsive mode of a synchronous generator, having its frequency response control system activated
HV-connected battery a large battery-storage unit connected to a distribution system or higher voltage system
HVDC link a direct current connection between two alternating current systems that utilizes an AC–DC converter at each end
inertia the inherent property of a mass that resists any change in its state of motion, whether an increase or decrease
inertia-less power a source of electrical power that does not possess any inertia
isochronous governor of an AC generator, a controller that is programmed to keep its frequency as close as possible to a preset frequency, usually the nominal frequency of the system
low-inertia system an alternating current system that has little inertia or insufficient inertia to operate in a stable fashion
nominal frequency the standard designed operating frequency of a power system
nonfrequency responsive mode of a synchronous generator, having its frequency response control system deactivated
nonisochronous governor of an AC generator, a controller that operates on a droop characteristic
ramp rate of a synchronous generator, the amount of responsive power generated per unit time
renewable energy resource an energy source that replaces itself, such as in wind or tidal power
responsive device a device capable of providing frequency response
slope characteristic the response characteristic of a nonisochronous governor
smart appliance an item of electrical hardware, such as a heating system, a refrigerator or an air-conditioning system that is an integral part of a smart grid
smart control system a control system that is part of a smart grid
smart grid a system of intelligent controls and equipment that optimizes the usage of power by consumers
smart meter a meter for measuring the consumption of energy by consumers that is part of a smart grid

solar PV a method of generating electrical power by converting solar energy into electrical current

speed governor of a synchronous generator, a control system that regulates the rotational velocity (frequency) of the machine

staggered tripping the tripping of demands (loads) at several different frequencies as part of a demand-tripping scheme

step-change response the kind of frequency response in which the change in responsive output is abrupt

synchronous generator an AC electrical generator whose rotor rotates at a mechanical frequency equal to the electrical frequency of the system divided by $p/2$, where p is the number of magnetic poles of the machine

synthetic inertia a power electronic method of simulating the effect of physical inertia in an inertia-less power source

system damping the phenomenon by which system oscillations in frequency, current, voltage, power, and so on are attenuated

system demand the system active load plus ohmic losses; alternatively, the same as active load when losses are disregarded

system frequency the average apparent frequency of oscillation of AC machines connected to the system; the power frequency

system inertia the equivalent H-constant for a bulk system

system operator (SO) an entity responsible for overseeing the day-to-day operation of a power system

trigger frequency the preset frequency at which a relay operates to jettison demand (load) in a demand (load) tripping scheme

trip frequency the same as "trigger frequency"

virtual synchronous machine (VSM) a power electronic device that aims to simulate all the physical parameters of a synchronous machine

well damped possessing good attenuation of oscillations

wind farm a purpose-built area, within which there are large numbers of wind turbine generators that are connected by a local grid and the whole is connected to a larger grid system for power generation purposes

wind turbine generator (WTG) an electrical generator that consists of a rotor with large vanes that are moved by the wind and drives an alternator that generates electrical power

Chapter 3: Per unit systems for frequency analysis

AC system alternating current system, a system in which currents, voltages, and so on vary cyclically over time

active power usable power that is produced by the in-phase component of current and voltage

average power factor for a system, a representative power factor that depends on the power factors of individual AC generators

base quantity in a per unit system, the fixed quantity by which a physical quantity is divided to produce a per unit quantity

dimensionless having no physical units, simply a number

equivalent system inertia a representative inertia constant H for a system

fast frequency response frequency response with very short response times produced by means of electronics
frequency stability the ability of the frequency of a system to return to a steady-state operating value following a system disturbance
***H*-constant** the inertia constant
inertia the innate property of a massive body to resist changes in its own motion
inertia constant (*H*) a per-unit quantity obtained by dividing the rated kinetic energy of a generator rotor by the MVA rating of the generator
megavoltampere base (*S*) a base quantity for a generator given by its rated active power in MW divided by its rated power factor
moment of inertia (*J*) the rotational equivalent of the mass of a body
MVA rating the megavoltampere rating of a generator
numerical stability the ability of a numerical calculation algorithm to converge to a solution without losing numerical accuracy in the process
per unit quantity a quantity obtained by dividing a physical quantity by a base quantity
per unit system a system made up of per unit quantities
physical quantity a quantity that is composed of a number followed by a physical unit, for example, 100 MW
rated kinetic energy the kinetic energy for which a machine is designed
rated power factor the designed lagging power factor limit for a generator, for example, 0.8
rotational kinetic energy (*K*) the energy possessed by a massive object due to its rotation
swing equation the equation that defines the dynamic behavior of the rotor of a generating machine when there is a mismatch between generation and demand on the system
system frequency the average apparent frequency of oscillation of AC machines connected to the system; the power frequency
system inertia the equivalent *H*-constant for a bulk system

Chapter 4: Initial analysis of the frequency control problem and a derivation of the swing equation

balanced system a power system in which the total generation equals the total demand
intact network the system in the absence of fault outages (operations); the system in the absence of any outages at all (long-term planning)
intact power system the same "intact network"
maintenance outages seasonal outages of lines, cables, transformers, generators, and other components for the purposes of cleaning, testing, refurbishment, and so on
normal operation relating to frequency, operation within the normal operational limits for the system, within a narrow band around the nominal frequency of the system
operational timescales roughly speaking, from 6 to 12 months ahead of real-time operation down to real-time operation
per unit obtained by dividing a physical quantity by a base quantity
planning timescales not real time (operations); roughly speaking, more than 6 months to a year ahead of real-time operation
prefault network the network before the onset of any fault condition
prime mover the input source of energy for a generator, for example, steam in the case of fossil fuel fired generation or water in the case of hydroelectric plants

rated frequency the frequency at which a machine or system is designed to operate

residual rotational power the difference between the mechanical power input and the electrical power output of a machine with rotating parts such as a synchronous generator

rotational kinetic energy the energy possessed by a massive object due to its rotation

swing equation the equation that defines the dynamic behavior of the rotor of a generating machine when there is a mismatch between generation and demand on the system

trip the loss of generation or demand from a system, whether planned or unplanned

unbalanced system a system for which the total generation does not equal the total demand

Chapter 5: Techniques available for calculating frequency response requirements

automatic frequency response frequency response produced by means of a preprogrammed control system that does not require manual intervention for its operation

break time following a successful relay operation, the time taken for a circuit breaker to successfully break a fault current

control diagram a logical or "block" diagram that represents in Laplace transform-space (s-space) the relationship between the input signal and the output signal of a process

demand tripping the preprogrammed jettisoning of demand (load) in response to a low-frequency signal for the purposes of restoring the frequency to normal levels

droop profile the response characteristic of a synchronous machine, having a negative slope or gradient

evolutionary timescales the time stages of development of the system frequency following a loss of generation or demand

frequency event a system event that causes the frequency to depart significantly from the normal operating frequency

frequency response requirements the frequency response needs of a system to cover for the largest losses of generation and demand

frequency sensitive mode of a synchronous generator capable of frequency response, the mode in which the ability to provide response is active

GRP a "general responsive profile," an arbitrary response profile represented by a mathematical power series with time t as the variable

***H*-constant** the inertia constant, either of an individual synchronous machine or an equivalent for the whole system

high-frequency requirements the frequency response requirements to cover for the loss of a block of load from the system

inertial response the natural response of a synchronous generator to a system power imbalance taken from the kinetic energy it has due its rotation and inertia

low-frequency requirements the frequency response requirements to cover for the loss of an amount of generation

major frequency event a frequency event where the deviation of frequency from its normal operating range is large

nominal frequency the standard designed operating frequency of a power system

normal operation the typical minute-by-minute operation of a power system in the absence of faults or any other unusual conditions

postevent following a loss of generation or demand or some other piece of equipment such as a line or transformer

primary response the response produced by automatic actions in the first 10 seconds or so following a system event

quiet period a period during the daily load cycle when the frequency is change relatively little

quiet system the same as "quiet period"

relay time the relay operational time, the time taken for a relay to operate measured from the initial detection of the fault to the reception of the signal by a circuit breaker

secondary response the response period following the primary response period when manual actions to correct any power balance that may exist

single busbar approximation the representation of a power system by a single busbar to which all generation and loads are attached for the purposes of frequency response analysis

spreadsheet method a method of calculating frequency response requirements using a spreadsheet

system frequency the average apparent frequency of oscillation of AC machines connected to the system; the power frequency

swing equation the equation that defines the dynamic behavior of the rotor of a generating machine when there is a mismatch between generation and demand on the system

tertiary response a response period sometimes defined in addition to primary and secondary response and following the secondary response period

total demand the sum of all demands of a system

total generation the sum of all generation of a system

total response requirements the sum of all response requirements of a system, consisting of low-frequency, high-frequency, and dynamic response in all time periods primary, secondary, and tertiary

Chapter 6: Analytical methods for solving the "swing equation"

complementary function the solution of a differential equation whose driving function is zero

concatenated forming a chain

contiguous sharing a common point or border

convolution theorem a function h combining two functions f and g according to Eq. (6.79)

driving function in a differential equation, a function that represents a source such as an applied force

first-order differential equation a differential equation in which the highest order derivative is order one

general responsive device any device that can provide frequency response

general solution a solution of a differential equation when it includes both the complementary solution and the driving function, the complete solution

governor response the changes in the active power output of a synchronous machine due to changes in the system frequency that are provided by a governor type control system

GRP general responsive profile, a general function of power output against time representing a "general responsive device"
Heaviside unit function the same as the unit step function (see next)
integrating factor a factor when applied to both sides of a differential equation enables the equation to be solved more easily
integration by parts a technique for integrating the product of two functions
inverse Laplace transform the reverse process to the Laplace transform (see next), obtaining $f(t)$ from $F(s)$
Laplace transform the mathematical process gone through to convert $f(t)$ to $F(s)$ as defined by Eq. (6.70)
Method 1 the "direct solution of the differential equation" method
Method 2 the "solution by Laplace transforms" method
particular integral an example solution of a differential equation whose driving function is not zero
piecewise linear function a single-valued function composed of a concatenation of linear sections
swing equation the equation that defines the dynamic behavior of the rotor of a generating machine when there is a mismatch between generation and demand on the system
system operator the entity responsible for overseeing the day-to-day operation of a power system
unit step function a function $u(t)$ which is 0 for $t \leq 0$ and 1 for $t > 0$

Chapter 7: Numerical methods for solving the "swing equation"

absolute error the difference between the numerical estimate at an iteration and the true value; the "global truncation error"
Adams–Bashforth–Moulton method a multistep predictor corrector numerical method
analytical method a method of solution that uses analysis of the differential equation rather than numerical approximations at successive steps
approximate solution an estimate of the true, analytical solution
backward difference the estimated solution at the $(n+1)$th step minus the estimated solution at the nth step
backward difference operator the operational form of the backward difference
Collatz Rule of Thumb a method for estimating the local truncation error of Runge–Kutta methods
convergent of a numerical scheme, the local truncation error decreases as the number of iterations increases
dependent variable the variable that depends on the independent variable, for example, y in the function $y(x)$ where x is the independent variable
error bounds the range of values within which the error lies
Euler's method a simple single-step method for solving first-order equations
exact solution the analytical solution
finite interpolative polynomial a polynomial consisting of a finite number of terms that passes through all the required values of the dependent variable
first-order equation the same as a first-order differential equation
first-order ordinary differential equation a differential equation that contains only the first derivative of an unknown variable

forward difference the estimated solution at the *n*th step minus the estimated solution at the $(n-1)$th step
forward difference operator the operational form of the forward difference
global truncation error the same as the "absolute error" (see earlier)
governor droop the negative slope of the governor control characteristic of a synchronous machine
independent variable the free variable upon which the dependent variable depends, for example, x in $y(x)$ where y is the dependent variable
initial condition the value of the dependent variable at a specified starting point of the independent variable
initial value problem (IVP) a differential equation whose initial condition is usually the zero of a time variable
interpolation method a method of approximating a function that connects a set of points by defining a function that passes through all the points
load damping coefficient the sensitivity parameter α that describes the dependency of the system load on the system frequency
local truncation error the error introduced in a single step of a numerical scheme
Milne–Hamming method a multistep predictor corrector numerical + method
modified Euler's method a modified version of Euler's method (see earlier) that uses a simple predictor corrector scheme
multistep integration method a numerical method that uses values of the dependent variable at more than two values of the independent variable in a single step
Newton's backward difference formula a numerical interpolation formula that uses backward differences (see earlier)
Newton's forward difference formula a numerical interpolation formula that uses forward differences (see earlier)
numerical instability a phenomenon in which a solution diverges or becomes oscillatory around the exact solution due to numerical rounding errors
numerical method a method of solving an equation by using successive approximations that are intended to converge to the exact solution
numerical stability a phenomenon in which the behavior of a numerical solution is independent of rounding errors
order of a method the value of the exponent of the step length h in the formula for the truncation error of a numerical scheme minus one
predictor–corrector method a numerical solution method in which each step consists of a "predictor" and a "corrector" stage, the "corrector" being a function of the "predictor," to increase the accuracy of the estimated solution at the step
Runge–Kutta method a single-step method in which interpolations are performed by subdividing each step
second-order method a method whose truncation error is $O(h^3)$
single-step integration method an integration method in which each iteration proceeds from the previous one by a single-step length h in the independent variable
standard form of the "swing equation" the form of the "swing equation" as given in Eq. (7.1)
swing equation the equation that defines the dynamic behavior of the rotor of a generating machine when there is a mismatch between generation and demand on the system
Taylor series expansion a function expressed as a power series in increasing powers of the independent variable

truncation error the difference between the numerical, approximate solution and the exact solution (see "local truncation error," "global truncation error," and "absolute error" above for categorizations)

weighting factor one of a set of modifying parameters in a Runge–Kutta scheme that are applied to the values in the interpolated subdivisions within a time step

Chapter 8: The "control diagram" method

analytical method a method of finding a solution purely by mathematical analysis rather than by numerical approximation

block diagram the same as a control diagram (see next)

control diagram a logical or "block" diagram that represents in Laplace transform space (s-space) the relationship between the input signal and the output signal of a process

control engineering the topic within engineering devoted to the study and representation of methods related to the control of physical processes

control system a means by which physical parameters such as frequency can be controlled or moderated

convergence the phenomenon whereby an approximate or predicted solution becomes closer to the real or analytical solution on successive iterations

convergence tolerance an acceptable difference between an approximate solution and the exact solution

feedback branch in a control system, the backward contribution of the output to the input

feedforward branch in a control system, the forward contribution of the input to the output

gain the ratio of the magnitude of the output to the magnitude of the input

general responsive profile an arbitrary response profile represented by a mathematical power series with time t as the variable

governor response the active power output of a synchronous machine having an automatic speed governor

initial condition the value of a variable such as frequency at time equals zero

input variable a variable that is fed into a control diagram

instantaneous trip a trip of generation or demand that is very fast and whose change is represented as taking place in zero time

Laplace transform space a function expressed in the space of the Laplace variable s

load damping the dependency of the system load on the system frequency

nominal frequency the standard designed operating frequency of a power system

numerical method a mathematical procedure by which a solution to a problem is found by employing a sequence of numerical approximations that it is hoped will converge to the solution

output data the numerical results of a scheme or process

piecewise linear composed of a concatenation of linear sections

primary response the automatically produced active power response of generators in the period approximately 10 seconds after a system fault

power balance the equalization of active power generation with demand on a power system or the mechanical input with electrical output for a synchronous machine

power ramp a responsive active power output in which the active power increases linearly with time

reference frequency the frequency against which a measured frequency is compared in a feedback control system
response provision the amount of response that is made available
state-space format a mathematical model of a linear physical system represented as a set of input, output, and state variables related by first-order differential equations
steady state a state that is independent of time, a nondynamic state
transfer function a function expressed in terms of the Laplace variable s that represents the ratio of the output to the input
standard swing equation the swing equation as represented by Eq. (8.1)
system demand (load) the active power drawn from a power system
system frequency the average apparent frequency of oscillation of AC machines connected to the system; the power frequency
system generation the production of active power in the system by synchronous and asynchronous sources
system variable the variable x in the state-space representation of a system
swing equation the equation that defines the dynamic behavior of the rotor of a generating machine when there is a mismatch between generation and demand on the system
time constant a number that represents the time rate of decay of a physical variable in a system

Chapter 9: Some important practical applications

credible loss a loss from the system of a generator, load, or item of equipment that has a relatively high probability of happening during operation
downloading for response operating a synchronous generator at less than full output to be able to provide response if there is a loss of generation
exact solution a solution using analytical theory only, not a sequence of numerical approximations
frequency-responsive mode of a synchronous generator, the mode of operation in which frequency response is enabled
frequency-responsive plant generators in frequency-responsive mode (see earlier)
fully loaded of a generator, running at the maximum permissible output
general ramped response a general response that is linear in the time t
generator governor droop the response characteristic of a synchronous machine, having a negative slope or gradient
generator response characteristic the graph of the responsive output of a generator, plotted against frequency or time
generator scheduling the process by which the list of generators that will run is chosen
generation to be tripped the generator that is assumed will be suddenly disconnected from the system in a frequency response study
governor action of a synchronous generator, the operation of a governor mechanism for regulating the system frequency
governor characteristic the graph of the responsive output against frequency of a speed governor
***H*-constant** the inertia constant in units of seconds
high-frequency excursion a fault condition in which the system frequency rises suddenly above the upper limit of its normal operating range due to the loss of demand (load) from the system

inertia the inherent property of a mass that resists any change in its state of motion, whether an increase or decrease
initial frequency the frequency at time equals zero
islanding protection the automatic tripping of a generator connected to the distribution system when the system frequency departs significantly from normal to protect the machine from damage
load damping the dependency of the system load on the system frequency
load-damping coefficient the sensitivity parameter α that describes the dependency of the system load on the system frequency
loading up the initial increase of the active power output of a generating machine to its scheduled operating level
loss of grid protection the same as islanding protection (see earlier)
loss of mains protection the same as islanding protection (see earlier)
low-frequency excursion a fault condition in which the system frequency rises suddenly above the upper limit of its normal operating range due to the loss of generation from the system
low-inertia system a system whose H-constant is near the minimum value for stable operation of that system
merit order the list of generators that are selected to run
nominal a designed or nameplate value
nominal frequency the standard designed operating frequency of a power system
nonresponsive generation generation that is not capable of providing frequency response
normal operation relating to frequency, operation within the normal operational limits for the system, within a narrow band around the nominal frequency of the system
normal operating range the limits for normal operation, a small band either side of the nominal frequency
normal operational frequency a frequency within the normal operating range (see earlier)
over-frequency condition an operating condition in which the frequency is too high, a high-frequency excursion (see earlier)
part-loaded of a generator, operating at less than full output, less than fully loaded (see earlier)
per unit a unit obtained by dividing a physical unit by a base value
ramped response profile a response profile whose graph of active power response against time is a straight line with a positive gradient
response requirements the frequency response needs of a system to cover for the largest losses of generation and demand
responsive generation generation that can provide frequency response
RoCoF rate of change of frequency
RoCoF procedure a procedure for calculating the rate of change of frequency
schedule (to) to select a generator to run
slope and setpoint the gradient of a straight line characteristic and the value of frequency for which the output is zero
statutory requirements the obligations put on operation by law
swing equation the equation that defines the dynamic behavior of the rotor of a generating machine when there is a mismatch between generation and demand on the system
system frequency the average apparent frequency of oscillation of AC machines connected to the system; the power frequency

system load the total active power drawn from a system, the total of generator outputs; the total of generator outputs less ohmic losses
system response requirements the response requirements (see earlier) of the system
total initial generation the total generation at time equals zero
total initial load the total load at time equals zero
tripped disconnected from the system, either intentionally or unintentionally
under-frequency condition an operating condition where the frequency is too low, a low-frequency excursion (see earlier)

Chapter 10: Challenges of operating systems with high penetrations of renewables (low-inertia systems)

AC alternating current
convertor based a wind turbine generator design that employs an AC–DC convertor or two back-to-back AC–DC convertors
doubly fed induction generator (DFIG) a wind turbine generator (WTG) whose rotor is connected directly to the stator and through a back-to-back HVDC converter
energy storage device any device that can store energy for future use, such as a battery or a capacitor
exporting power is leaving the system
fast frequency response a method of frequency response that relies on very fast switching of a power electronic device for its output
fault level the magnitude of the fault current multiplied by voltage, using given in units of megavoltamperes (MVA); a measure of the corresponding inherent strength of a power system at a point
fossil fuel a fuel, such as coal, oil, or gas, that is extracted from the ground
frequency response the change in the active power output of a source in response to changes in the system frequency
full converter a design of wind turbine generator (WTG) in which the power generator on the LV side is connected in series through a back-to-back converter to the HV side
HVDC high-voltage direct current
importing power is entering the system
inertia erosion the loss of inertia from a system
intact system the system in the absence of fault outages (operations); the system in the absence of any outages at all (long-term planning)
interconnector a connector usually between two countries and usually employing an HVDC link
nonconverter based not employing an HVDC converter
PID controller a device employing proportional, integral, and derivative controllers
PV photovoltaic
smart grid a system of intelligent controls and equipment that optimizes the usage of power by consumers
solar PV farm a concentrated location of many solar photovoltaic panels for power export to the grid
steady state a situation where system variables such as frequency and voltage are not changing with time

storage typically a battery, capacitor, or similar device for holding back energy to be used at a future time if needed

synthetic inertia simulated inertia provided by a power electronic device

system inertia the total inertia H of a system

Type 1 WTG the turbine blades are connected via a gear box to a simple induction generator and then connection is directly to the grid

Type 2 WTG this type is the same as the type 1 WTG except for the addition of a variable resistor in the rotor circuit

Type 3 WTG this type is the doubly fed induction generator (DFIG). The rotor circuit is connected to the stator circuit both directly and through a back-to-back HVDC converter

Type 4 WTG this type is connected entirely via back-to-back converters to the grid

Type 5 WTG this type employs a variable speed drive train in series with a speed/torque converter to deliver a constant output shaft to a synchronous generator that is then connected to the grid

virtual synchronous machine (VSM) a power electronics device that simulates the behavior of a synchronous machine

voltage regulation the modulation of a system voltage so that it lies in a narrow band around a specified target voltage

wind farm an array of wind turbine generators (WTGs) that are interconnected with each other and then feed their combined output via a connection to the grid

wind turbine generator (WTG) a method of generating electricity using large turbine blades and the wind as the primary energy source

Chapter 11: "Smart grids" and the "system frequency"

agent one of the modules of a "multiagent system" (see later)

arbitrator agent one of the agents of a "multiagent system"

artificial neural network (ANN) a learning algorithm that is based on the theory of how the human brain functions

biomass plant a generating station that burns waste products

cogeneration recycling the otherwise unused heat output of a power station

combined heat and power (CHP) the same as cogeneration (see earlier)

control the maintaining of a system variable, for example, frequency or voltage, to within acceptable levels

crossover one of the standard features of a genetic algorithm (GA)

daily load profile the pattern of variation of the system loads throughout the day

DC link the same as an HVDC link, an interconnector that is composed of a DC circuit or circuits that have an AC–DC converter at each end

diagnosis agent one of the agents of a "multiagent system"

distribution system the lowest level of a grid system that feeds the loads

EV electric vehicle

expert system (ES) a decision-making algorithm that is composed of a "knowledge base" and an "inference engine"

fault detection agent one of the agents of a "multiagent system"

genetic algorithm a learning algorithm that is based on the laws of genetic selection

global warming the increase in the temperature of the earth's atmosphere that is attributed to the overproduction of "greenhouse gases" by humans

greenhouse gases those gases that have been identified as contributing to "global warming"

heuristic relying on learning for its function, as, for example, in a heuristic algorithm

high voltage (HV) the highest level or levels in a system, for example, the 275 and 400 kV networks in the British grid system

judgment index agent one of the agents of a "multiagent system"

knowledge base a database of information for use by an expert system (ES)

Kyoto Protocol the international agreement made to reduce the levels of "greenhouse gases" in the earth's atmosphere

inference engine the part of an expert system (ES) that makes decisions

low voltage (LV) the lowest voltages of a grid system, such as 33 and 11 kV on the British grid system

medium voltage (MV) the voltage levels in between low voltage (LV) and high voltage (HV) in a grid system

microgrid a very detailed grid system that includes all the consumer voltage levels

mixed integer programming a mathematical optimization routine that allows both integer and real variables

multiagent system (MAS) a large-scale control system that is composed of modules called "agents," each of which fulfills a specific function

mutation one of the standard features of a genetic algorithm (GA)

network analysis the analysis of power flows, bus voltages, frequency, etc., in a power system using computer-based tools

optimal power flow a computer algorithm that selects the best values for system variables such as bus voltages and power outputs of generators

principle of linear superposition in a linear system, the rule which allows the addition of two solutions to produce a third solution

PV farm a solar photovoltaic farm

scheduling agent (SA) one of the agents of a "multiagent system"

simplex method a linear mathematical routine used widely in optimization problems

smart grid a system of intelligent controls and equipment that optimizes the usage of power by consumers

solar PV solar photovoltaic plant

stability the ability of a system to return to its normal operating condition following a disturbance

supergrid another name for the highest operating voltages of a power system

supervisory control and data acquisition (SCADA) traditionally, a monitoring and data gathering system used in the operation of a power system

system frequency the apparent average speed of rotation of synchronous generators in a power system

system monitoring agent one of the agents of a "multiagent system"

training an activity needed to initiate the operation of a genetic algorithm (GA)

transmission operator (TO) the entity that oversees the day-to-day operation of a power system

Chapter 12: Conclusions

abnormal operation operation during a fault condition such as the unplanned loss of generation or load
accuracy closeness to the analytical or ideal solution
Adams–Bashforth–Moulton methods a set of *predictor-corrector* numerical methods in which an Adams–Bashforth *predictor* is used in combination with an Adams–Moulton *corrector*
artificial neural network (ANN) a heuristic algorithm that is modeled on the functioning of the human brain
backward difference the value at the current step minus the value at the previous step
biomass plant generators that recycle waste products for combustion
block diagram a diagram using pictorial blocks to show logical connections between successive variables
bulk systems whole or extensive systems
combined heat and power scheme a generation method that recycles heat
continuous response response that is smooth, that is whose derivative is continuous
control diagram the same as a block diagram
controllable domestic load domestic load that is controlled as part of a *smart* network
converter based employing one or more *HVDC* convertors
daily load cycle the variation of demand (load) during the day
demand (load) the consumption of power that must be provided by generators
demand (load) shedding the deliberate jettisoning of demand (load) triggered by falling frequency
demand-side management ways of controlling usage of electricity by the consumer
direct solution of the differential equation a straightforward, analytical method of solving a differential equation
energy storage device a device for storing electricity, such as a battery or a capacitor
Euler's method the simplest numerical method for solving a differential equation
exporting sending power outward from a system
fast frequency response the rapid response provided by power electronics
forward difference the value at the next step minus the value at the current step
frequency control the control of the local or *system frequency*
frequency events system events that have a significant effect on the *system frequency*
frequency response erosion a reduction in the *frequency response* available on a system
frequency response requirements the amount of power that must be held back on generators for use during periods of *abnormal operation*
frequency stability the ability of a system to withstand and control changes in the *system frequency*
generator (speed) governor the device on synchronous machines that controls changes in frequency
general responsive device a responsive device other than a synchronous machine
general responsive profile the graph of responsive power output against time of a *general responsive device*
genetic algorithm a heuristic algorithm that uses the laws of genetics for its operation
***H*-constant** the *inertia constant* of a machine or system

HVDC high-voltage direct current

HVDC link a direct current (DC) link that has an AC–DC convertor at each end

high-frequency event a system event which involves a significant rise in the *system frequency*

high-frequency excursion the same as a *high-frequency event*

importing drawing power in from outside the system

inertia the innate property of matter that resists changes to its state of motion

initial value problem (IVP) a mathematical term to describe a differential equation whose solution passes through a single point known as the "initial value"

intact network a network with no fault outages (operations) or with no outages at all (long-term planning)

integrating factor a multiplicative term for simplifying the solution of a differential equation

Kyoto Protocol the international agreement to reduce the level of greenhouse gases in the Earth's atmosphere

linear obeying the principles of superposition, etc.

local truncation error the change in error in a numerical solution from one step to the next

low-frequency excursion a system event where the frequency falls significantly below the normal operating range

maintenance outages removal or isolation of elements of the system so that checking or repairs can be done

manual actions changes to system plant or equipment that require human intervention for their completion

Milne–Hamming methods a set of *predictor–corrector* methods that use a *predictor* due to *Milne* in conjunction with a *corrector* due to *Hamming*

modified Euler's method a *predictor–corrector* improvement to *Euler's method*

moment of inertia the equivalent of mass in a rotating system

multiagent system a control system made up of many generalized components known as "agents"

multistep method a numerical method in which function evaluations are made at two or more steps at each iteration

MVA rating the megavoltampere rating of a machine

nonconverter based a machine not employing a power electronic convertor

nonresponsive generation generation not activated to produce response

normal operation operation of the power system during periods in which faults are absent

numerical accuracy the closeness of a numerical solution to the analytical solution

numerical methods methods that rely on replacement of the derivatives of a differential equation with finite differences

numerical stability the ability of a sequence of numerical operations to maintain accuracy

order of a numerical method the exponent of the order of the local truncation error minus one

Paris Agreement an international agreement to reduce the level of "greenhouse gases" in the Earth's atmosphere

per unit a system of representing quantities by dividing a "physical quantity" by a "base quantity"

physical units non-per-unit units such as megawatts
predictor–corrector method a two-step numerical method in which a *predictor* stage is followed by a *corrector* stage at each iteration
quiet system a system in which the *frequency* is changing relatively slowly
rate of change of frequency (RoCoF) the derivative of frequency with respect to time immediately following a *system event*
reserve a holding back of generating output to cover system events
response the power held on deloaded generators for the purposes of opposing frequency changes due to the sudden loss of generation or demand from the system
responsive generation generation that is preset to produce *response*
rotational kinetic energy the energy of motion of a rotating body
Runge–Kutta methods numerical methods in which each *single-step* interval is divided into subintervals that are used in the calculation at each iteration
single-step method a method that only requires variables to be evaluated at the current step at each iteration
solar PV solar photovoltaic plant
solar PV farm a large array of *solar PV* panels
smart grids a system of intelligent controls and equipment that optimizes the usage of power by consumers
solution by Laplace transforms the solution of a differential equation such as the *swing equation* by using Laplace transforms
spreadsheet method a method of programming the solution of a mathematical problem by means of a spreadsheet such as Excel
stability the ability to recover normal operation following changes on the system
state space the space of physical variables rather than Laplace transformed variables
swing equation the fundamental equation governing the motion of synchronous machine rotors following the loss of generation or demand from the system
synchronous machines electrical machines whose rotors rotate at the synchronous speed divided by the number of pole pairs
synthetic inertia the effect of having inertia synthesized by power electronic devices
system damping the property which governs how quickly a system disturbance dies down following a disturbance
system frequency the average apparent frequency of oscillation of AC machines connected to the system; the power frequency
system inertia the property of a system that opposes change; the average H-*constant* of a system
system requirements the needs of the system to enable correction following disturbances
Taylor series expansion the expression of a function by a power series
transfer function the function in s-space that relates the output to the input of a system
tripped lost from the system due to a fault
truncation error the difference between the true solution and that provided by a truncated power series representation
unit step function a function $u(t)$ which is 0 for $t \leq 0$ and 1 for $t > 0$
variation of demand with frequency the phenomenon whereby the amount of power load (demand) varies with system frequency
virtual synchronous machine (VSM) the semblance of the properties of a synchronous machine produced by means of power electronics

List of symbols used

Chapter 1: The need for frequency control

α	a positive constant
$D(\%)$	the droop of a generator governor in percent
f_N	the nominal frequency of the system in per unit
k	a positive constant
$P_{\mathbf{LOAD}}$	the active power load in per unit
$P_{\mathbf{RESP}}$	the additional active power generated by response
pu	per unit

Chapter 2: What can provide frequency control?

No symbols are used in this chapter.

Chapter 3: Per unit systems for frequency analysis

H	machine inertia constant
$J_{\mathbf{ind}}$	individual machine's moment of inertia
K	rotational kinetic energy
$KE_{\mathbf{rated}}$	rated rotational kinetic energy
$KE_{\mathbf{total}}$	total rotational kinetic energy
MVA	megavoltamperes
$\mathbf{MVA}_{\mathbf{base}}$	base megavoltamperes
MW	megawatts
$\omega_{\mathbf{0\ m}}$	rated mechanical speed in radians per second
$\omega_{\mathbf{rated}}$	rated speed in radians per second
$\mathbf{pf}_{\mathbf{rated}}$	rated power factor
$P_{\mathbf{total}}$	total active power output of all machines
S	megavoltampere rating
s	seconds

Chapter 4: Initial analysis of the frequency control problem and a derivation of the swing equation

J	moment of inertia
$K_{\mathbf{rot}}$	rotational kinetic energy
ω	rotational speed in radians per second

ω_m mechanical speed in radians per second
ω_{mrated} rated mechanical speed in radians per second
$\omega_{0\ \mathrm{m}}$ initial mechanical speed in radians per second
P_e electrical power output in megawatts
P_{epu} electrical power output in per unit
P_m mechanical power input in MW
P_{mpu} mechanical power input in per unit
P_{res} residual power
S_n nominal megavoltamperes, megavoltampere base

Chapter 5: Techniques available for calculating frequency response requirements

α a positive constant
a_0, a_1, a_2, a_3 constants
A_1, B_1, C_1, D_1 constants
c_1, c_2, c_3 constants
$D(\%)$ droop of responsive machine in percent
Δf difference of the "system frequency" from the set point frequency
f the "system frequency"
f_n the nominal frequency
GRP general responsive profile
H_{eq} the equivalent inertia constant following a loss
k a positive constant
ω system speed in radians per second
P_e electrical power output
P_{GEN} active power generation
$P_{\mathrm{GEN(NR)}}$ nonresponsive power generation
$P_{\mathrm{GEN}n\mathrm{(R)}}$ responsive power at nominal frequency
P_{LOAD} active power load
$P_{\mathrm{LOAD}n}$ active power load at nominal frequency
P_m mechanical power input
$P_{\mathrm{non\text{-}resp}}$ nonresponsive active power generation
P_{resp} responsive power generation
P_{resp1} primary period responsive power generation
P_{resp2} secondary period responsive power generation
P_{resp3} tertiary period responsive power generation
P_{resp1tot} total primary period responsive power generation
P_{resp2tot} total secondary period responsive power generation
P_{trip} active power generation tripped
t time in seconds
t_{trip} trip time
$u(t - t_{\mathrm{trip}})$ unit step function, stepped at $t = t_{\mathrm{trip}}$

Chapter 6: Analytical methods for solving the "swing equation"

α a constant
α_{Hz} the constant α, in Hz

α_{pu}	the constant α, in pu
A_1, B_1, C_1, D_1	constants for the "direct solution of the differential equation" method
A_2, B_2, C_2, D_2	constants for the "Laplace transform" solution method
a_0, a_1	constants
b_1, b_2, b_3, b_4	constants
c	a constant
c_0, c_1, c_2, c_3	constants
c_4, c_5, c_6	constants
Δf	difference of the "system frequency" from the set point frequency
f_0	initial frequency, frequency at $t = 0$
$f(0)$	initial frequency, frequency at $t = 0$
f_n	the nominal frequency
$(f \times g)(t)$	the convolution function of $f(t)$ and $g(t)$
$F(s)$	the Laplace transform of $f(t) \equiv \mathcal{L}\{f(t)\}$
$g(t)$	a general function of time t
H	inertia constant in seconds
k_1	a constant
$\mathcal{L}\{f(t)\}$	the Laplace transform of $f(t) \equiv F(s)$
$n!$	the factorial of n
$P_{GEN(NR)}$	nonresponsive power generation
$P_{GENn(R)}$	responsive power generation at nominal frequency
P_{LOADn}	active power load at nominal frequency
P_{TRIP}	active power generation tripped
S	the variable in 'Laplace transform space'
t	time in seconds
t_{TRIP}	the time at which active power generation is tripped
u	a "dummy variable," used in integration
$u(t), v(t)$	variables used in "integrating by parts"
$u(t - t_{trip})$	unit step function, stepped at $t = t_{trip}$
x	the independent variable
y	the dependent variable

Chapter 7: Numerical methods for solving the "swing equation"

α	a constant
a_0, a_1	constants
Δt	forward difference
∇t	backward difference
$D(\%)$	droop of responsive machine in percent
$\lvert\varepsilon_n\rvert$	absolute error at the "n^{th}" iteration
$\phi(x,y)$	general function of x and y
f	the "power system frequency"
f_0	the initial frequency, frequency at $t = t_0$
$f(t_0)$	the initial frequency, frequency at $t = t_0$
$f'(t_0)$	the first derivative of "f" with respect to "t" at $t = t_0$
$f''(t_0)$	the second derivative of "f" with respect to "t" at $t = t_0$
f_n	the nominal "power system frequency"
k	a constant

k_1	a constant
k_2	a constant
H	the "inertia constant" in seconds
h	step length, time step
$P_{GEN(NR)}$	nonresponsive power generation
$P_{GENn(R)}$	responsive power generation at nominal frequency
P_{LOAD}	active power load
P_{LOADn}	active power load at nominal frequency
P_{TRIP}	active power generation tripped
P	signifying a "predicted value"
T_α	"truncation error" at some value of $x = \alpha$
$T(x,h)$	"truncation error" expressed in terms of x and h
t	time in seconds
t_0	the initial time, usually chosen to be zero
$u(t - t_{trip})$	unit step function, stepped at $t = t_{trip}$
w_i	weighting factor in a Runge–Kutta scheme
x	independent variable
y	dependent variable
y_n	the value of "y" at the "nth" iteration
y_{n+1}	the value of "y" at the "$(n + 1)$th" iteration
y_0	value of y at $x = x_0$
$y(x_0)$	value of y at $x = x_0$
$y'(x)$	first derivative of y with respect to x
ζ	some value of x within a specified range

Chapter 8: The "control diagram" method

A, B, C, D	constants
a_0, a_1	constants
b_1	a constant
Δf	difference of the "system frequency" from the "set point frequency"
Δu	change in the input
Δy	change in the output
f	the "power system frequency"
f_0	the "initial frequency"
f_n	the "nominal frequency"
G	the "feed-forward" constant
H	(1) the "inertia constant" in seconds; (2) the "feedback" constant
K	the gain
k_1	a constant
P_G	active power generation
$P_{GEN(NR)}$	nonresponsive power generation
$P_{GENn(R)}$	responsive power generation at nominal frequency
P_L	active power load
P_{LOADn}	active power load at nominal frequency
P_{TRIP}	active power generation tripped
R	the ramp rate of generation

RoCoF	rate of change of frequency
T	the time constant
t	time in seconds
$U(s)$	the denominator
u	the "input variable"
$u(t - t_{trip})$	unit step function, stepped at $t = t_{trip}$
x	the "system variable"
$\dot{x}$	the same as "dx/dt"
$Y(s)$	the numerator
y	the "output variable"

Chapter 9: Some important practical applications

α	a constant
A_1, B_1, C_1, D_1	constants
a_0, a_1	constants
Δf	difference of frequency from nominal
ΔP_{RESP}	change in responsive active power output
$D(\%)$	responsive generator droop constant in percent
df/dt	rate of change of frequency≡RoCoF
f	power system frequency
f_0	initial frequency
f_n	nominal power system frequency
GW	gigawatts
H	inertia constant in seconds
Hz	Hertz
k	a constant
k_1	a constant
J	moment of inertia
MW	megawatts
ω_n	nominal speed in radians per second
pf_n	nominal power factor
$P_{GEN(NR)}$	nonresponsive active power generation
$P_{GENn(R)}$	responsive power generation at nominal frequency
$P_{G(TOT)}$	total active power generation
P_{LOAD}	active power load
$P_{LOAD\ n}$	active power load at nominal frequency
$P_{L\ (TOT)}$	total active power load
P_n	nominal active power base
P_{TRIP}	active power generation tripped
RoCoF	rate of change of frequency≡df/dt
S_n	nominal MVA base
Sys. Dem.	active power system demand
t	time in seconds
$u(t - t_{trip})$	unit step function, stepped at $t = t_{trip}$
$t = 0^+$	an infinitesimal amount of time after time equals zero

Chapter 10: The challenges of operating systems with high penetrations of renewables (low-inertia systems)

"AC"	alternating current
"DC"	direct current
DFIG	doubly fed induction generator
FFR	fast frequency response(?)
HVDC	high-voltage direct current
PV	photovoltaic
VSM	virtual synchronous machine
WTG	wind turbine generator

Chapter 11: "Smart grids" and the "system frequency"

AA	arbitrator agent
ANN	artificial neural network
CHP	combined heat and power
DA	diagnosis agent
DOE	Department of Energy (United States)
DS	distribution system
DSO	distribution system operator
ES	expert system
EV	electric vehicle
FDA	fault detection agent
HV	high voltage
JIA	judgment index agent
LV	low voltage
MAS	multiagent system
MV	medium voltage
PV	photovoltaic
SCADA	supervisory control and data acquisition
SA	scheduling agent
SMA	system monitoring agent
TO	transmission operator

Chapter 12: Conclusions

No symbols used in this chapter.

Chapter 1

The need for frequency control

1.1 An introduction to frequency control

1.1.1 The subject of the book

The subject of our book is the stability and control of the power system frequency in high-voltage alternating current ("AC") systems. The focus will be on operational issues at the national transmission system level. The author's personal experience is with the Great Britain national transmission system, which is fundamentally an island "AC" system with several direct current ("DC") connections, both to offshore wind farms and to mainland Europe. There are also some internal "DC" connections that make up part of the Great Britain "national grid" system (Diagram 1.1).

1.1.2 Frequency stability and control as a system requirement

The stability and control of system frequency is an essential requirement of system operation that must be attended to extremely carefully, particularly in more recent times, when the composition of both generation and demand profiles has been changing rapidly, to include technological developments such as wind generation, solar photovoltaic (PV) systems, demand-side management, and "smart grids." All these things will be covered to a greater or lesser degree in this chapter and in the later chapters of this book.

1.1.3 Combining the old with the new

We should never forget the basic principles upon which the transmission and distribution systems operate. Therefore in this book, we shall attempt to cover some of the more "traditional" aspects of frequency stability and control, but we shall also try to cover some aspects of the more "modern" challenges faced increasingly by control engineers in large and small power systems throughout the world, focusing particularly on the issue of falling inertia levels that are due to the increasing presence of inertia-less plant.

Modern Aspects of Power System Frequency Stability and Control.
DOI: https://doi.org/10.1016/B978-0-12-816139-5.00001-1

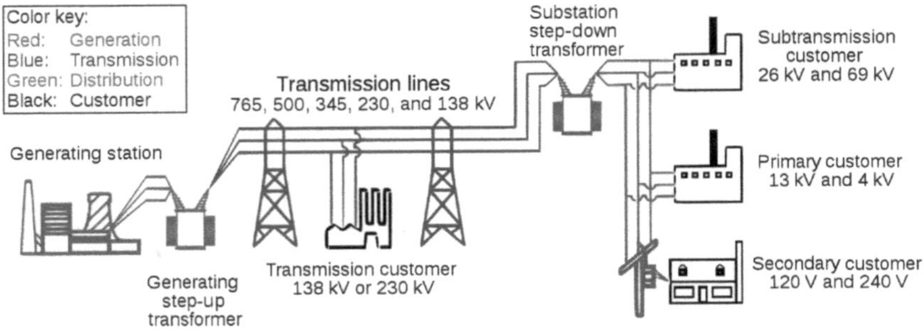

DIAGRAM 1.1 Schematic of a typical national transmission system. *US Department of Energy.*

1.1.4 Nomenclature: what is meant by "the frequency" in this book?

1. *The "power system frequency"*

   When we talk about "the frequency" in this book, we shall always mean the "power system frequency." This may be thought of as the apparent, measured average speed of rotation of those AC machines actively connected to the electrical network. It is, with few exceptions, always close to the system power frequency, for example, 50 Hz in the United Kingdom or 60 Hz in the United States. It is the frequency often seen in electronic displays in electricity control rooms and in the lobbies or foyers of electricity control buildings.
2. *Frequencies other than the "power system frequency"*

   In this book, we are not considering the various subharmonic frequencies (frequencies less than the power frequency) or superharmonic frequencies (frequencies greater than the power frequency) frequencies that are known to exist on the power system, for example, those assorted frequencies that appear following a trip of generation or load from the system. We shall consider all these, for the purposes of our analyses here, to be negligible.
3. *The justification for only analyzing the "power system frequency"*

   The assumption of only considering the power frequency is justified because, although observed to be present following a system fault, supersynchronous frequencies form a comparatively small proportion of the power and decay quickly. Meanwhile, subsynchronous frequencies are a different subject altogether and may or may not be present following a system fault depending on system conditions. For more details on this subject, the reader is directed, for example, to Ref. [1].
4. *The control of the "power system frequency" within specified limits*

   The frequency of an AC power system needs to be controlled within specified limits to maintain the stability of the system, for reasons which we now proceed to explain.

1.1.5 Satisfactory normal operation of the power system

1. *A general statement*

   To ensure the continued normal operation of a power system, the system frequency needs to be maintained close to the nominal frequency for that system (e.g., 50 Hz in the United Kingdom and 60 Hz in the United States). This is a general statement that is applicable to the operation of all AC power systems. See Ref. [2], Chapter 11, for further details and discussion on this subject.
2. *The assumption of an AC system*

   Note that we are making an implicit assumption here that the system we are discussing is "AC," not "DC." However, in many countries (including the United Kingdom), there is an increasing change to a more "mixed" system with significant amounts of both "AC" and "DC" components and circuitry present.
3. *The issue of system inertia*

   One of the main issues developing here is one of inertia, which, traditionally, in purely "AC" systems in the past was not an issue because there used to be a large enough fleet of traditional synchronous machines connected to the system not to have to be too much concerned with the issue of the possible shortfall of inertia and its consequences. However, the issue is becoming increasingly relevant in the "mixed AC and DC" systems that we have in modern times.

1.1.6 The operating condition of synchronous machines

The rotational speed of the "AC" machines connected to the system must not be too low or too high. If the rotational speed (the "speed") is too low, the machines will cut out and the power allocated to feed the loads connected to the system will be lost. If the "speed" is too high, the rotating machines could "pole slip," that is the traction between the magnetic fields of the rotor and stator could be lost, and the machines may disconnect from the system due to the consequent triggering of the automatic pole slip protection. During overspeed, if the currents and voltages within the machines exceed their ratings, the machines may also sustain damage and consequently require costly repairs even if the "pole slipping" condition is not reached.

1.1.7 Deviations of the system frequency from normal ("nominal")

More generally, if the system frequency deviates too much from the "nominal" (i.e., from the "rated" frequency), the following undesirable effects may occur:

1. *Consequences of a significant drop in frequency*

   These include the stalling of generators and motors, high magnetizing currents in induction motors and transformers, the loss of auxiliary motor drives in power stations, the development of errors in equipment that utilize electric timers ("electric clocks go slow"), the reduction in the values of the reactances of circuit elements, and the loss of blocks of demand through underfrequency, including those controlled by frequency-dependent load-tripping relays.

2. *Consequences of a significant rise in frequency*

   These include the tripping on overspeed of synchronous generators and motors, errors developing in equipment utilizing electric timers ("electric clocks go fast"), the increase in the reactances of circuit elements, and the loss of blocks of demand through overfrequency.

   Either low- or high-frequency deviations can ultimately lead to instability, which in turn can lead to demand control and hence loss of supply.

1.1.8 A further comment about system impedances

Since the values of impedances within the network depend upon the system frequency being considered, if the frequency deviates too much from the nominal frequency than other system, effects may come into play, such as a change in real system and reactive losses, which can potentially affect power loadings on transmission and distribution transformers and therefore on voltages throughout the system.

1.1.9 The continuous requirement to regulate the system frequency

For the reasons outlined in the earlier sections, we must always ensure that the "power system frequency" is as close as possible to the "nominal frequency." In a national control room, this is indeed one of the primary tasks of the control engineer, and in the past in Great Britain in addition great pride was taken in keeping the "frequency error," that is, the integral of the deviation of the system frequency from nominal frequency with time, as small as possible.

1.1.10 Methods of regulating the system frequency

There are various traditional (and some not-so-traditional) methods of controlling the system frequency that will be mentioned in this book. In the case of synchronous generators, these machines are invariably fitted with a "speed governor" system that regulates the frequency by adjusting the power output of the generator, either to the rated frequency ("isochronous governors") or operating on a "slope" profile ("nonisochronous governors") in addition to having an "automatic voltage regulation" device to regulate the voltage [2].

1.1.11 Control of frequency in the modern era

1. *Electronic control of frequency*

   In the case of more modern, "environment-friendly," generating devices, such as wind turbines and solar PV systems, regulation of the local frequency by purely electronic means has now been made technologically possible and is becoming increasingly more prevalent in power systems throughout the world.
2. *The increasing speed of response*

   Typical of modern responsive devices are their extremely fast response times, which are of the order of milliseconds or, in some cases, even faster. These magnitudes of rapid response reaction times can be achieved now from, among other technologies, large batteries and high-voltage direct current (HVDC) links.
3. *Limiting the speed of response*

   One major principle that should always be borne in mind when selecting the speed of response is that it is not always a good thing to have arbitrarily fast response times. This is because *too* fast a response may introduce new instabilities, such as small-signal instabilities, into the network. Therefore judicious selection and coordination of response times always remain necessary.

1.1.12 The wider system potentially at risk

The importance of maintaining an accurate power balance on the power system at large should never be underestimated. If the system frequency is not maintained within the prescribed statutory limits, and machines become disconnected from the system as a result, the whole system may become unstable, and in the absence of appropriate remedial actions, the entire system may be at risk of collapse.

1.2 An introduction to the "system requirements"

1.2.1 A definition of the "system requirements"

The "system requirements" for the purposes of frequency stability and control may be described as the power that must be held in reserve on, for example, thermal generators, DC links, and other more modern sources of power, such as wind generators, to cover for the potential losses of generating sets, blocks of load, lines, transformers, and other equipment on the power system.

1.2.2 The variation of the "system requirements" with the "daily load cycle"

The "system requirements" for a power system vary with the time of day (the "daily load cycle"), since the demand on the system is continuously changing and indeed the "daily load cycle" itself varies with the day of the week and with the week of the year.

1.2.3 The degree of integrity of a power system and the "system requirements"

The "system requirements" need to be calculated separately for each of several different possible system states. By this we mean that the system infrastructure with its connected generation plant and loads is ever changing, for a variety of different reasons. Consequently, there is not just one set of calculations to be done, but many.

There are different maintenance outage patterns to contend with different generation and load patterns. These will include many planned scenarios. However, there are also unplanned events that can change the structure of the power system and suddenly change the balance of power within it. For example, a fault may result in the loss of a generating set, or a block of load or a DC link. All these possibilities, and more, must be evaluated in calculating the "system requirements."

In the next section, we begin to examine the range of possibilities that exist and must be studied.

1.3 The system requirements for the "intact" ("prefault") system

1.3.1 A definition of the "intact system"

In the context of frequency stability and control, when we speak of the "intact system," we usually mean a system in which there have been no unplanned losses of generation, demand, transmission or distribution lines, transformers, capacitors, reactors, or anything else. This definition does not exclude the possibility that there may be planned maintenance outages present, or planned outages taken to allow the connection of new equipment to the system. These two things are always present in a real system, although their numbers and extent vary seasonally and with the level of activity of the construction market.

1.3.2 The approximate balance of power

Apart from relatively small minute-to-minute variations, for most of the day, the "system demand" usually does not change very quickly. The generation and demand connected to the system are roughly in balance, and so the system frequency is at or near the nominal frequency for the system in question (e.g., 60 Hz in the United States and 50 Hz in the United Kingdom). Provided all the conditions described earlier are maintained, the frequency will remain at, or close to, the nominal frequency of the system.

Relatively little change to the profile of generation synchronized to the system is required during these relatively quiet times, when "load following" is achieved relatively effortlessly by automatic governor action and relatively small changes to the total generation scheduled on the system.

1.3.3 The daily load cycle

However, during a typical day, there are certain times at which the system demand changes relatively quickly. Such times include the time in the morning, usually around breakfast time, when domestic and commercial lights are switched on, kettles are plugged in, and the working day begins (the "morning pickup"). Here, the demand on the system is increasing relatively rapidly (see Diagram 1.2 for the graph of a typical daily load cycle).

During the working week, when the morning pickup is over, the demand tends to level out and remain relatively constant for some time (the "morning plateau").

Another time when the demand can be changing relatively quickly is the time in the afternoon around which office hours come to an end. During the working week, this time can have the highest demand of the day (the "evening peak").

When the evening peak is over, the system demand will decrease again moving into the overnight period when the demand is relatively low, until the following morning, when the demand begins to pick up again in the daily cycle.

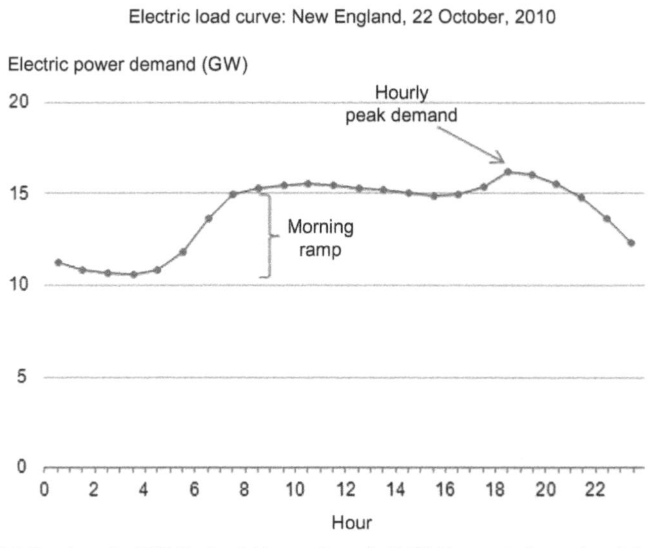

DIAGRAM 1.2 A typical "daily load (demand) cycle." *US Energy Information Administration.*

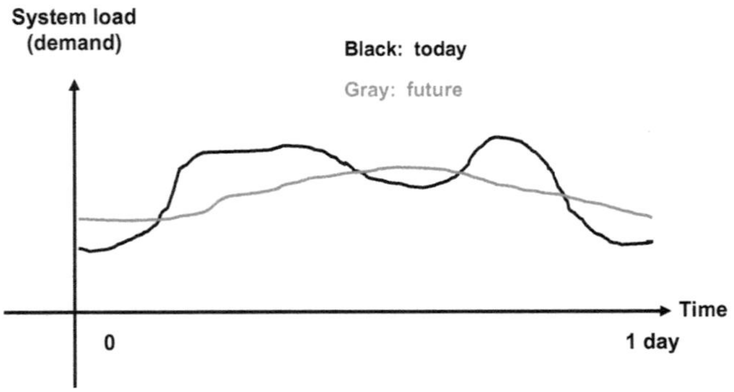

DIAGRAM 1.3 Changes to the shape of the daily load (demand) profile. *Author's own sketch.*

During such times of relatively high rate of change of the system demand, since the active power balance is also changing rapidly, the scheduling engineer needs to authorize generating units ("sets") to increase or decrease their outputs to the power system.

1.3.4 Modern changes to the shape of the daily load profile

Relatively recently, with the advent of ever-increasing amounts of solar power to the generation mix of many electric power systems, and also with the advent of large batteries and more "intelligent controls" on commercial, industrial, and domestic demand ("smart grids"), the shape of the daily load profile has been changing. Instead of the traditional routine of "morning pickup," "morning plateau," "afternoon plateau," "evening peak," "overnight minimum," and so on, we are beginning to experience things such as "afternoon peaks" and not insignificant demands overnight as well (Diagram 1.3).

In other words, in some countries such as the United Kingdom, the daily load cycle has become not as predictable as it once was, and this has presented new and difficult challenges to control engineers in their daily task of following the demand cycle and keeping the power system frequency in balance.

In addition, due to the intermittency of wind generation, the generation profile is also becoming increasingly unpredictable. We are surely entering a new era in frequency control.

1.4 System requirements following the loss of a generating set

1.4.1 Introduction

1. *The loss of a generating unit from the system*

   The most important potential change to the intact system described earlier that we should consider is the unplanned disconnection ("loss") of a generating unit. This event is sometimes referred to in system operations circles as a "generator trip."

2. *The effect on the system frequency*

   The loss of a generating set causes the system frequency to fall, because now there is a deficit of power feeding into the system. Of all the potential changes to the intact system, this event has the capacity to affect the system frequency the most, because of the large size of some generating units compared with the system size, especially during times of low demand. It is also one of the most frequent events, so must be studied and planned for very carefully.

1.4.2 The primary causes of the loss of generation from the system

A generator may become disconnected unexpectedly from the system for a variety of reasons, including, but not limited to the following:

1. An electrical fault may develop on an item of equipment, such as a ground fault may appear on a terminal busbar caused, for example, by a lightning strike on it.
2. A mechanical fault may develop, such as the interruption of fluid flow in a turbine system, leading to the partial or total loss of power output from the generator.
3. The generator may become unstable and trip due to the presence of instability in a nearby generating unit. This kind of chain reaction is sometimes referred to as "cascade tripping."
4. If the generating unit in question is a wind turbine, it may cut out due to the wind speed being either too low or too high. Wind turbines can only function in a specific range of wind speeds, outside which they cut out completely.
5. If a generating unit is on a spur feeding through an HVDC link, the link, and therefore the infeed from the generator into the main system, could be lost due to the possibility of the failure of the link's control system. If this happens then the output from the generating unit could also be lost as seen from the standpoint of the main system.

Typically, the loss of a large block of generation from the system has the following kind of effect in terms of the profile of the power frequency plotted against time (Diagram 1.4).

On losing the block of generation, the first thing that happens is the frequency begins to fall. The *rate of change of frequency* here is usually referred to as the "RoCoF". The magnitude of the *RoCoF* is a very important quantity in system control and is considered in more detail later in Chapter 9, Some important practical applications, of this book.

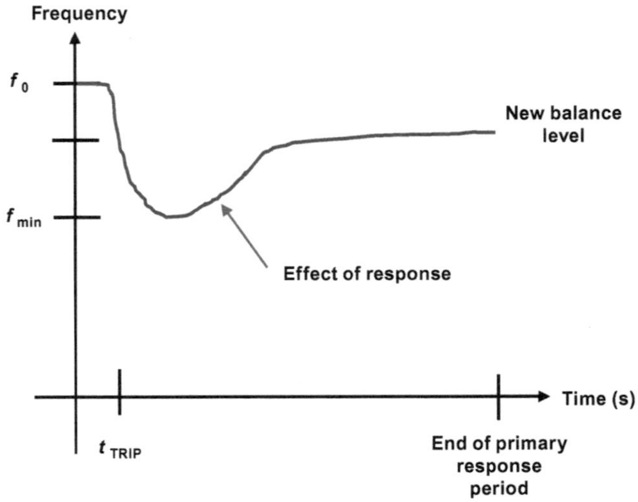

DIAGRAM 1.4 The effect on the system frequency of a generation loss. *Author's own sketch.*

1.4.3 Rate of change of frequency relays

In the Great Britain distribution systems, "RoCoF" relays are installed on small generating plant to detect the "loss of mains" and thereby protect the generating plant against damage when the consequent frequency unbalance occurs. They can be activated on "underfrequency," "overfrequency," or on the "rate of change of frequency."

Consequently, one of the main principles of system frequency control is to avoid "hitting" one of the *frequency disconnection trigger levels* selected for these devices, or to exceed the *RoCoF limits* for either "underfrequency" or "overfrequency" situations.

1.4.4 Loss of generation: following the initial fall in frequency

After the first second or two following the generation trip, the "RoCoF phase," the system encounters four controlling effects: *(1) automatic governor action, (2) the effect of frequency-dependent loads, (3) automatic demand-tripping schemes, and (4) response from nonsynchronous sources*.

We shall consider each of these effects in turn.

1.4.5 Loss of generation: automatic governor action

Most synchronous generators are equipped with automatic speed governing systems. These detect and react to the system frequency in such a way as to counter the detected deviation of the system frequency. In this respect, they

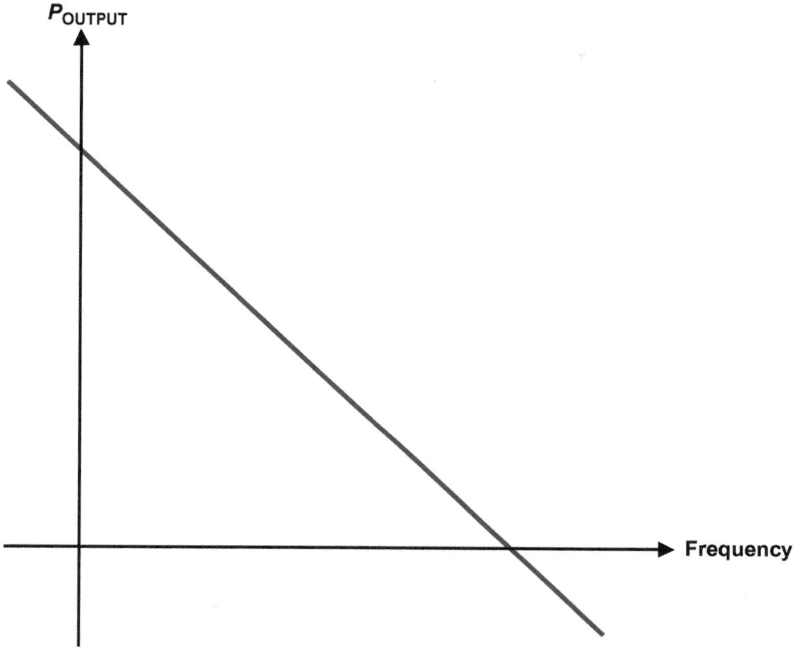

DIAGRAM 1.5 The speed governing droop characteristic of a generator. *Author's own sketch.*

are "negative feedback" systems, and their effect is to stabilize the frequency (see, for example, Ref. [1], Chapter 11, p. 582, for further details).

As a "negative feedback" system, the speed governor of a synchronous machine operates on a so-called droop characteristic, which is roughly a straight line in the central part of its operating range (see Diagram 1.5 for a sketch of this).

The slope, or gradient, of the speed governor characteristic is called the "droop" and is measured in percentage points according to how much the frequency changes when the machine changes from "no output" to "full output."

For example, if the speed of the machine reduces by 4% when the power output changes from "no load" to "full output," then the "droop" is therefore 4%.

The action of the *droop characteristic* can be expressed in the form of an equation as follows:

$$\Delta P_{RESP}(pu) = -k(pu)\cdot\{f(pu) - f_N(pu)\} \tag{1.1}$$

where f_N is the nominal frequency and k is a positive constant. Hence, we see that when the frequency is less than the nominal frequency, the responsive power is positive, and conversely when the frequency is greater than the nominal frequency, the responsive power is negative.

The constant k of (Eq. 1.1) is related to the "droop D in %" of the governor as follows:

$$D(\%) = 100/k(pu) \tag{1.2}$$

For a step-by-step derivation of this result, see Section 5.3.7.2 later in this book.

1.4.6 The selection or otherwise of governing action

1. *Responsive and unresponsive modes*

   It is the usual practice for the speed governing system ("speed governor") of a synchronous generator to be made able by design to be selected to either *active* (in "frequency-responsive mode") or *inactive* (in "frequency unresponsive mode").
2. *Inertial response only*

   A synchronous machine whose speed governor is *inactive* (in "frequency unresponsive mode") will still react to changes in system frequency, but only by means of its natural, so-called inertial response which is always present because of the inertia ("H-constant") possessed by the machine. The machine has a store of energy due to its rotating inertia: its "rotating kinetic energy." We remind the reader that "inertial response" is the involuntary and uncontrolled reaction of the generating machine to any imbalance between the load and generation on the system.
3. *Inertial response plus governor response*

   By contrast with the preceding section, a synchronous machine whose governing system is *active* (in *frequency-responsive* mode) will react to changes in the system frequency by both (a) *governor* and (b) *inertial response.* By reason of this double capability, such a generating set is therefore more effective at responding to and limiting system frequency changes than one in frequency unresponsive mode and whose response is only inertial. The improvement is principally because "inertial response" by its nature dies away quickly, since it by its own action seeks to neutralize any power imbalance, whereas if governor response with a "droop" characteristic is installed, the response characteristic is not transitory but persists well beyond the time when inertial response has died away.

1.4.7 Loss of generation: the effect of frequency-dependent loads

The presence of induction motors, which are present as part of the national demand (load) of a transmission system, causes the overall demand (load) to be dependent on the power frequency. Ordinary resistive

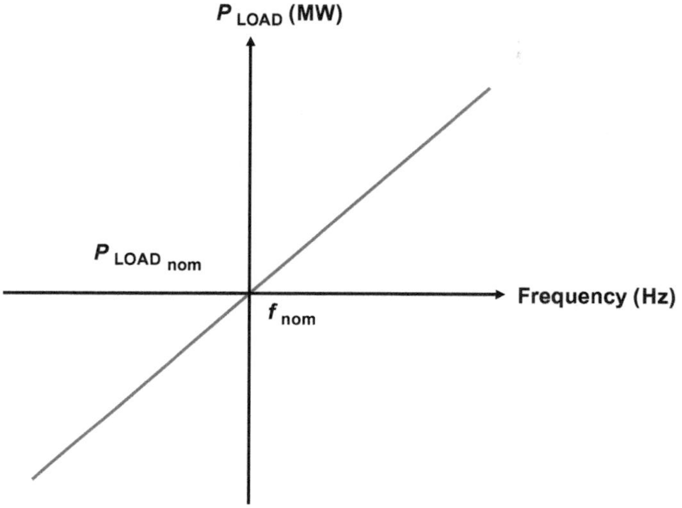

DIAGRAM 1.6 The form of the dependency of the load on the frequency. *Author's own sketch.*

loads such as lighting and heating are not frequency dependent (see, for example, Ref. [2], p. 584, for a discussion of this) and so do not have any effect in this respect.

The mathematical dependency of demand on frequency is roughly linear and may therefore be expressed in equation form as follows:

$$P_{\mathrm{LOAD}}(f) = P_{\mathrm{LOAD}}(f_N) \cdot \{1 + \alpha(f - f_N)\} \tag{1.3}$$

where f_N is the nominal frequency of the system and α is a positive constant. This relationship is depicted in Diagram 1.6.

As the frequency *increases*, the load also *increases*; conversely, as the frequency *decreases*, the load also *decreases*. As such, in contrast to the "droop" characteristic of the generator response covered earlier, it is a "positive feedback" system.

The effect of having frequency-dependent loads as part of the system demand is *always to stabilize the frequency*. This result is shown in Diagram 1.7, where a comparison is made on the development of the system frequency between (1) having frequency-responsive loads and (2) not having frequency-responsive loads.

1.4.8 Partial loading of generating plant

Traditionally (in the United Kingdom at least) generating plant has in the past been partially loaded for two distinct, but related, purposes. These are referred

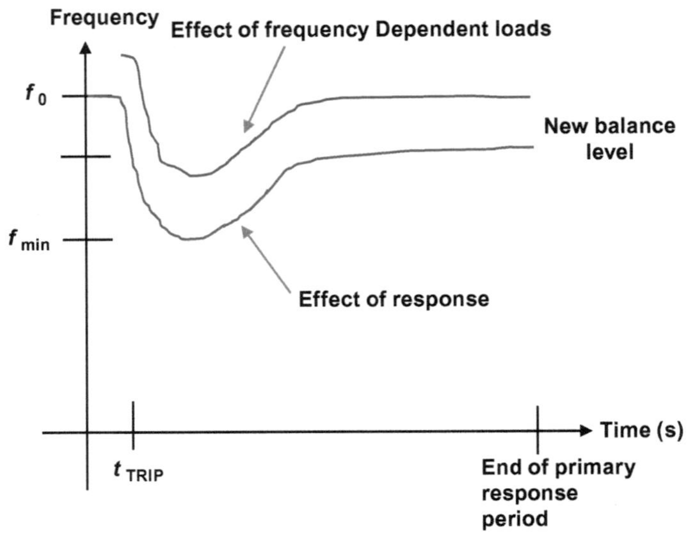

DIAGRAM 1.7 The effect on the system frequency of having frequency-dependent loads. *Author's own sketch.*

to in the United Kingdom as "response" and "reserve." These are summarized here as follows:

1. Response
   a. *The question*

      So that a generating unit that is controlled by a "nonisochronous" speed governor (i.e., one with a "droop" characteristic) in active mode can generate more power following a fall in the system frequency, it must first be operating in a "de-loaded" state, that is, it must initially not be operating at its full output. With this rather obvious statement in mind, clearly a decision must be made regarding by how much the generator must be de-loaded in its initial state.

   b. *The answer*

      The approach that is usually used is to first, for a given set of predicted demand levels during the day, to select a group of generators (potential "responsive plant") for the task of providing response when needed and consequently to de-load all of these by about the same amount, say roughly 10% (this amount of de-loading is allocated at the nominal system frequency). Then when the need arises, these generators will operate on governor action to respond to a fall in frequency as desired.

2. *Frequency response "erosion"*
   a. *What is frequency response "erosion"?*

      If the system frequency at the onset of the loss of an amount of generation is less than the nominal frequency, then *less than* 10% (or whatever percentage has been allocated) of that available at *nominal frequency* will be available for use. This is known as frequency response "erosion."

b. *The explanation*

The cause of frequency response "erosion" is as follows. If a responsive generator has been set up to be de-loaded by, say, 10% at the "nominal" system frequency, then at any frequency *below* the nominal frequency, it will be de-loaded by *less than 10%* because of the machine's power output being constrained by the so-called droop profile (see Diagram 1.8 for an illustration of this).

Therefore if the power system starts out at a frequency less than the nominal frequency, the whole of its allocated response holding will not be available to it, because some of its response capability has already been used up.

3. *Reserve*

By "reserve" we usually mean a quantity of generation set aside for use if necessary. Reserve can be held for several different reasons. Reserve can be held, for example, to cover for any "demand forecasting errors" that may arise at some point during the day. As another example, reserve can be held to cover for unexpected losses of generating plant. And, of course, somewhat paradoxically, reserve can be held for response.

1.4.9 Loss of generation and generation rescheduling

1. *Manual actions*

After some time has elapsed following the loss of a block of generation, and all automatic actions have been exhausted, a period follows when "manual actions" to complete the restoration of the system power

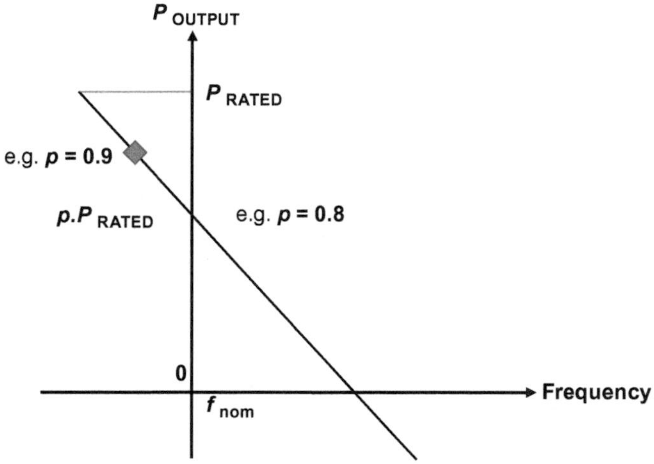

DIAGRAM 1.8 The phenomenon of frequency response erosion. *Author's own sketch.*

frequency (if necessary) to within operational limits are undertaken. This period begins, typically, about a few minutes after the original event that disturbed the frequency.

2. *What are "manual actions"?*

   "Manual actions" consist in the main part of issuing instructions to increase or decrease the power outputs of selected generating plant, or perhaps to alter the network configuration in some way. A generator operator will respond to the instruction (or request—it is a response to an energy trade in Great Britain) by changing the set point of the generator. In exceptional circumstances, additional machines may need to be synchronized to the system or, on the other hand, desynchronized from it. In either case, this process is known as "generator rescheduling."

3. *"Manual actions" and "reserve"*

   Usually, on a power system, as mentioned in the previous section, an amount of generation is held "in reserve" for times including these when extra generation is required at short notice. This can be called upon, for example, if the amount of "de-loaded plant" scheduled for frequency control purposes proves insufficient for needs during real-time operation. This can occur during the period when all automatic response has been exhausted and therefore "manual actions" are the only remaining possibility to restore the power system frequency to within acceptable levels.

1.5 System requirements following the loss of a demand block

1.5.1 Demand (load) trips

The second most important potential change to the intact system that should be considered in frequency planning studies is the loss of demand (load), known as a "demand (load) trip." A loss of demand causes the system frequency to rise, contrary to a loss of generation, which causes the frequency to fall.

1.5.2 Loss size

Usually, but not always, a typical demand loss during the daily operation of a bulk power system is of a smaller magnitude than a typical generation loss. Its effect is therefore correspondingly smaller, but the change to the system frequency is in the opposite direction so that the mitigating actions required to bring the system frequency back to near nominal are of a different nature.

1.5.3 The causes of "demand (load) trips"

The reasons for "demand (load) trips" include, but are not limited to, faults on distribution lines and faults on supply transformers. In fact the largest loss of demand on the Great Britain system is currently the loss of an exporting interconnector. In general, demand is usually lost as a block, since groups of demand are attached from the transmission system to the distribution system by individual supply transformers, each of which supplies its own block of demand.

1.5.4 The task of the control engineer

The task of the control engineer during operational timescales following a demand loss is to prevent the frequency from going too high and causing problems. As such, enough responsive generation must be scheduled to ensure that this does not happen.

1.5.5 Actions available to the control engineer during a "high-frequency event"

1. *Automatic response*

   Following an unplanned loss of demand, those generators selected to be in "responsive mode" will automatically reduce their frequency along the "droop characteristic" as the power system frequency rises following the fault, that is, the loss of some of the demand from the system. This will occur after a time delay of a few seconds, the exact magnitude of the day varying from one generator to another according to their individual design.
2. *Generator rescheduling*

   After the end of the "automatic response period," generators can be rescheduled by having their mega-watt (MW) outputs reduced or by being taken off the system to allow the system frequency to come down further if necessary at this stage. This is a "manual action" performed under the guidance of the control engineer at the control center.

1.5.6 Demand (load) shedding

As a parting remark to close this section with, demand or load is sometimes tripped deliberately. This is done at a preset "trigger frequency" that is numerically less than the nominal frequency and is employed to assist in the recovery of the system frequency during a *low-frequency* event. It is then known as "demand (load) shedding." We emphasize that this is done deliberately and is not a "fault." It is an essential mechanism to protect the majority of electrical consumers at the expense of the loss of a minority.

1.6 The system requirements following the loss of lines, transformers, busbars, and other transmission equipment

1.6.1 The loss in bulk of generation and demand

Of course, in relation to the earlier discussion, *the loss of an individual item of equipment from a power system can lead consequentially to the loss of an amount of generation or demand from the power system.* However, in operational timescales, in the context of frequency planning studies, generation and demand are usually treated in bulk as individual potential losses.

1.6.2 The largest losses

Of special significance to the safe operation of the system are the largest potential losses of generation and demand from the power system at each moment in time throughout the operational day. *These are very important because they determine the largest response requirements at any time.*

1.6.3 The loss of equipment other than generators or loads

The loss of components other than generators and loads can also affect the system frequency, but usually the effects of losing one of these are much less severe than losing either generators or loads. Nevertheless, in planning or operating a large power system, these events may also need to be taken into consideration individually, for example, in studying the feasibility of a potential new generation connection.

1.6.4 The loss of lines

The primary effect of losing a transmission or distribution line is to *interrupt the flow of power through the system.* The power that was earlier being transmitted through the line must now be diverted elsewhere. One effect of this redistribution of the power flows in the system is to reassign the voltages in the locality of the fault.

However, another effect is possible. If the line in question had generation or load teed off from it, then the loss of the line would automatically trigger the consequent loss of that generation or load. In this case there would be a significant effect on the system frequency.

Another possibility might be the consequence of a "protection maloperation." For example, protection outside the zone of the line could erroneously detect a fault in its zone and trip equipment, including transformers connecting generation or demand to the system, thereby again disturbing the system frequency.

1.6.5 The loss of transformers

As hinted earlier, the loss of a transformer or transformers connecting generation to the system or feeding loads could have the secondary effect of altering the system frequency by reason of the associated loss of generation and/or demand. The magnitude of the frequency change would clearly be dependent on the size of the generation or demand lost in each individual case and may need to be examined separately.

1.6.6 The loss of busbars

If busbars, in substations or otherwise, develop faults and are tripped off the system as a result of the busbar protection, clearly whatever is connected to the busbar in question would probably also be tripped off the system. Therefore the loss of a busbar could indeed also involve the loss of one or more transformers, generating sets, blocks of demand, and so on, with associated effects on the system frequency, which needs to be assessed separately.

1.6.7 Summary

There are therefore many potential *scenarios* on the power system that could lead as a result to a change in the balance of power generation and consumption and so lead to a change in the system frequency.

In theory, all these need to be studied by the control engineer, but in practice due to practical constraints only the cases with the most severe effects can be examined.

1.7 The monitoring of the system frequency in real time

1.7.1 Introduction

The frequency of a large interconnected power system such as a national transmission network is monitored on a continuous basis, usually from the location of a central control center of some kind. This close tracking of the frequency is essential for ensuring that the power frequency always remains within the required limits designed for the system.

1.7.2 Daily demand variations

During normal, steady-state operation of an intact system, as noted earlier, we do not expect the power system frequency to change very much from minute to minute.

However, at certain times in the daily load cycle, changes in demand will be relatively high, for example, traditionally, in the early morning during the working week when the load increases rather quickly ("morning pickup") and in the early evening during the working week when the load drops off significantly.

At these times it is necessary for the *system operator* to manage actively the changes in frequency resulting from the changes in demand by instructing generators to increase or decrease their outputs to match the changes in demand. This of course is a planned process.

1.7.3 Unplanned changes

At other times, there may be unplanned changes to the system. These are handled rather differently. In the case of the sudden loss of generation or load, for example, certain items of plant connected to the system (usually generators), which have been designed to change their outputs in response to changes in system frequency, will increase or decrease their power outputs automatically.

These changes in output will oppose the changes in frequency that they detect. Hence, when the power frequency decreases, their power output will increase, and when the power frequency increases, their power output will decrease.

How much a given "responsive" plant changes its power output to a given change in frequency is dependent upon the individual settings chosen for its control system. With the advent of "green" energy sources, increasingly power frequency control is being provided by, for example, wind turbines, HVDC links, batteries, and demand-side management.

1.8 Modern challenges in frequency control

1.8.1 Introduction

In view of the rapidly changing composition of modern power networks, both in the generation mix and in the type of transmission circuits employed (both "AC" and "DC" rather than just "AC"), the modern power system engineer is facing new challenges.

1.8.2 Hybrid systems

Whereas in the past large power systems have been almost completely based on AC and the sources of energy for generating stations have been almost entirely composed of fossil fuels and nuclear fuel, now the nature of generation is becoming very different from this in many parts of the world. Today, as mentioned earlier, the power system engineer must contend with a very broad mix of generation types, both old and new.

We now have coal-fired, oil-fired, and gas-fired units, hydro plant, wind turbines, DC links, batteries, and new demand types such as solar farms and charging stations for electric cars all connected to the same state or national systems. What difference does this new composition make to frequency control?

The short answer to this question is that many of these new, more ecological-friendly types of power generation are very unlike their older counterparts in the way that they affect the power system to which they are connected, and in how they provide frequency response to the system.

1.8.3 Newer types of generation

Whereas the traditional rotating generators are, as it were, ideally suited to an AC system, having both electrical and mechanical inertia and a means of varying quickly their power outputs to counteract changes to the power balance on the system, the same cannot be said for many of the newer types of generation.

For example, DC links do not have any mechanical inertia, although their control systems can be programmed to provide power response. Even wind turbines, although constructed with rotating mechanical machinery, are nowadays mostly connected to power systems by means of back-to-back HVDC converters so that any inertia they have is not transmitted to the power system to which they are connected.

A decrease in available inertia in the case of some power systems can have negative consequences for the stability of these systems.

1.9 Modeling of the power system for analyzing frequency behavior

1.9.1 Introduction

As part of this book, we shall be considering different ways of modeling the "power system frequency problem" by constructing mathematical models of the system for the purposes of analyzing its frequency behavior. We shall try to find the solutions of these models as far as we can.

1.9.2 Aims

One of our main aims in this book will be to try to calculate the frequency response requirements of the power system for scenarios, that is, different levels of generation, demand, and inertia. We shall also consider the assessment of the RoCoF immediately following a system incident involving the loss of generation or demand.

1.9.3 The "swing equation"

With this primary objective in mind, we first develop an approach to the "swing equation" and investigate the different possible ways to solve it. We shall give examples using *MATLAB* throughout the book to illustrate our ideas.

1.9.4 Low inertia systems

In Chapter 10, The challenges of operating systems with high penetrations of renewables ("low inertia systems"), of this book, we shall be examining closely the challenge of obtaining and maintaining control of the frequency in systems that are low in inertia, because, as we shall demonstrate, inertia is a vital ingredient in the ability of a power system to respond successfully to events that trigger sudden deviations in the system frequency.

Therefore it is essential that some reliable means of replacing the inertia lost from the closure of traditional power plants is found as a matter of urgency.

1.10 The next chapter: what can provide frequency control?

In the next chapter, we shall be covering in some detail the following topics: (1) traditional providers of frequency control; (2) frequency response: clarification of terminology; (3) continuous response; (4) step-change response; (5) new providers of frequency control; and (6) the issue of system inertia.

References

Books

[1] P.M. Anderson, B.L. Agrawal, J.E. Van Ness, Sub-synchronous Resonance in Power Systems, IEEE Press, 1990. Under the sponsorship of the IEEE Power Engineering Society, ISBN 0-87942-258-0.

[2] P. Kundur, Power System Stability and Control, Power System Engineering Series, McGraw-Hill Education, 2017. 22nd reprint, Electric Power Research Institute, ISBN-13: 978-0-07-063515-9; ISBN-10: 0-07-063515-3.

Chapter 2

What can provide frequency control?

2.1 Traditional providers of frequency control

2.1.1 Introduction

In this section, we discuss the sources of "frequency response" that have been used for controlling the "system frequency" in the past. These sources of frequency response consist of two main kinds: (1) response from synchronous generator sets regulated by means of a "speed governor" and (2) response that is achieved by tripping blocks of demand.

2.1.2 Governor response: synchronous generators

1. *The use of synchronous generators for response*

   Most types of *synchronous generators* can be set up to provide frequency control for the system to which they are connected. The well-established examples are coal-, oil-, gas-fired and hydroelectric generators. Nuclear generators can also be set up for frequency control, but widespread use of them for this purpose is not made for commercial reasons. To maximize their income from the sale of active power, they usually need to be run at full output outside maintenance periods.
2. *The form of the response characteristic of synchronous generators*

   Synchronous generators can be run in two modes: (1) "frequency responsive" and (2) "nonfrequency responsive."

   When selected to be in "frequency responsive" mode, synchronous generators operate on a negative feedback type, approximately linear, "slope" characteristic which was depicted in Diagram 1.5.

   When the system frequency falls, the power output of the generator rises. Conversely, when the system frequency rises, the power output of the generator falls. This principle was introduced in Chapter 1, The need for frequency control, and is described further in Section 2.2.

   When selected to be in "nonfrequency responsive mode," synchronous generators do not operate on a "slope" characteristic. Their output

Modern Aspects of Power System Frequency Stability and Control.
DOI: https://doi.org/10.1016/B978-0-12-816139-5.00002-3

changes dynamically to imbalances in system generation and demand due to "inertial response" only.

3. *A note on "isochronous" and "nonisochronous" governors*

   As mentioned in Chapter 1, The need for frequency control, traditional generator speed governors are of "isochronous" or "nonisochronous" type. In the case of the "isochronous" type the controller is programmed to keep the frequency as close as possible to a preset frequency, usually the "nominal frequency" for the system. On the other hand, governors with a "droop characteristic" are "nonisochronous" and are programmed such that when the generator speed falls, the MW output increases, as previously described in Chapter 1, The need for frequency control. See Ref. [1], pp. 587 and 589, for more details on these two types of governor.

2.1.3 The use of the jettisoning of blocks of demand (load) for response

1. *Demand tripping*

   During a system frequency fall a block of demand can be set to "trip" to help return the frequency back to normal levels. It is traditionally used as a contingency measure. This method cannot, of course, be used following a system frequency rise, because in that case tripping the demand would cause the frequency to rise further, contrary to what we want to happen. This method of tripping blocks of demand in the "demand tripping" technique is therefore only useful following a fall in the system frequency.

2. *Trigger frequencies*

   In keeping with traditional operational practice, some demand blocks are typically preselected to be tripped when the system frequency falls below a predefined "trigger frequency" level, rather than at a predefined time. Multiple "trigger frequency" levels can be selected whereby different generators are preset at different trigger levels for a "staggered tripping" kind of arrangement.

2.2 Frequency response: a clarification of terminology

The term "frequency response" can be used in either one of two different, but related ways:

1. To describe the *commercial service* provided by generators to help ensure that the system frequency is maintained within operational limits. This will typically involve the commercial details of the price structure applying to different output ranges of a machine.
2. To describe the technical means by which such a commercial service can be provided in practice.

The latter possibility is the sense in which we shall use the term in this book.

In the next two sections we shall discuss two, qualitatively different ways of providing such a frequency response service: "continuous response" and "step-change response." This is related to two different basic *shapes* that response profiles can take.

2.3 Continuous response

2.3.1 A definition of "continuous response"

By "continuous response," we mean that the response service is provided continuously in time, rather than being suddenly turned on, then suddenly turned off. Synchronous generators have in the past typically provided this kind of response.

2.3.2 An example of "continuous response"

Let us imagine that our synchronous generator incorporates, for example, a coal-fired steam turbine. Then a "speed governor" can be included for the purposes of controlling the amount of steam entering the turbine, and therefore controlling also the power output of the machine. Subject to a short time delay, this governor acts continuously (the intervening gap is covered by the natural inertial response of the machine.).

2.3.3 Measurement of the system frequency

The speed governor typically includes a means of measuring the local system frequency at a given time, such that the amount of steam entering the turbine is higher when the frequency is lower and is lower when the frequency is higher, thus regulating the frequency in the desired way. Such an arrangement acts continuously through time, apart from a time lag of typically a second or two (sometimes more depending on the specific design of the machine), and so the frequency response it provides can therefore with some justification be called "continuous response."

2.3.4 The effect of continuous response on the "system frequency"

The power balance of the system is affected relatively gradually via a negative feedback characteristic, and the resulting changes in frequency are relatively smooth. Here presented graphically is what the effect of a continuous response characteristic on the "system frequency" typically looks like Diagram 2.1.

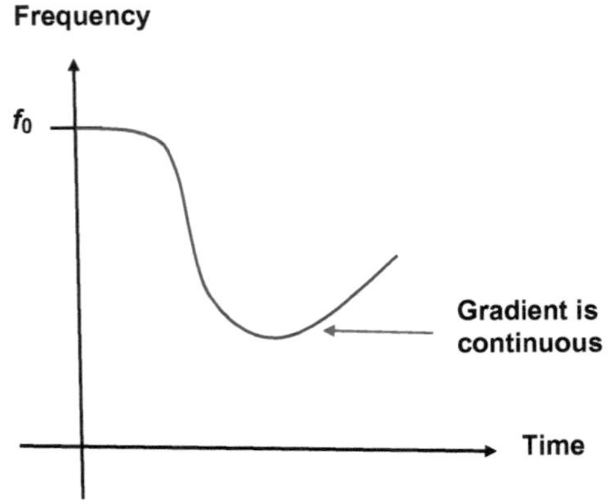

DIAGRAM 2.1 A typical graph of the effect of a continuous response characteristic on the "system frequency."

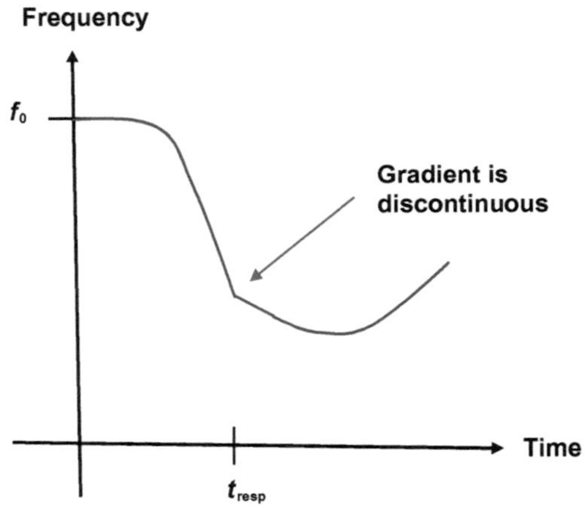

DIAGRAM 2.2 Typical graph of what the effect a step-change response characteristic on the "system frequency" looks like.

2.4 "Step-change" response

This kind of response has, in the past, been provided by the tripping of blocks of demand. (This method applies to low-frequency excursions only as mentioned in Chapter 1, The need for frequency control, and earlier in this chapter.) The power balance of the system is disrupted relatively suddenly, almost instantaneously, and so the change in frequency is angular, not smooth. In other words, although the change on frequency is continuous, the rate of change of frequency (RoCoF) (gradient) is discontinuous.

Here presented graphically is what the effect of a step-change response characteristic on the "system frequency" typically looks like Diagram 2.2.

2.5 New providers of frequency control

Today, of course, there exists a whole range of different technologies, newer than synchronous generator governors or demand tripping relays, which can potentially provide frequency response.

We mention the main ones here. They include the following:

1. high-voltage direct current (HVDC) links;
2. Wind turbine generators (WTGs);
3. Solar photovoltaic (PV);
4. Batteries;
5. Demand-side management (DSM); and
6. Smart grids.

2.5.1 High-voltage direct current links

1. *The step-change response of "HVDC links"*

   "HVDC links" can act like the kind of frequency control achieved by demand shedding, in the sense that the control systems of "HVDC links" can be designed to increase suddenly, or decrease suddenly (in millisecond timescales or faster) their power infeed into a power system, thus providing some measure of frequency control to that system.
2. *Continuous response from "HVDC links"*

   HVDC links have, in fact, been employed in many countries for some time now to provide continuous frequency control and have the capability to provide a droop frequency response like that of synchronous machines.

   However, for the current generation of HVDC devices, the response typically differs in two key respects. First, it is usually faster, taking 1–2 seconds. Second, there is (in the current generation) no inertial response such that in the intervening delay the power output remains constant.
3. *"HVDC links" as a proportion of total system response capability*

   HVDC links usually represent a relatively small proportion of the total system response capability. They are normally employed in conjunction with a much larger proportion that is made up of the more traditional categories of synchronous generator governor response and demand (load) shedding. The relative proportions will vary from system to system.

   See Diagram 2.3 for a simplified line diagram of an "HVDC link."

2.5.2 "Wind turbines" (wind farms)

1. *"Wind turbine" controls*

   Controls can be added to *wind turbines* to enable their power outputs to be varied with changes in the system frequency. This, therefore, enables them to provide "frequency control" in a manner like that of traditional synchronous generators but usually nowadays without the

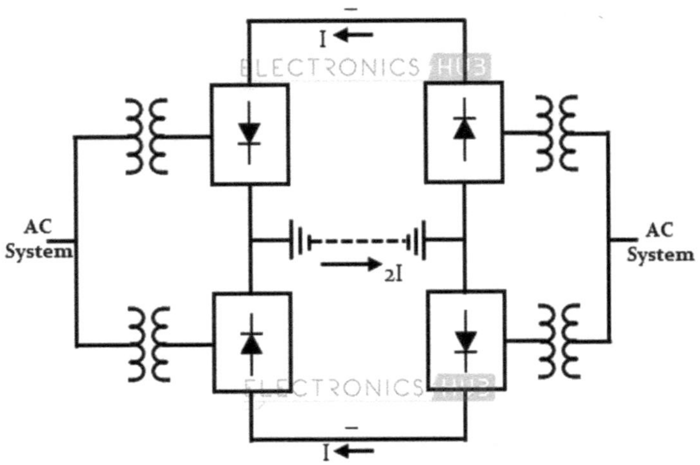

DIAGRAM 2.3 Typical simplified line diagram of an HVDC link. *Reproduced with permission from https://www.electronicshub.org/wp-content/uploads/2015/10/Homopolar-Link-HVDC.jpg.*

corresponding inertia. This feature of modern wind generators is discussed in more detail in the next paragraph.

2. *"Inertia-less power"*

   A salient feature of the great majority, if not all, of modern wind generators is that they *cannot provide any inertia to the power system*, as noted earlier. This point will be expanded upon further later in this book, but the main point here is that only active power, and not inertia, can be made available to the system by these devices for the purposes of providing frequency control. Therefore they are almost by definition less effective than synchronous generators for the purposes of controlling the "system frequency."

3. *The line diagram of a typical "DFIG (doubly fed induction generator)" WTG*

   See Diagram 2.4 for a simplified line diagram of a typical "DFIG" *WTG*.

2.5.3 "Solar photovoltaic" (solar farms)

"Solar PV" systems can also be used as a method of frequency control in a modern, high-voltage (HV) power system. A "solar PV" system consists of an array of solar panels that absorb the light from the sun during sunny periods and return electricity back into the grid during nonsunny periods.

"Solar PV" units can be tripped in response to a frequency monitoring input signal in a similar way to demand blocks. The important difference to bear in mind is that tripping "solar PV" is tripping an amount of generation,

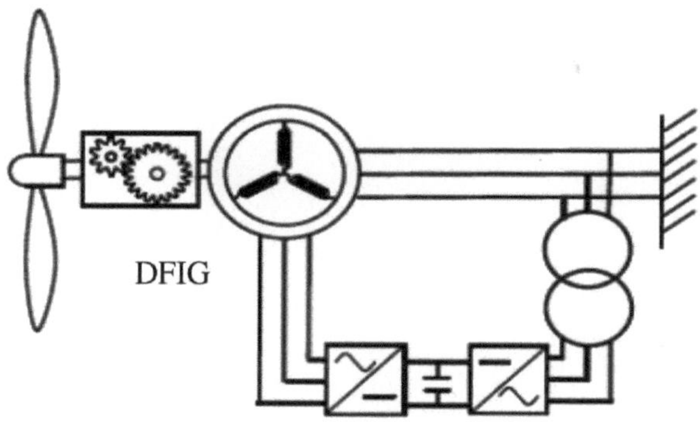

DIAGRAM 2.4 Typical simplified line diagram of a "DFIG" wind turbine generator. *DFIG*, Doubly fed induction generator. *Courtesy kalyan07wordpress.com.*

not tripping a demand; so the resulting system frequency change is in the opposite direction, that is, down not up.

An additional possibility for a "solar PV" system is that of providing "droop control" like that of the governor system of a synchronous generator. In this case the economic viability of this option would have to be explored to ensure that the product was cost effective.

2.5.4 High-voltage-connected batteries

Various types of *large batteries*, in conjunction with suitable control systems, can be used to provide frequency control. Moreover, when coupled with an inverter, the inverter can provide reactive power (voltage control) to the system. In this way, both "real" and "reactive" power regulation capabilities can be provided, giving an overall control capability like that of a synchronous generator. See Diagram 2.5 for a diagram of a large battery in conjunction with an inverter.

2.5.5 Demand-side management

2.5.5.1 What is "demand-side management"?

"Demand-side management" (DSM), also known as "demand-side response," is a term used to describe the regulation of consumer energy demand patterns, by techniques such as financial incentives and educational initiatives (see, for example, Ref. [2] for more details). The idea is to provide incentives at specific times which modify the demand in real time.

2.5.5.2 The objectives of "demand-side management"

One objective of "DSM" is to encourage less usage of electrical energy during peak hours by shifting the distribution of demand toward off-peak times such as overnight and at the weekends.

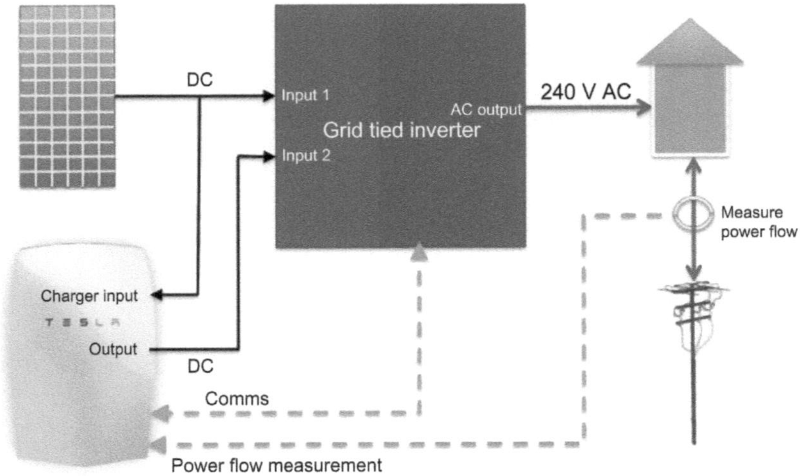

DIAGRAM 2.5 Schematic diagram of a large battery in conjunction with an inverter and solar PV installation. *PV*, Photovoltaic. *Courtesy www.solarquotes.com.au.*

Such a process does not necessarily decrease the total energy consumption but can be expected to reduce the investment needed in transmission networks and the running costs of power plant needed for times of peak demand.

It can also make the task of controlling the "system frequency" easier for the control engineers in national control centers.

2.5.5.3 Some examples of "demand-side management"

1. *Energy storage*

   An example of "DSM" is the use of energy storage units (e.g., batteries) to store energy during off-peak hours and discharge them later, during peak hours.
2. *Electric vehicles (EV)*

   Another example of "DSM" is in the use of rechargeable "EV." Thus the EV are charged at purpose-built charging stations, disperse their charge during usage of the vehicle, and then charge up again at the charging station and the cycle continues.

2.5.5.4 Some applications of "demand-side management"

1. *Compensating for intermittency*

   A recent application for "DSM" is to aid grid operators in balancing intermittent generation from wind and solar units, particularly when the timing and magnitude of energy demands does not coincide with those of renewable generation.

2. *"Smart" appliances*

    Another application of "DSM" involves the remote control of domestic appliances such as refrigerators. These can be switched on or off remotely to increase or decrease the power system demand as required.

2.5.6 Smart grids

2.5.6.1 What is a "smart grid"?

A "smart grid" may be described as an electrical supply network that uses digital communications technology to detect and react to local changes in usage. "Smart grids" can in principle be used by the "system operator" to engineer part of the system demand pattern for the purposes of improved frequency control.

2.5.6.2 Why are "smart grids" important?

"Smart," controllable grids can be used to optimize the use of available infrastructure. Provided all the technological challenges can be overcome satisfactorily, the real potential benefit of "smart grids" should be the greater controllability over the generation to demand balance, and hence over the system frequency, than has been possible to date.

2.5.6.3 How do "smart grids" work?

1. *The hardware*

    A "smart grid" employs a variety of operational and energy-related measures such as "smart meters," "smart appliances," "renewable energy resources," and "energy-efficient resources."

2. *The software*

    "Smart grids" also use what may be termed "smart control systems" that make use of up-to-date programming tools such as "artificial intelligence" techniques such as "expert systems" to optimize the usage of the grid.

3. *References*

    The reader is directed to, for example, Refs. [3,4] at the end of this chapter for more detailed information about "smart grids."

4. *The connection of "smart grids" with frequency control*

    The impact that "smart grids" can potentially have on frequency control is covered in more detail later in Chapter 11, "Smart grids" and the "system frequency." The basic idea is that by optimizing all available resources the daily "demand cycle" may be smoothed out; and hence, the "system frequency" may be controlled more easily.

2.6 The issue of system inertia

2.6.1 The factors necessary for achieving effective "frequency control"

There are several factors that work together to produce effective "frequency control," which, as we have mentioned, depends fundamentally on the balance between active power generation and demand.

These factors may be considered under the following headings:

2.6.1.1 Regulation of the output power of generators

One of these factors is the capability to regulate the *active power output* of generating machines (including nonsynchronous machines). This means the ability of a power source to counteract any disturbance in the power system, usually caused by either a loss of generation or load elsewhere.

2.6.1.2 Modification of the system demand

Another factor in the ability to control the power system frequency effectively is the ability to modify the *system demand.*

As discussed earlier, one way to help to restore the system frequency following a low-frequency event, such as the loss of a generating set, is to reduce the system demand by "demand (load) tripping." By tripping preselected blocks of demand at defined "trip frequencies," the system power balance may be, at least partially, restored to its normal operating level. This method, as previously mentioned, is only effective against the loss of generation.

2.6.1.3 Use of "modern technologies" to alter the power balance

As discussed earlier, "modern technologies" such as "HVDC links," "WTGs," and solar "PV" technology may be employed, with suitable controllers, to regulate the system frequency.

2.6.1.4 The time factor

1. *The rate of change of power output*

   Yet another factor important for producing effective frequency control is the *time factor*. As hinted at earlier, the rate at which the power output of a "responsive device" is changing is also important, both in terms of the effectiveness of the response provided and for the subsequent stability of the power system.
2. *Example: the "synchronous generator"*

   For example, in the case of a "synchronous generator" with an active speed governor, its effect on the power system frequency is determined by the gradient of its "droop profile" and by its "ramp rate," that is, by how many *MW per second* it can increase its active power output. Both these factors are important.

3. *Example: "demand tripping"*

   In the case of "demand tripping" the *precise timing of the trip event* as well as the *magnitude of the demand tripped* are critical to the subsequent behavior of the power system. If a demand block is tripped too early, instability may ensue. On the other hand, if a demand block is tripped too late, it may not be possible to bring the frequency back to its "prefault" level in sufficient time to meet requirements.

2.6.1.5 The "inertia" factor

A fourth factor to be considered is one which is, perhaps, sometimes taken for granted in systems dominated largely by generators that have adequate amounts of rotating machinery in them (i.e., synchronous machines), but it needs to be considered carefully in systems that have not (e.g., systems with large amounts of "solar PV"), that is, the "inertia" of the system.

We shall now go on to examine in more detail the question of inertia: what is it, and what role does it play in the stability of the power system?

2.6.2 A short discussion on the need for "inertia" in power systems

2.6.2.1 What is "inertia"?

"Inertia" is commonly understood to be an innate resistance to the change of motion possessed by a physical object. In the case of a "traditional" alternating current (AC) generator, which has solid rotating components (the rotor, alternator, turbines, and so on), the rotational speed of which determines the electrical power frequency, this "inertia" is embodied in the mass of its rotating parts. Due to their "inertia," these rotating parts exhibit a resistance to any change in their speed of rotation, whether this be an increase or a decrease. There is an electromagnetic coupling between the rotating parts of the machine and the electrical network that is also stabilizing.

2.6.2.2 *Generators* with *massive, rotating parts*

As mentioned earlier, a generator whose construction is based on rotating parts resists any change to the power frequency, whether positive or negative. In other words, such a generator, inherent in the way it is designed, provides frequency control in a natural way to the power system to which it is connected.

2.6.2.3 *Generators* without *massive, rotating parts*

By comparison, consider the case of an electrical generator that does not have any rotating parts. It has only one way of controlling the system frequency, that is, by changing its power output. Power sources in this category include "solar PV," "HV batteries," and "HVDC." We can say that because they have no rotating parts, they are inherently less stable than power sources that do and require additional means of stabilization to function properly in an "AC" system.

2.6.2.4 *Generators with massive, rotating parts whose inertia is inaccessible to the grid*

From the point of view of the wider system, for example, a modern wind generator has no inertia even though it has massive, rotating parts. This is because a modern wind generator is connected to the power system by a back-to-back "HVDC" convertor. Because of this, power mismatches on the grid system do not affect the speed of the rotor, and we can therefore say that its inertia cannot be "seen" by the system to which it is connected. This situation would apply to any rotating plant separated from the main system by back-to-back convertors, such as the popular "DFIG" design mentioned earlier in the chapter.

2.6.2.5 *The requirement for a minimum system inertia*

1. We can say that a certain minimum amount of inertia is required to support any given power system so that it has the capacity to remain stable after it becomes unbalanced in some way, for example, when a generating set is suddenly disconnected from the system following a fault.
2. Some work has already been done in this area for the British system (see Ref. [8]). The results of these studies show that, for a variety of system loading conditions, below a certain value for the "system inertia" the system is transiently unstable.

2.6.2.6 *Synthetic inertia*

"Synthetic inertia" is a relatively recent concept. The idea behind it is that by generating an appropriate amount of power from a nonsynchronous source, the effect of inertia can be simulated during the "RoCoF" period following a system power imbalance event (see Ref. [3] for a discussion on this concept).

However, there are limitations to the practicability of this concept in its original form. It has been discovered that the idea is valid for the "RoCoF" phase but needs modification in the form of the addition of a feedback loop for later timescales. Unfortunately, such a feedback loop incurs unwanted delays that are destabilizing.

An important concept of a real synchronous machine that needs to be included in any simulation is the "inertial response," that is, the immediate, natural response of a synchronous machine to a power imbalance before the governor response takes over after a few seconds when the "inertial response" has died away.

The ultimate argument for "synthetic inertia" being equivalent to actual physical inertia is that it is possible to synthesize a power waveform that resembles to all intents and purposes the power waveform produced by an actual synchronous machine following a fault, namely a "virtual synchronous machine (VSM)."

Research work is currently being done in this area to ascertain whether the "VSM" can really double as a "synchronous machine" where frequency stability is concerned (see, for example, Ref. [9] at the end of this chapter for more details on this).

2.6.2.7 Fast frequency response

However, you may ask, what if we can have arbitrarily fast-acting electronic controls on, say, an "HVDC link" or a "back-to-back convertor"? Would this capability then solve the problem of low-inertia systems? (For a discussion of the difference in definitions between the terms "synthetic inertia" and "fast frequency response," consult Refs. [4], [6] and [7] for more details.)

2.6.2.8 Possible drawbacks of "fast frequency response"

As we know historically, in the case of generator "automatic voltage regulators" that act to modulate power system voltage, acting quickly is effective in countering transient changes, but acting too quickly can trigger oscillatory (small signal) instabilities.

Therefore we must tune the speed of response "appropriately to the situation." Does this solve our problems? Well, perhaps, provided we can simulate damping electronically.

2.6.2.9 The insufficiency of power changes on their own

From the earlier discussion, it is suggested that it is clearly not enough to counter frequency changes solely using power output changes on their own. This is, fundamentally speaking, because power output changes on their own without the assistance of enough system inertia (actual or otherwise) could possibly lead to system-wide instabilities.

2.6.2.10 System damping requirements

A related issue to "system inertia" is *system damping*. If an oscillation is *poorly damped*, even if the power frequency returns eventually to normal operational levels, it may do so in a highly unstable way, for example, with prolonged oscillations in between, which is clearly unacceptable.

Any oscillations which arise must be sufficiently *well damped* so that good control of the power system is always maintained. See Ref. [5] for a discussion of system damping in relation to fast frequency response.

2.6.2.11 Some further potential drawbacks of "purely electronic" response

Let us assume that we can achieve, by purely electronic means, a satisfactory degree of damping for all possible unwanted oscillations that may follow a planned power injection for the purposes of regulating the system frequency. Is there anything left that we should consider?

The answer is: well, yes, at least two things:

1. The power response on HV electronic equipment has a *sharp cutoff* at each end of its range, such that once the range of power is exhausted it can "go no further," unlike the case with AC generators, for example, which are capable of considerable operation beyond its normal range.
2. During the operation of HV electronic equipment such as HVDC links, back-to-back converters, and so on, there is a risk of rectifier or inverter faults such as *commutation failure*. These can cause a unit to shut down completely, thus removing it, and consequently its effect, from the system.

2.6.2.12 Current research activity

Since much research is still going on in the field of power electronics and its use to regulate the system frequency (see, for example, Ref. [9]) and other system parameters in new ways, particularly in the rapidly expanding field of "smart grids," it remains to be seen just how far a system *can* remain stable following an event on an AC system that is deficient of inertia (a "low-inertia system"). This is currently a very interesting area of research whose success will be critical for the satisfactory operation of power systems large and small.

2.7 The next chapter: per unit systems for frequency analysis

In the next chapter, we "come back down to earth again" to begin to look at how to model power systems for the purposes of frequency stability and control. The following topics will be covered:

1. per unit systems for individual machines;
2. per unit systems and the power system at large;
3. how the foregoing analysis may be used in practice; and
4. the fundamental relationship between frequency stability and inertia.

References

Books

[1] Power system stability and control, in: P. Kundur, Power System Engineering Series, McGraw-Hill Education, Electric Power Research Institute, 22nd Reprint, 2017. 978-0-07-063515-9; ISBN: 10: 0-07-063515-3.

[2] S.F. Bush, Smart grid: communication-enabled intelligence for the electric power grid, Demand-Response and the Advanced Metering Infrastructure, Wiley-IEEE Press, 20139781118820216.

[3] J. Zhu, Optimization of power system operation, Operation of Smart Grid, Wiley-IEEE Press, 20159781118887004.

[4] M. La Scala (Ed.), From Smart Grids to Smart Cities, Vol. 2 of Electrical Engineering Series, Advanced Smart Grids Set, Wiley/iSTE, 2017. ISBN 978-1-84821-749-2.

Articles and Papers

[5] P. Fairley, Can synthetic inertia from wind power stabilize grids?, IEEE Spectr. November 7, 2016.

[6] R. Eriksson, N. Modig, K. Elkington, Synthetic inertia versus fast frequency response: a definition, IET Renewable Gener. 12 (5) (2018) 507–514.

[7] K. Dallmer-Zerbe, E. Spahic, G. Kuhn, R. Morgenstern, G. Beck, Fast frequency response in the UK grid—challenges and solution, in: 13th IET International Conference on AC and DC Power Transmission (AC DC 2017), 2017.

[8] M. Yu, A. Dyśko, A. Roscoe, C. Booth, R. Ierna, H. Urdal, J. Zhu, Effects of swing equation-based inertial response (SEBIR) control on penetration limits of non-synchronous generation in the GB power system, in: International Conference on Renewable Power Generation, 2015 (RPG 2015), 2015.

[9] J. Zhu, J. Hu, W. Hung, C. Wang, X. Zhang, S. Bu, et al., Synthetic inertia control strategy for doubly-fed induction generator wind turbine generators using lithium-ion supercapacitors, IEEE Trans. Energy Convers. 33 (2) (2018) 773–783.

Chapter 3

Per unit systems for frequency analysis

3.1 Per unit systems: an introduction

3.1.1 What is a per unit system?

A "per unit system" is a system that expresses each of the parameters or variables of a system in terms of the ratio of a "physical quantity" to a "base quantity" to produce a "per unit quantity." Thus, for example, if a "physical quantity" happens to be 250 MW (megawatts) and the "base quantity" for the system is 100 MW, then the quantity can, alternatively, be expressed as being 250 MW/100 MW = 2.5 per unit.

3.1.2 Dimensions of "per unit" quantities

Usually, but not always, "per unit quantities" are "dimensionless," that is, they have no physical units; they are simply a number by itself. In frequency stability studies a common exception to this rule is the "inertia constant," usually designated by the symbol H, which has units of *seconds (s)*, which is sometimes alternatively, but equivalently, expressed in the alternative units of *MW s/MVA*. Another example is time, usually represented by the symbol "t," which is almost invariably left in units of seconds.

3.1.3 Why do we need per unit quantities?

1. *Removal of physical units*

   Per unit quantities can be useful to us in two main ways. First, by using per unit quantities, equations are usually simplified, including *removing most of the physical units from them.*
2. *Numerical stability considerations*

   Second, a parameter or variable expressed in per unit form is *often close in value to 1*. As one consequence of this, the "numerical stability" of any subsequent calculations we may do using per unit quantities is increased significantly compared with using physical units where

Modern Aspects of Power System Frequency Stability and Control.
DOI: https://doi.org/10.1016/B978-0-12-816139-5.00003-5

numbers are often of greatly varying magnitudes attracting numerical instability, either because values of quantities are very small, or very large, or a mixture of the two.

3.2 Per unit systems: individual machines

3.2.1 The individual machine

1. *Available per unit systems for the individual machine*

   When we are considering only an *individual machine* (e.g., a synchronous generator), there are several well-established systems available for defining per unit quantities for the physical quantities involved in the technical description of the machine (see, for example, please see Ref. [1]).
2. *The physical quantities covered by existing per unit systems*

   The physical quantities covered by existing per unit systems typically cover a wide variety of electromagnetic and mechanical quantities, for example, moments of inertia, powers (real and reactive), torques, currents, voltages, inductances, reactances, resistances, magnetic fluxes, speeds, and time constants (for more details on this please see, for example, see Ref. [1], Section 3.4).
3. *Nonsuitability of individual machine quantities for whole system studies*

   Most of these quantities, however, are only meaningful for modeling the individual machine in question and are not suitable for application to the wider system. There are, however, ways of doing this as well (see Section 3.3, which describes one method of doing this). This latter approach makes it possible to treat the whole system as a single entity for the purposes of frequency response studies.

3.2.2 An example of defining a "per unit" quantity for frequency studies

As an example of defining an "individual machine" per unit system, to convert the "moment of inertia" of a single machine from physical units to "per unit" quantities, the standard way to proceed is to divide the *rotational kinetic energy K* of the machine at rated speed by its *voltampere base (S)* to form the *inertia constant (H)*:

$$H \equiv \frac{K}{S} = \frac{(1/2)J\omega_{0m}^2}{\mathrm{MVA_{base}}} \tag{3.1}$$

in which ω_{0m} denotes the rated rotor speed in *mechanical radians per second* (see, for example, Ref. [1], Section 3.9.2). As remarked on earlier, the resulting per unit quantity *H* has the units of *MW s/MVA* or, equivalently, *seconds*.

3.3 Per unit systems and the power system at large

When we are considering a "bulk system," that is, one that has many machines connected to it, we can still define per unit quantities for this system, in addition to the per unit quantities for the individual machines connected to it. In the context of analyzing such a system for frequency stability and control analyses, we are primarily interested in the *active power outputs* (MW generated) and *inertias* (i.e., the "moments of inertia") making up the system. We shall look first at the topic of the active power outputs.

3.3.1 Active power outputs

1. *A bulk generator base quantity*

   In the case of active power output, we can add together the active power outputs of individual generating machines to produce a total generated power output for the system, which can then be used legitimately as a "base quantity" that is representative of the bulk system. There is no problem with that.
2. On the other hand, we could equally well add together the individual blocks of active *power demands* connected to the system to produce another base quantity that is also representative of the bulk system. Either would do since, discounting power transfers to and from adjoining systems, these are roughly the same and therefore a good representation of our system.

3.3.2 System mega-voltampere rating

Each synchronous machine connected to the system has its own, individual *MVA rating*, equal to its *rated power output* divided by its *rated power factor (lagging)*. While it is not correct to define a system *MVA rating* by adding together the *MVA ratings* of the individual machines connected to it, one can nevertheless, for example, define a *system MVA rating* by dividing the sum of MW outputs of individual machines in it by a kind of "average" system power factor (see Section 3.3.3 for more details on this).

3.3.3 Inertia ("moment of inertia") J_{ind}

The rotor assembly of each individual synchronous generating machine connected to the system, each one consisting of the alternator, turbines, connecting sections, and so on and has its own moment of inertia which we can call, for our purposes, J_{ind}. The whole assembly looks something like that depicted in Diagram 3.1, which is of a hydroelectric plant, and so the rotor assembly is configured vertically. Most other plants have the rotor assembly configured horizontally.

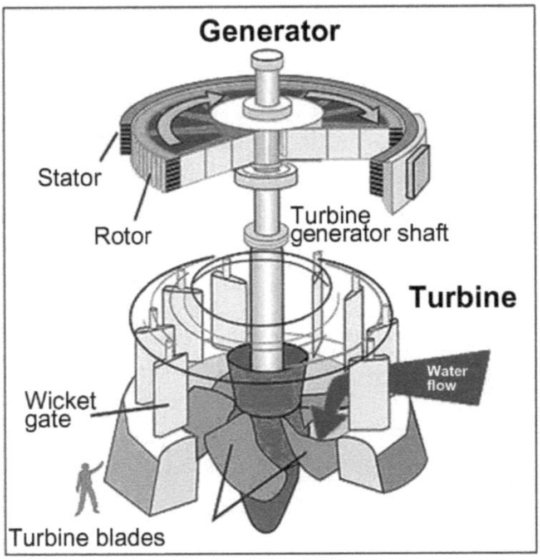

DIAGRAM 3.1 An "AC" generating machine rotor assembly (hydroelectric plant). *US Department of Energy.*

The "moment of inertia" of the rotor assembly of a generating machine is embodied in the definition of the "rated kinetic energy" of its rotor assembly, as follows:

$$\mathrm{KE_{rated}} = \frac{1}{2} \cdot \sum J_{\mathrm{ind}} \omega_{\mathrm{rated}}{}^2 \tag{3.2}$$

in which ω_{rated} is the machine's rated "speed" in radians per second.

3.3.4 The incorrectness of adding together the inertias

It is *not correct* to add together the moments of inertia of the individual machines connected to the system to produce an "equivalent system inertia."

However, it *is quite correct* to add together the rated (rotational) kinetic energies of all the synchronous machines in the system to obtain a *total system rated rotational kinetic energy*. (This is sometimes referred to as the "MVA × *H*" of the system.) Therefore from the earlier argument, a valid inertia representing the whole system may be defined as follows.

3.3.5 Adding together the kinetic energies of the individual machines

As an example, suppose we have three synchronous generators connected to a system whose individual kinetic energies are $\mathrm{KE_1}$, $\mathrm{KE_2}$, and $\mathrm{KE_3}$. Then the total kinetic energy of these generators is given by the expression:

$$\begin{aligned} \mathrm{KE_{total}} &= \mathrm{KE_1} + \mathrm{KE_2} + \mathrm{KE_3} \\ &= \frac{1}{2} J_{\mathrm{ind1}} \omega_{\mathrm{rated1}}{}^2 + \frac{1}{2} J_{\mathrm{ind2}} \omega_{\mathrm{rated2}}{}^2 + \frac{1}{2} J_{\mathrm{ind3}} \omega_{\mathrm{rated3}}{}^2 \end{aligned} \tag{3.3}$$

We have assumed here that the system is in the "steady state," that is, in rotational equilibrium, so that all the rated rotational speeds are the same, hence

$$KE_{total} = \frac{1}{2}\omega_{rated}(J_{ind1} + J_{ind2} + J_{ind3}) \tag{3.4}$$

To define an *H*-constant for the whole system in a similar way to that for an individual generating machine, we can now proceed in the following way.

3.3.6 Calculating an "average power factor"

First, we can define an "average power factor" for the system, as follows:

The rated power factor pf_{rated} for an *individual machine* is given by

$$pf_{rated} = \frac{P_{rated}}{S_{rated}} \tag{3.5}$$

where P_{rated} denotes the rated MW power output, and S_{rated} the *MVA* rating of each machine. So now, for our three-machine system we can write

$$P_{total} = P_1 + P_2 + P_3 = pf_1 \cdot S_1 + pf_2 \cdot S_2 + pf_3 \cdot S_3 \tag{3.6}$$

$$= pf_{average} \cdot S_{average} \tag{3.7}$$

The "average power factor" can be defined as the sum of the power factors of the individual machines, weighted by their individual MW power ratings. Thus

$$pf_{average} = \frac{pf_1 \cdot P_1 + pf_2 \cdot P_2 + pf_3 \cdot P_3}{P_{total}} \tag{3.8}$$

Combining Eqs. (3.7) and (3.8), by eliminating $pf_{average}$, and rearranging the resulting expression, we obtain an expression for the "total voltampere rating" of the system:

$$S_{total} = \frac{P_{total}^2}{pf_1 \cdot P_1 + pf_2 \cdot P_2 + pf_3 \cdot P_3} \tag{3.9}$$

We note that the right-hand side of Eq. (3.9) is made up entirely of known parameters.

3.3.7 Defining an "*H*-constant" for the system as a whole

The final step in defining a per unit inertia constant (*H*-constant) for the *whole system* is to divide Eq. (3.4) by Eq. (3.9) to give

$$\begin{aligned} H_{system} &= \frac{KE_{total}}{S_{total}} \\ &= \frac{1}{2}\omega_{rated}(J_{ind1} + J_{ind2} + J_{ind3}) \cdot \frac{(pf_1 \cdot P_1 + pf_2 \cdot P_2 + pf_3 \cdot P_3)}{P_{total}^2} \end{aligned} \tag{3.10}$$

Notice how Eq. (3.10), which represents multiple machines, has the same general form as Eq. (3.1), which represents a single machine.

3.4 What is the practical use of the above analysis?

3.4.1 The answer to the question

What is the practical use of the above analysis, you may well ask. The answer to the question is that the point of all the above analysis is that it enables us to assign inertia constants H to bulk systems and hence to compare directly their inertias. By this means, we can tell at a glance if a power system we are analyzing has a high or low inertia compared with that of another system, or with the same system with a different scenario of generation selection or loading level, or both.

3.4.2 Looking further ahead in the book

In Chapters 6–10 of this book, we shall be analyzing the frequency stability of power systems in different ways; the inertia constant H of the system we are analyzing at any time will be an essential and critical parameter in all our calculations.

3.4.3 In the next section

In the next section, we shall be considering the question of the fundamental relationship between "frequency stability" and "inertia."

3.5 The fundamental relationship between "frequency stability" and "system inertia"

3.5.1 An intuitive definition

The relationship between "system inertia" and "frequency stability" is a fundamental one. In basic terms the more inertia that a system has the more stable to changes to frequency it will be. Conversely, if we do not have enough inertia on the system then we are at risk of the system becoming unstable. All this may seem rather obvious, but, nevertheless, we think this should be stated "for the record."

3.5.2 Can we run an "AC" system without any inertia?

As discussed in Chapter 2, What can provide frequency control?, much research has been undertaken over the last few years into ways of electronically simulating the effects of inertia. However, as we have seen in that chapter that the employment of what may be called "fast frequency response" techniques is unlikely to replace *completely* the need for inertia in "AC" systems; for the reason stated there, *inertia* inherently resists any change in frequency, whether an increase or decrease, whereas a *power injection* can only correct the frequency one way or the other, not both ways at the same time.

3.5.3 The "system frequency" in "inertia-less systems"

If there is no inertia at all in the system, how can we define a "system frequency"? Even if we were able to simulate a "system frequency" by electronic means, what value should we select it to have?

3.6 In the next chapter: the initial analysis of the frequency stability problem and a derivation of the "swing equation"

3.6.1 The fundamental system power balance

In the next chapter, we shall begin by looking at the fundamental balance on the power system: what is in balance and what happens when this balance is disturbed? We shall look at this in different timescales, from long-range planning to second-by-second operation. We shall take a close look at what is happening physically inside a synchronous machine during an imbalance of power on the system.

3.6.2 Analysis of the frequency stability problem "from first principles"

From this point of focus, we shall proceed to analyze the frequency stability problem "from first principles." This will entail starting from the basic physics underlying the problem, and as a result we shall derive the most fundamental form of the "swing equation," which will be at the core of the investigative analyses that we shall perform in the later chapters of this book.

Reference

Book

[1] Power system stability and control, in: P. Kundur, Electric Power Research Institute Power System Engineering Series, McGraw Hill Education, 22nd Reprint, 2017.

Chapter 4

Initial analysis of the frequency control problem and a derivation of the swing equation

4.1 Introduction

4.1.1 The ultimate aim of this chapter: deriving the "swing equation"

In this chapter, we shall ultimately be looking at how to derive the "swing equation" from "first principles." We shall be employing a "rate of change of energy" approach and applying it to our problem of power input and power output with a rotating mass in between.

4.1.2 In the next sections

In the next section, but one, we shall be looking at the elements of the fundamental power balance in the intact system. After that we shall be looking at the power balance in the system following a loss of generation or demand.

Before these topics, however, in the next section, we shall be clearing up a possible ambiguity in terminology before going on to discuss power balances both at the individual machine level and in the wider system.

Finally, we end the chapter by deriving the "swing equation" itself from "first principles."

4.2 The clarification of some terminology

4.2.1 The "intact power system"

Let us first clarify some terminologies that are often used, perhaps sometimes a little loosely: what do we mean by the "intact power system"? There are two main answers to this question: one relating to longer term (planning) timescales and the second to shorter term (operational) timescales.

Modern Aspects of Power System Frequency Stability and Control.
DOI: https://doi.org/10.1016/B978-0-12-816139-5.00004-7

It is important to be clear about which one we are referring to at any given time, since they have potentially different meanings depending on the specific context.

4.2.2 Longer term (planning) timescales

In longer term timescales, say more than a year out from actual operation, when the details of the specific planned (maintenance) outage schedule of an electrical network are unknown, we usually mean by "intact network" the complete network, with *neither planned outages* (maintenance outages) *nor unplanned (fault) outages* (also often referred to as "trips" in colloquial system operations usage).

4.2.3 Shorter term (operational) timescales

In shorter term timescales, say less than a year out from actual operation, when the network model already contains the details of any proposed planned (maintenance) outage patterns or schedules, by "intact network" is normally meant the same as "prefault network," that is, "maintenance outages" have been included in the model, but "trips" have not.

4.2.4 In the next section

In the next section, we shall discuss the elements making up the active power balance in a system, first for the wider system and then for individual machines.

4.3 Elements of the fundamental balance in the intact power system

4.3.1 The power balance across the wider system

1. *The power balance for the wider system*

   The wider system can be "balanced" or "unbalanced" at any chosen point in time. Ignoring internal machine dynamics and any individual phase imbalances that may exist, the system is "balanced" if the *power generated equals the loads connected to it plus an allowance for losses.* The system is "unbalanced" if this is not the case.
2. *Dynamic power changes on the wider system*

   Due to relatively small, dynamic changes, even in the absence of faults, the balance of a power system is constantly changing, with the generated power rarely *exactly* equaling the demand (i.e., the load plus losses). This is evidenced in real time by the system power frequency wandering in a relatively small band around the *rated frequency* for the system.

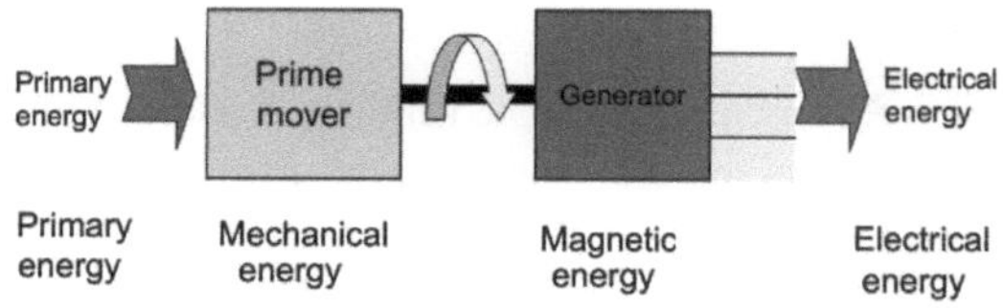

DIAGRAM 4.1 The individual machine power balance. *US Department of Energy.*

4.3.2 The individual machine power balance

1. *The balance of electrical power output and mechanical power input*

   On the level of an individual machine, there is of course more going on inside it than is evident in the wider system beyond its terminals. However, within a synchronous machine, there exists the basic balance between *electrical power output and mechanical power input.* In the absence of faults, these are approximately equal in magnitude (see Diagram 4.1).

   On the left-hand side of Diagram 4.1, we have the mechanical input power generated by the *prime mover*, and on the right-hand side the output power (the load) associated with the *electrical output* of the machine. For the purposes of our further analysis, let us call these two powers P_m and P_e, respectively.
2. *Machine torques*

   In terms of the *torques* $\omega \cdot P_m$ and $\omega \cdot P_e$ acting on the machine shaft, these are equal and opposite, resulting in a nonaccelerated motion of the shaft in accordance with "Newton's second law of motion" applied in a rotating system such as we have here.

4.4 The imbalance following a system loss

4.4.1 Internal power balance

Using the terminology used in the previous paragraph, whenever P_m equals P_e, the rotating parts of the generator (considered a single mass) will move at a constant angular speed, that is, there will be no angular acceleration. Let us call the standard (rated) speed ω_{0m} (for "omega-zero mechanical"). This speed is measured in "mechanical radians per second." This is the speed for which the machine is designed. (It should be noted that there can be a power balance at other speeds, either greater or less than the design speed.)

If, on the other hand, P_m does not equal P_e, then the rotational speed of the rotor—alternator assembly will not be constant, but the assembly will experience an *acceleration* or *deceleration*, which of these occurs will be depend on the relative magnitude of the two powers.

This situation of imbalance occurs whenever there is an unplanned loss of either generation or demand (load) from the system.

4.4.2 Rotational kinetic energy

The kinetic energy K_{rot} contained in a rotating mass having a moment of inertia J, and a mechanical angular speed ω_m is given by the well-established formula:

$$K_{\mathrm{rot}} = \frac{1}{2} \cdot J \cdot \omega_m^2 \tag{4.1}$$

(e.g., see Ref. [1]).

4.4.3 Residual rotational power

An imbalance between the mechanical power input P_m and the electrical power output P_e of a synchronous machine results in a residual rotational power, which we can call P_{res}, which is the difference of these two powers. Hence

$$P_{\mathrm{res}} = P_m - P_e \tag{4.2}$$

This residual power P_{res} goes into accelerating or decelerating the rotor assembly and is equal to the rate of change of the kinetic energy K_{rot} with time of the rotor assembly as defined by Eq. (4.1).

We now move to a derivation of the basic equation that governs rotor oscillations in an electrical generator, namely the "swing equation."

4.4.4 A step by step derivation of the "swing equation"

We know from our knowledge of basic physics that the difference between the mechanical input power and the electrical output power of the machine is equal to the rate of change of the kinetic energy K_{rot} of the rotor assembly (treated as a single mass) with time.

Hence, using Eqs. (4.1) and (4.2) and the symbols we defined earlier, we have

$$\begin{aligned} P_m - P_e &= \frac{\mathrm{d}}{\mathrm{d}t}\left(\frac{1}{2} \cdot J \cdot \omega_m^2\right) \\ &= \frac{1}{2} \cdot J \cdot \frac{\mathrm{d}\left(\omega_m^2\right)}{\mathrm{d}t} \\ &= \frac{1}{2} \cdot J \cdot 2\omega_m \cdot \frac{\mathrm{d}\omega_m}{\mathrm{d}t} \end{aligned} \tag{4.3}$$

Hence,

$$P_m - P_e = J \cdot \omega_m \cdot \left(\frac{\mathrm{d}\omega_m}{\mathrm{d}t}\right) \tag{4.4}$$

4.4.5 A common assumption, but is it really needed?

If you were familiar the *swing equation* already, you may have noticed that there appears to be an "extra" ω_m just in front of the derivative term. This is not a mistake. To simplify Eq. (4.4) from a nonlinear differential equation into a linear differential equation that is easier to solve, the approximation is usually made that this "extra" ω_m is assumed to be equal to $\omega_{m \cdot \text{rated}}$, so that in per unit terms it will be equal to 1 and so apparently "disappear" from the equation.

As far as the author is aware, in applications the foregoing assumption has always been made in the past; but in Chapter 6, Analytical solutions, of this book, we shall attempt to retain the original form given by Eq. (4.4) to try to gain more accuracy in our calculations. How successful we shall be in this objective we shall find out later in the book.

4.4.6 In the next section

In the next section, we shall move on to deriving a "per unit" form of the "swing equation" that will be useful to us in our numerical calculations later in this book.

4.4.7 The "per unit" form of the "swing equation"

Let us now take Eq. (4.4) and convert it into a useful "per unit" version to facilitate our later calculations. First, we need to define some base values.

4.4.7.1 Base values

If we are considering a single machine, by tradition we normally select the nominal voltampere rating of the machine, S_n (*n* for "nominal") for the base value for the power terms. (It is possible to choose the rated MW output instead.)

Hence, we have

$$P_{m\ \text{pu}} = \frac{P_m}{S_n} \tag{4.5}$$

and

$$P_{e\ \text{pu}} = \frac{P_e}{S_n} \tag{4.6}$$

For the angular speed ω_m, we select for the base value the rated speed ω_n. Hence, we also have

$$\omega_{\text{pu}} = \frac{\omega_m}{\omega_n} \tag{4.7}$$

We convert the moment of inertia J to the H *constant* in the usual way (see, for example, Ref. [1], Section 3.9.2), using S_n for the voltampere base:

$$H = \frac{1}{2} J\, \omega_n^2 / S_n \tag{4.8}$$

4.4.7.2 The step by step conversion of Eq. (4.4) into "per unit" form

Substituting for P_m, P_e, ω_m, and J from Eqs. (4.5)–(4.8) into Eq. (4.4) we get

$$S_n\left(P_{m\,\mathrm{pu}} - P_{e\,\mathrm{pu}}\right) = \left(\frac{2 \cdot H \cdot S_n}{\omega_n^2}\right) \cdot \left(\omega_n \cdot \omega_{\mathrm{pu}}\right) \cdot \left(\omega_n \cdot \frac{\mathrm{d}\omega_{\mathrm{pu}}}{\mathrm{d}t}\right) \tag{4.9}$$

Canceling out the occurrences of S_n and ω_n and dropping the "pu" subscript (i.e., understanding now all quantities to be in "per unit"), we obtain finally:

$$P_m - P_e = 2 \cdot H \cdot \omega \cdot \frac{\mathrm{d}\omega}{\mathrm{d}t} \tag{4.10}$$

Thus, in Eq. (4.10), we have the basic, unlinearized, "per unit" form of the *swing equation* which we shall use as the basis of our investigations in all subsequent chapters of this book.

4.5 The next chapter: available techniques for calculating frequency response requirements

4.5.1 Approaching the calculation of the frequency response requirements

In the next chapter, we shall be looking first at what is happening, if anything, to the power system frequency during "normal operation," we shall then look at what happens following the loss of some generation or a block of demand. We then begin to look at how we may be able to calculate the frequency response requirements for the system beginning with the "swing equation." For this, we need to be able to model the multimachine system at different periods following the loss of generation or demand.

4.5.2 Some modeling details

To calculate accurately the frequency response requirements of the system, we need to be able to model the effect of the variation of demand with frequency and to be able to model the effect of generator speed governor action. In the next chapter, we shall derive a mathematical form to be solved that models a general case, that is, which can be applied to the different time periods we are interested in.

4.5.3 Available solution techniques

At the end of the next chapter, we shall introduce some of the possible mathematical and other analytical methods available to us for solving the general form of the "swing equation" that we have derived in this chapter.

Reference

Book

[1] Power system stability and control, in: P. Kundur, Electric Power Research Institute, McGraw Hill Education, 22nd Reprint, 2017.

Chapter 5

Techniques available for calculating frequency response requirements

5.1 Introduction to techniques available for calculating frequency response requirements

5.1.1 A description of the objective of this chapter

In this chapter we shall be building up in logical steps to a format for the "swing equation" that we shall be able to use later in the book for calculating the "frequency response requirements" in the three major situations operationally: during "normal operation," that is, "prefault" ("preevent"); "postfault" ("postevent") "low-frequency excursions," and "postfault" ("postevent") "high frequency excursions."

5.1.2 Requirements during "normal operation"

We shall look in more detail later at the earlier sorts of "frequency event" that could occur during "normal operation," what could happen to disturb the system frequency and what this entails for the "system frequency" and consequently what correctional measures need to be put in place by the "system operator."

5.1.3 Evolutionary time periods "postevent"

We shall then describe the different "postevent" time periods that the system goes through following a loss of generation or demand from the system. The former causes the "low-frequency excursion" type of event while the latter causes the "high frequency excursion" type of event.

5.1.4 Putting together a usable form of the "swing equation"

We shall be starting with the basic form of the "swing equation" that we derived in Chapter 4, Initial analysis of the frequency control problem and a

Modern Aspects of Power System Frequency Stability and Control.
DOI: https://doi.org/10.1016/B978-0-12-816139-5.00005-9

derivation of the swing equation, with the objective of building up a robust, standardized version of the "swing equation" that we shall be able to use more easily and refer to for solving practical examples in the future chapters.

5.1.5 The available techniques for solving the "swing equation"

Finally, at the end of this chapter, we shall describe briefly the different techniques available to us for solving our standardized form of the "swing equation" in preparation for the following chapters, in which we shall be giving examples using MATLAB of calculating "frequency response requirements" for the different situations mentioned earlier.

5.2 The power system frequency and frequency requirements during "normal operation"

5.2.1 A description of the "quiet" system

Let us first discuss those instances during the operation of the power system when the frequency is changing relatively little. At such times, when there are no unplanned outages from the system or any faults on it, changes to the system frequency are mostly due to system demand changes and other relatively minor changes on the system, such as transformer tap-changing and circuit switching.

During these "quiet" periods on the system, it is possible to control relatively easily the system frequency to within a narrow band defined around the nominal frequency. The actual size of this band varies from system to system. Relatively small power changes from the available sources (generators) connected to the system can be used to correct the frequency here and there, as required. No major power changes are required to maintain the system balance.

5.2.2 "Major frequency events" on the system

From the point of view of frequency control, usually the most important imbalance that can occur on the system results from the unplanned disconnection (loss) of a large generating unit, because the resulting type of imbalance will affect the power frequency the most due to its size. Such a large generation loss may, for example, be caused by a fault occurring on the output terminals of a large generating unit.

On some systems, particularly those with interconnections to other systems, it may be the case that the loss of an "exporting interconnector" is numerically larger than the loss of the largest generator on the system, including the loss of the largest "importing interconnector." In this case the worst "high frequency event" will be more onerous than the worst "low-frequency event."

Following the loss of a large generating unit from the system, the frequency *falls* and this change in the frequency must be reversed to recover the system frequency to within its normal operating range. As described in Chapter 1, The need for frequency control, of this book, the system and some of its major components are subject to various risks during the time that the system frequency has departed significantly from its normal operating region around the nominal frequency.

Another kind of major loss from the system is the loss of a block of demand (load). We recall from Chapter 1, The need for frequency control, that the system frequency *rises* immediately following a loss of load.

In most power systems, the loss of a block of demand represents a smaller amount of power than the loss of a large generating unit. Therefore the frequency deviation caused by a loss of load is usually less than that caused by a loss of a large generating set, albeit it is in the opposite direction. Again, we cite the possible case of the loss of a large exporting interconnector being more onerous than the loss of the largest generator.

5.2.3 "Minor frequency events" on the system

Some power imbalances with less severe associated power frequency changes can come from other kinds of disturbances, such as faults that result in the loss of one or more transmission circuits, distribution circuits, or transformers from the system. These, although having less impact on the frequency than generation losses, should also be studied to ensure that the integrity of the system can nevertheless still be maintained.

One may also put into the category of "minor frequency events" the loss of "small" generating sets and the smaller loads from the system. It should, however, be borne in mind that it is possible to lose several generating sets in quick succession and therefore that the loss of several smaller units may be equivalent to the loss of one, much larger, unit.

5.2.4 The need for control during "quiet periods"

The control engineer, as discussed earlier, in addition to providing restitution of the frequency following major and minor frequency events, also needs to be aware of any frequency response requirements that could arise during normal operation, that is, during operation of the intact (unfaulted) network. These requirements exist because, as mentioned earlier, the power frequency constantly changes due to the ever-changing background of generation and demand, and therefore there is a practical need to maintain the system frequency within relatively narrow prefault operational limits.

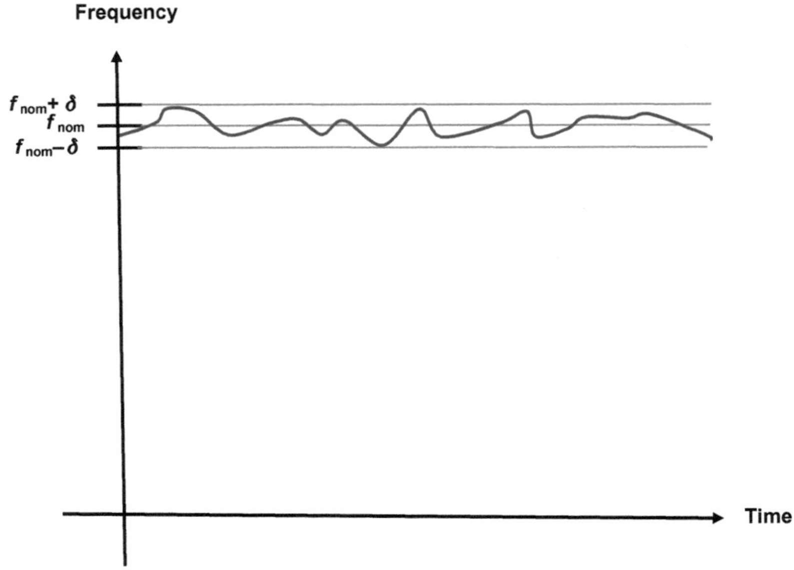

DIAGRAM 5.1 The variation of frequency during "quiet periods."

Please see Diagram 5.1 for a graphical depiction of the relatively small changes of frequency around the nominal frequency during "quiet periods."

As mentioned before, keeping these dynamic changes in the frequency in check requires relatively little response to be made available on the system, but this amount, strictly speaking, needs to be considered a part of the *total response requirements* for it.

In the next section, we will examine the time periods into which the evolution of the system frequency is conventionally divided following the loss of a large generating set or a large block of demand. These time periods are defined according to which, if any, controls are currently available on the system to restore the frequency.

5.3 The time periods of frequency evolution following a system loss

5.3.1 Immediately postfault

Immediately following the unplanned outage of a large generating unit or a large demand block from the power system, generation and demand are no longer in balance in the system at large. Putting it another way, the power system is no longer in a steady-state condition and the system frequency is changing, as the system searches for a new balance point at a new frequency.

We shall now examine in more detail the two most frequent kinds of system loss that can substantially disturb the system frequency. These, to reiterate, are a major loss of generation and a major loss of demand.

First, let us consider generation:

5.3.1.1 A major loss of generation

Let us look first at the case of a major (large) loss of generation from an initially balanced system, that was, until just recently prior to the event, in a "steady-state" operating condition.

In this immediately postfault situation, the generation and demand connected to the bulk system are now no longer in balance. Indeed, *the total demand is now larger than the total generation*. Therefore power must come from some additional source to redress the power imbalance. This power comes from the kinetic energy of the rotors of synchronous generators. The rotors lose kinetic energy to convert to electrical energy to feed the surplus load that has just appeared on the system.

Since the generator rotors have lost some of their kinetic energy, their rotational speed falls, as described by Eq. (4.1) of the previous chapter. Please see Diagram 5.2 for a sketch of the system frequency immediately following a major loss of generation.

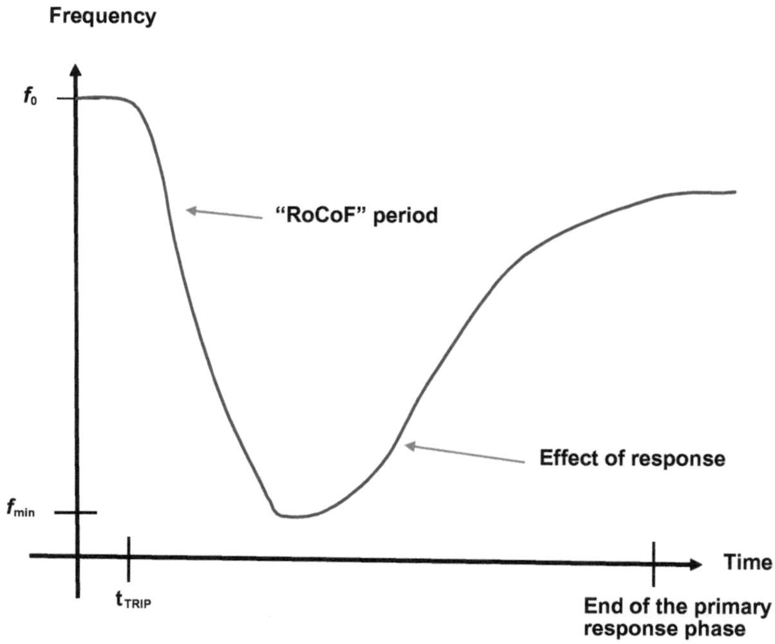

DIAGRAM 5.2 The frequency following a major loss of generation from the system.

We recall that we have said earlier that that power must come from some source so that the power balance may be redressed. We shall now look at the possible sources of such power.

5.3.1.2 Inertial response

At first, before any sort of governor or other response is becomes activate, some extra power is provided from the remaining generating machines on the system.

As mentioned earlier, this power is converted from the kinetic energy of the rotating parts of the generators and other rotating machinery still connected to the system. This "rotating kinetic energy" is converted into some of the extra electrical energy needed to rebalance the system immediately following the initial power imbalance that arose from the initial generation trip.

The extra power generated by rotating machines, at this stage following the loss, is known as *inertial response*, because it derives solely from the *inertia* (more strictly, the *kinetic energy*) of the rotating machines. It is important to emphasize that this initial "inertial" response is not associated with any governor response or any other kind of response (this will come at a slightly later timescale of the evolution of the frequency).

In the next section, we shall look at the role played by the speed governor and other kinds of response and ask the question of when this becomes active following a major loss of generation from the system.

5.3.1.3 The timescales for the activation of speed governor response and other kinds of response following a system incident

1. *The speed governors of synchronous machines*

   As introduced in Section 1.2.2.5, the speed governors of synchronous machines (when selected to be in "frequency responsive mode") operate on a "droop profile." In practical terms, this means that when the system frequency falls, the power output of the machine will rise.

   Conversely, when the system frequency rises, the power output of the machine will fall.

   The above is true as far as it goes, but, however, it is not the complete story of what happens. This is because the response from the speed governors of synchronous machines is *not instantaneous* following a speed change, but there is a *time delay* before any power increase or decrease from the machine can be realized.

   The origin of the time delay is the finite time needed for certain processes to be activated and be completed before any response starts to appear.

The processes referred to are

- **a.** detection of the speed change on the system;
- **b.** transmittal of the signal to open the valve controlling the output power of the prime mover;
- **c.** receipt of the signal to open the valve controlling the output power of the prime mover;
- **d.** physical opening of the valve controlling the output power of the prime mover.

The time required for these processes to complete varies from machine type to machine type but, in total, it is of the order of one second for fossil fuel machines. We recall from our discussion earlier that during this time delay, only the inertial response is active in the machine.

2. *Other kinds of response*

a. *Demand tripping*

As described earlier in the book, response can also be provided by *demand tripping*, when blocks of demand are deliberately jettisoned to raise the frequency of the system. This kind of response is typically achieved by arming relays to trip certain circuit breakers controlling blocks of demand at a preset *trip frequency*, so that demand is automatically tripped when the system frequency falls below a certain level. By its nature, this sort of tripping is activated on the timescales required to operate the breakers involved, that is, of the order of 100–150 ms after the signal to trip is sent from the detecting relay.

b. *Other kinds of controlled tripping*

In general, any kind of generation or demand (e.g., wind power, solar power, various demand-side sources, and sinks) can in principle be set up to be tripped by a relay whose activation is based on a measurement of the system frequency. Thus the timescales involved for tripping will in general consist of a "relay time" plus a "break time."

In the case of purely electronically controlled equipment, such as DC links, their response activation times are much faster, of the order of milliseconds or even microseconds, since no relatively slow mechanical systems are involved here: it is all power electronics.

5.3.1.4 Loss of demand (load)

Following a loss of demand, on the other hand, we are presented with the opposite situation to that with which we were presented following a loss of generation. Instead of the frequency falling postevent, it rises, and it therefore must be brought back down again to restore a "normal operating frequency" (Diagram 5.3).

Since the tripping of demand cannot be used to restore the system frequency following a loss of demand, since the frequency would go "the wrong way," any restorative action to correct the frequency in this case must

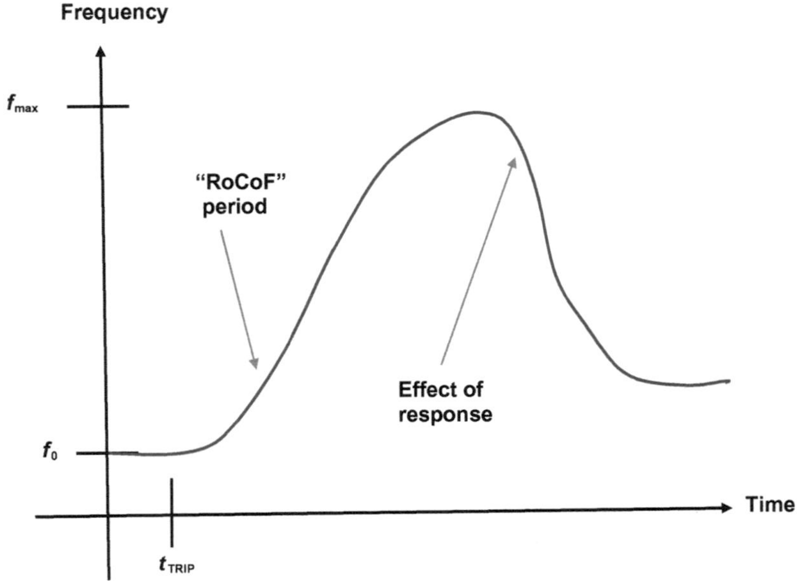

DIAGRAM 5.3 The system frequency following a loss of demand from it.

be to reduce the output from generators, including synchronous generators, wind farms, and DC links.

Clearly, in the case where we have lost some demand from the system, the speed governors of synchronous generators still operate on their "droop" profiles, but this time they will act to *reduce* both the speed and the power output of the machine, rather than the opposite (that is, to *increase* both the speed and the power output of the machine).

5.3.2 The time period from 1 second up to about a minute postfault

5.3.2.1 The definition of the "primary response" period

The time period later than about 1 second, and in some cases up to about 1 minute, dependent upon the actual system, postfault is usually called the "primary response" time period. In this time period we typically see the actions of automatic frequency correction devices such as generator speed governors.

There are additional ways available to the control engineer to enable the postevent frequency to recover satisfactorily. These include the tripping of demand at selected, preset trigger frequencies, as mentioned earlier. This method is, of course, only effective for the case when the system frequency is falling. When the system frequency is rising, such an action would cause the system frequency to rise further and so cannot be used, also as mentioned earlier.

5.3.2.2 The nature of the "primary response" period

The "primary response" period is, first and foremost, the time during which we have *automatic frequency response* from power system devices operating.

Manual actions are performed during a later period, namely, during the "secondary response" period (see later for the definition of this).

5.3.2.3 Typical actions during the "primary response" period

Typically, during the "primary response" period, we see a ramping up of power from those synchronous generators that are active response mode following an incident that has involved a loss of generation, or a ramping down following an incident involving a loss of demand.

In addition, as described earlier, we would expect to see response from such facilities as demand tripping (for low-frequency incidents only) and response being generated from a variety of other sources, as described earlier. No manual actions are normally performed during this period.

5.3.3 The response time period of more than about a minute postfault

The response time period of more than about a minute postfault is usually called the "secondary response" time period. During this time period, instructions can typically be issued "manually" to generating units of various kinds to increase or decrease their outputs, with a view to further restoring the system frequency to normal operating levels, if necessary, beyond the "primary response" period.

Some utilities define a further, "tertiary response," time period beyond the "secondary response" time period. This additional categorization is in some ways arbitrary, since relatively "slow" actions are typically performed within it, such as plant rescheduling and other, smaller, adjustments, in a similar way to within the "secondary response" time-period.

In the next section, we shall begin our analysis in detail of the "swing equation" as we begin to think about how to solve it for subsequent use in some practical examples that will appear later in the book. In general, we shall be developing the capability to predict the evolution of the frequency and to be subsequently able to calculate the frequency response requirements for the system.

5.4 Approaching the solution of the "swing equation"

5.4.1 The general aims of solving the "swing equation"

5.4.1.1 The reasons for solving the "swing equation"

Why do we want to solve the "swing equation" in the first place? The basic answer is that, although the "swing equation" provides us with some useful information, such as whether the rotor of a machine will accelerate or decelerate in response to a power imbalance, it does not in its present form tell us anything about how the system frequency changes with time following such an imbalance.

5.4.1.2 The structure of the "swing equation"

Structurally, the "swing equation" is of the form "$\mathrm{d}f/\mathrm{d}t = g(f,t)$" in which f denotes the system frequency and g is some general function of f and t. What we should ideally like to do is, by some convenient method, to "integrate" the equation so that it reemerges in the more useful form "$f = f(t)$," so that we have an expression for the system frequency as a function of time and certain other known parameters such as power injections and system loads.

5.4.1.3 The ability to calculate the frequency response requirements

Having solved the "swing equation," we shall have obtained a usable "evolution" function for the system frequency, that is, we shall be able to *predict* the system frequency following a system power disturbance. This will be of great use to us, since having this capability we shall be more able to plan and control the subsequent recovery of the system frequency. Indeed, of the greatest value to us will be the means to calculate the *frequency response requirements for our system* at each stage following the disturbance.

5.4.1.4 The need to be able to model the system

Before we can solve the "swing equation," however, we need to be able to *model* the power system during the various successive time periods following the system disturbance that have been discussed earlier in the previous sections.

Therefore in the next section, we shall, beginning with the basic form of the "swing equation" given at the end of Chapter 4, Initial analysis of the frequency control problem and a derivation of the swing equation, look at how the "swing equation" changes in the different time periods following a system disturbance, with an emphasis on what happens as different kinds of response appear on the system.

5.4.2 The "evolution" of the power system frequency: the basics

For ease of reference, we have repeated below the "per unit" form of the *swing equation* for a *single synchronous machine of inertia constant H* in the form which was given at the end of the last chapter as Eq. (4.10)

$$P_m - P_e = 2 \cdot H \cdot \omega \cdot \frac{d\omega}{dt} \tag{5.1}$$

As mentioned earlier, for a given power network, it would be helpful to use a form of this equation to *predict the future behavior* of the system's angular frequency ω, measured in the units "radians per second" ($=2\pi f$ where f is the power frequency in Hz), following a system power imbalance event.

We recall that we would like to be able to predict the future behavior of the system frequency because we want to take the optimal actions to enable the system frequency to be restored (if necessary) to within statutory operating limits.

Once we have the capability to be able to predict the frequency "evolution" following a system incident, we shall be in a much better position to be able to calculate the *response requirements* for a list of "the most likely" system scenarios that involve the loss of major amounts of generation or demand.

5.4.3 The frequency response requirements: a definition

When we say that we would like to find the *total frequency response requirements* for a system, we mean that we want to be able to answer the question, "how much power is it necessary to hold in reserve on the available generators to cater for all credible scenarios (that is, to support the most likely system configurations, faults, generation and/or load losses)?" In the next section, we discuss how this task may be accomplished.

5.4.4 How do we go about calculating the system frequency response requirements?

5.4.4.1 Using a power system analysis tool

For a given system and system configuration, one way to estimate the required frequency response requirements is first to model the system in one of the many commercially available power system analysis suites such as "PSSE" or "DIgSILENT Power Factory." Then, several further steps need to be accomplished to produce a result for the frequency response requirements of the system.

5.4.4.2 The general procedure for calculating frequency response requirements using a network analysis tool

In general, the following stages need to be gone through in calculating the system frequency response requirements using a commercial analysis tool:

1. *Build a model of the power system in the chosen tool*
   a. *Network items to be represented*
      Building a model of the power system in the chosen tool involves, in general, representing overhead lines, underground cables, transformers,

generators, loads, and other items of system equipment such as busbars and possibly even switches depending on the tool selected.

b. *Justifiable approximations that can be made*

However, in the case of frequency calculations, some approximations may be justifiably made to make the calculations easier. For example, it is not necessary for frequency response calculations to represent the system in all its three phases (plus neutral lines) or to model individually all the circuit breakers in the system.

c. *The "single busbar" approximation*

One could even argue that, for frequency response calculation purposes, the entire system could be represented by a single busbar, to which all generation and loads could be attached. This is indeed one approach that could be used but seems to be a waste of the modeling capabilities of a typical commercial tool.

d. *Associated consequences of making approximations*

One main consequence connected with approaches involving modeling the generation and load on a single busbar is that power system active losses are not represented. Another consequence is that voltage and consequently reactive power are also not represented.

2. *Choose a set of times throughout the day for which the frequency response requirements are to be calculated*

It is common practice to select a set of times throughout the day for which one wishes to calculate the frequency response requirements. The frequency response requirements are calculated at intervals like this because it is not necessary or perhaps even possible to calculate the frequency response requirements on a continuous basis. Therefore this is done at a manageable number of times during the day, for example, every half-hour. This schedule will also typically coordinate in some way with the schedule for allocating generation to the system and fundamentally to the load forecasting intervals during the day.

3. *Determine, for each of the above set of times, the expected greatest loss of generation and the expected greatest loss of load*

For each of the "critical times" during the day as described in the preceding section, the generation and demand for the system are predicted using information known about the generation profile and demand forecasting results. From this information, the largest potential losses of generation and demand can be determined.

4. *Decide, for each of the above set of times, which generating sets and demand blocks are to be used for providing frequency response*

a. *Selecting generation sets for response*

Having determined the total requirements for generation to be deloaded at each of the set times during the day, it is necessary to decide which generating sets will provide response for the loss of the largest set predicted to be running for each of these times.

Which generating sets to de-load will typically be decided according to their relative prices as submitted through a suitable commercial bidding mechanism.

An alternative method, applicable in the case of power systems in which generators do not make bids or offers for response is to make the selection of which responsive sets to use based on some other criterion such as fuel efficiency.

b. *Selecting demand blocks for response*

Demand tripping can be used in conjunction with de-loading of generators for response. As described earlier, when used to assist frequency response, blocks of demand are usually set to trip when the system frequency falls below a certain predefined level. In this respect, the trip is not dependent upon time but upon frequency.

Consequently, how demand tripping, if selected, complements the response selected from synchronous (and other) generators need to be investigated with the chosen analysis tool, largely by trial and error, until an acceptable solution combination is obtained.

5. *By a process of trial and error determine, for each of the above set of times, the optimal de-loadings of generating plant and tripping of demand blocks required*

Unless the search process is automated in some way, the required amount of response holding on generators and the correct level of arming of demand tripping schemes must be determined by a process of trial and error. This process essentially consists of the following algorithmic steps:

For each of the set times

a. Begin with an estimated amount of response from generators and demand tripping schemes, noting the relative effectiveness of these two different sources of response.

b. Using the estimated amounts of response, run a simulation in the chosen analysis tool and note the course of the system frequency following the greatest losses of both generation and demand (not that demand that is part of the demand tripping schemes). This will require two separate simulations per set time.

c. Compare the frequency evolution following the largest generation and largest demand losses separately with the appropriate standard defined for the system being studied.

d. If the standard has not been reached, iterate the above process until it has, for both largest generation and largest demand losses separately.

e. If the standard has been reached but with response to spare, remove some of the response and recalculate, for both generation and demand losses separately.

f. Iterate the above process until the response allocated just pulls the frequency to within the required standard. Again, the largest generation loss and the largest demand loss must always be studied separately.
g. At the end of the process detailed earlier, the required amounts of response required for each of the largest generation loss and the largest demand loss will have been determined.
h. Repeat the whole process for the next set time and so on until the requirements for all set times have been calculated. Then stop.

In the next section, we shall study an alternative method to the use of a commercial power system analysis tool as described earlier. The spreadsheet method has the advantage over the commercial analysis tool method of ease and speed of calculation.

5.4.4.3 The "spreadsheet method"

1. Description of the "spreadsheet method"
   a. *The rationale behind the "spreadsheet method"*

      Since the problem we are trying to solve, namely the calculation of frequency response requirements, is not strongly dependent (at its most basic level, at least) on the *spatial* representation of the power network, but mainly on the total amount of generation and demand in it, one can argue that we are justified in approaching the problem simply by employing a (suitably programmed) spreadsheet. In other words, we are using a kind of "single-bus" approximation to model the network.
   b. *How the "spreadsheet method" should be programmed*

      The spreadsheet should be programmed with a suitable solution of the "swing equation" that can be used to calculate the frequency response requirements for both the "low frequency" (loss of generation) and the "high frequency" requirements of the system.

      We shall be considering later in this book the various possibilities available to us for solving the "swing equation." Having chosen a solution method, one of them can be programed into the spreadsheet.

      One can then build round the chosen method of solving the "swing equation" an iterative loop that enables the frequency response requirements to be calculated using the "general procedure" outlined in Section 5.4.4.2 (5) earlier.
2. *Advantages of the "spreadsheet method"*

   The main advantages of using a spreadsheet approach are those of simplicity, convenience, and ease of maintenance. They are also relatively easy to program using the inbuilt spreadsheet functions.
3. *Disadvantages of the "spreadsheet method"*

   One obvious drawback to the spreadsheet method is that the detailed modeling of power system components available in a typical commercial power system suite is not easily duplicated. However, modeling of power system components is possible to some extent, and this allows the adequate modeling of such features as generator governors and load-tripping schemes that can be used for the calculation of the frequency response requirements.

In the next section, we shall examine the possibility of programming the code necessary to calculate the frequency response requirements in a language such as C++, R, or some other currently available high-level computing language.

5.4.4.4 Program coding in a high-level computer language

1. *Description*

   Of course, it is possible instead of using a simple spreadsheet for our programming, to use one of the high-level computer programming languages available. All calculations that are representable in a spreadsheet such as Excel are representable instead in C++ or R, to give two examples.
2. *Advantages of program coding in a high-level computer language*

   One of the main advantages of using a program language such as C++ or R to code the solution of the "swing equation" and to calculate the system's frequency response requirements is that such features as "do loops" and other iterative techniques are made available, thus allowing the solution code to be quite compact.
3. *Disadvantages of program coding in a high-level computer language*

   A possible disadvantage of using a high-level programming language is that to avoid making errors the code maintenance ideally needs to be taken on by someone who has a good grasp of the language.

   Also, certain platforms and operating systems must be available to be able to compile and run the program, which may involve a certain cost outlay. In addition, during upgrades of the platform or operating system modifications may also need to be made to the code by an experienced user who must, therefore, be made available.

In the next section, we look finally at a further possibility, that of using a control diagram type approach to "program" the solution of the "swing equation" and to calculate the frequency response requirements of the system.

5.4.4.5 "Programming" by building a control diagram representative of the problem

1. *Description of the "control diagram" approach*

   As is well known (see, for example, see Ref. [1]), it is possible to represent the "swing equation" in the form of a "control diagram" with a feedback loop as in Diagram 5.4.

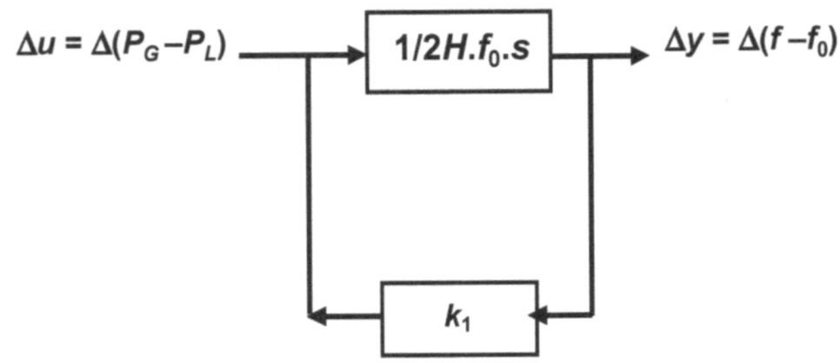

DIAGRAM 5.4 Control diagram representation of the "swing equation."

In Diagram 5.4 the "feedback arm" represents the action of "load damping." To this could in principle be added further blocks to represent, for example, automatic governor response.

2. *Advantages of the "control diagram" approach*

   The "control diagram" approach is in an especially convenient form for many engineers since it mirrors the form of many common feedback systems. Also, in a general kind of a way, it is clear what is going on because it is a pictorial, not a symbolic, representation. In this sense it is easier to grasp its physical meaning than with, say, an algebraic representation.

3. *Disadvantages of the "control diagram" approach*

   Some of the disadvantages of the "control diagram" approach are as follows:

   First, although the diagrammatic representation is simple enough for the "swing equation" in its simplest, most basic, form, it becomes quite a lot more complex if the extra features to model such things as the generator governor profile and demand tripping schemes are to be included as well.

   Second, the diagrammatic representation of the "swing equation" still needs to be "iterated," with successive values of the response available, to be able to calculate the frequency response requirements of the system according to the general procedure laid down in Section 5.4.4.2 (3) earlier.

4. *In the next section*

   Having spent some time discussing some aspects of the possible tools at our disposal for solving the "swing equation" and for calculating the frequency response requirements of the system, in the next section we shall now turn back to the mathematical modeling aspect of the solution process to build up a representation of the system at various stages of the frequency evolution following the system incident that first disturbed the frequency.

5.4.5 The modeling of a "multimachine" system

Returning once again to our mathematical analysis, we observe that Eq. (5.1) describes how a *single machine* behaves in isolation. Can we derive equations similar in form to Eq. (5.1) to describe the behavior of a *multimachine system*?

Let us examine again the various "evolutionary timescales" of the system frequency that we discussed in Section 5.2 and try to construct equations to model them.

5.4.5.1 The power balance in the preevent system

Let the total power generation equal the total power demand exactly at some initial power frequency, say f_0 Hz at some time t_0 seconds. For this brief instant, the power system will therefore be in balance. So, we can write

$$\Sigma P_{\text{GEN}} - \Sigma P_{\text{LOAD}} = 0 \tag{5.2}$$

evaluated at the assumed initial power frequency f_0. As usual, the "Σ" sign denotes the arithmetic sum of the quantities placed in front of it.

For the purposes of analyzing the subsequent evolution of the power system frequency, it is helpful now to divide all the system generators into four broad categories, based on their functions on the network:

1. *"Responsive" generators*

   This categorization applies to a generator that is in "frequency sensitive mode," that is, its power output operates on a "droop" profile and is therefore "frequency sensitive."

   We can write

$$P_{\text{resp}} = \Sigma^n P_{i\ \text{resp}} \tag{5.3}$$

   where $\mathbf{\Sigma^n}$ denotes that we are adding together the outputs of a total of "n" frequency responsive machines, each having a subscript i, where i increments from 1 to n in steps of 1.

   At any moment in time during the operation of a power system, we know that certain generating machines will have been selected to be operating in *frequency sensitive mode* so that they can compensate for changes in the system frequency.

   Such a kind of response is automatic, that is, it does not require manual intervention of any sort for its normal operation.

   The "droop profile" of a responsive synchronous machine can be selected to suit the characteristics of the system. The "droop" profile determines how much the power output of the machine varies with a given change in the system frequency. The "droop profile" is discussed in further detail in Section 3.5.1.2 later in this chapter.

   In the case of other kinds of frequency responsive generators, such as wind turbines, the flexibility of the electronic control systems of these devices enables them to *simulate* a "droop profile" (or a selection of other profiles as required), if necessary.

2. *Nonresponsive generators*

   The power output of this kind of generator does not respond to changes in the system frequency by use of a governor type or any other type of frequency responsive system. Such a generator is in termed as being in "frequency insensitive" mode. However, "inertial response" is still present in synchronous machines, as always.

To represent several nonresponsive generating sets, we can write

$$P_{\text{non-resp}} = \Sigma^n P_{i\ \text{non-resp}} \tag{5.4}$$

where $\boldsymbol{\Sigma}^n$ denotes that we are adding together the outputs of a total of "n" nonfrequency responsive machines, each having a subscript i, where i increments from 1 to n in steps of 1.

Eq. (5.4) represents the adding together the power outputs of all the nonresponsive (frequency insensitive) generating sets on the system.

3. *"Trip" machine(s)*

In our modeling this kind of generator (or group of generators) will refer to those which, due to some unplanned event, are about to be lost from the system, as part of some system event.

So, we can write in this case

$$P_{\text{TRIPS}} = \Sigma^n P_{i\ \text{TRIP}} \tag{5.5}$$

where Σ^n denotes that we are adding together the outputs of a total of "n" "trip" machines, each having a subscript i, where i increments from 1 to n in steps of 1.

Note: It is quite possible that more than one generating set may be "lost" at the same time. However, in the practical experience of the author it is far more likely that one trip is followed by another trip with a small time-interval in between the two events.

4. *Combining the above*

We can now rewrite Eq. (5.2) as follows:

$$P_{\text{resp}} + P_{\text{non-resp}} + P_{\text{TRIPS}} - \Sigma P_{\text{LOAD}} = 0 \tag{5.6}$$

which now shows individually the contributions of the different categories of generation making up the "generation" part of an initial power balance

5.4.5.2 The system power balance immediately postevent

Let us now look at how the power balance changes following a system event.

Let's say our event is to completely disconnect (lose) the generation P_{trips} instantaneously from the system. Then, immediately following this event we will have the following adjusted power balance:

$$P_{\text{resp}} + P_{\text{non-resp}} + 0 - \Sigma P_{\text{LOAD}} = 2 \cdot H_{\text{eq}} \cdot \omega \cdot \frac{d\omega}{dt} \tag{5.7}$$

where H_{eq} represents the equivalent *H-constant* for the postevent system, that is, the *H-constant* that no longer includes the tripped machine(s). Such an *H-constant* can be defined as in Eq. (3.10) at the end of Chapter 3, Per unit systems for frequency analysis.

The power system is no longer balanced and consequently the frequency is now changing. This condition is modeled by the right-hand side of Eq. (5.7) not equaling zero.

Because we have just lost some *generation*, the frequency is in fact *falling*, that is, *dω/dt* is *negative*. (If instead we had just lost some *demand*, the frequency would have been *rising*, that is, *dω/dt* would have been *positive*.)

In the next section, we move on to describe the power system balance during the primary response period.

5.4.5.3 The system power balance during the primary response period

In the primary response period response from synchronous generation with a "continuous" characteristic like the one depicted in Diagram 2.1, and response from batteries, solar farms, demand blocks, and other sources with a "step-change" characteristic like the one depicted in Diagram 2.2 come into play.

During this time period, we need to add on a term to Eq. (5.7) to represent response, so that it becomes

$$P_{\text{resp}} + P_{\text{non-resp}} + 0 - \Sigma P_{\text{LOAD}} + \Delta P_{\text{resp 1}} = 2 \cdot H_{\text{eq}} \cdot \omega \cdot \frac{\mathrm{d}\omega}{\mathrm{d}t} \tag{5.8}$$

where $\Delta P_{\text{resp 1}}$ represents all the kinds of response active during the primary response period taken together at any point in time during this response time period.

The total response at time t may be written as

$$P_{\text{resp 1}} = \Sigma^{n} P_{\text{resp 1 i}} \tag{5.8a}$$

where $\boldsymbol{\Sigma^n}$ denotes that we are adding together the outputs of a total of "n" sources of response, each having a subscript i, where i increments from 1 to n in steps of 1.

Note: $P_{\text{resp 1}}$ can be modeled as some function of f and t.

5.4.5.4 The system power balance during the secondary and tertiary response periods

During the secondary and tertiary response periods, we typically see generator active power (MW) outputs being rescheduled. We recall that the secondary and tertiary time periods are the time periods during which the "manual" actions to adjust generator outputs are added to already-existing "automatic" responsive actions that are provided by such devices as generator speed governors.

At a specific time "t" during the secondary response period we can write

$$P_{\text{resp}} + P_{\text{non-resp}} + 0 - \Sigma P_{\text{LOAD}} + P_{\text{resp 1 tot}} + \Delta P_{\text{resp 2}} = 2 \cdot H_{\text{eq}} \cdot \omega \cdot \frac{\mathrm{d}\omega}{\mathrm{d}t} \tag{5.9}$$

where $P_{resp\ 1\ tot}$ represents the total response delivered by the end of the primary response period and $\Delta P_{resp\ 2}$ represents the current power generation output changes in the secondary response time period measured from the beginning of the secondary time period.

Similarly, at a time t during the tertiary response period (if any) we can write

$$P_{resp} + P_{non\text{-}resp} + 0 - \Sigma P_{LOAD} + P_{resp\ 1\ tot} + P_{resp\ 2\ tot} + \Delta P_{resp\ 3} = 2 \cdot H_{eq} \cdot \omega \cdot \frac{d\omega}{dt} \tag{5.9a}$$

where in addition to the foregoing $P_{resp\ 2\ tot}$ represents the total response produced during the secondary response time period and $\Delta P_{resp\ 3}$ represents the response produced so far in the tertiary response period (if any).

5.4.6 The variation of demand with frequency

Due to the characteristics of frequency-dependent loads, such as induction motors, that are present in the system, the magnitude of the total system demand varies with the system frequency (please see Ref. [1], Chapter 11, p. 584 for a more detailed description of this effect).

This effect can be modeled in the following way

$$\Sigma P_{LOAD} = \Sigma P_{LOAD\ n}\{1 + \alpha \cdot (f - f_n)\} \tag{5.10}$$

where α is some constant and the subscript "n" denotes that the quantity concerned is evaluated at the *nominal frequency* of the system.

The "variation of demand with frequency" effect is very important because in fact it contributes measurably to the recovery of the system frequency following an event on the system. Therefore to model the loads correctly in Eqs. (5.2)–(5.9) inclusive by including this important effect, in each instance the term "ΣP_{LOAD}" should be replaced by the expression given in the right-hand side of Eq. (5.10).

The numerical size of the effect varies from system to system. On the British system, α is approximately *2% per Hz*, that is for every 1 Hz fall in system frequency, the active power load falls by 2%. Conversely, for every 1 Hz rise in system frequency, the load rises by 2%.

5.4.7 Forms of the variation of responsive power with frequency

5.4.7.1 The variation of response with frequency in synchronous machines

As referred to earlier, when operating in its normal range, a synchronous machine speed governor operates on a "droop" profile, that is, the responsive power output rises as the frequency falls in a linear manner. This can be expressed as follows:

$$\Delta P_{\mathrm{RESP}} = -P_{\mathrm{RESP}} \cdot k\Delta f \qquad (5.11)$$

where f represents the system frequency and k is a positive constant. Thus in form the variation of responsive power with change in frequency is similar to the variation of demand with frequency, except for the minus sign in Eq. (5.11) which indicates a "negative slope" rather than the "positive slope" of Eq. (5.10).

In diagram form the relation given by Eq. (5.11) looks like Diagram 5.5.

Notice that the "slope" (gradient) of the profile may be adjusted (not during operation) to suit the requirements of the local system.

For an in-depth description of the functioning of the speed governing system of a synchronous machine, see, for example, Ref. [2], Chapter 3, Section 3.2.

5.4.7.2 *The relationship between the constant* k *and the droop* D(%) *of the machine*

When the speed (or frequency) of a turbo-alternator drops by D(%) between no load and full load on the turbine, it is said to have a "droop" of D(%) (see Ref. [2], p. 113 for more details).

By contrast, the definition of k in Eq. (5.11) says that when the power output of the machine increases by ΔP_{RESP} pu, the frequency drops by k pu.

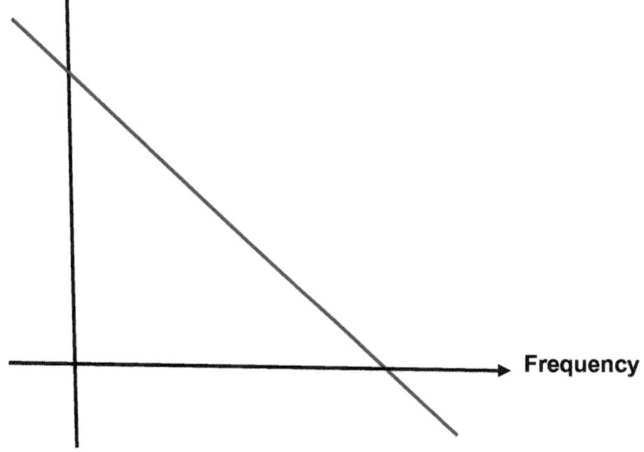

DIAGRAM 5.5 The "droop" profile of a synchronous machine.

How can we reconcile these two definitions? Starting from Eq. (5.11) and writing out the terms explicitly in per unit, we have

$$\Delta P_{RESP}(\text{pu}) = -k(\text{pu})\cdot\Delta f(\text{pu}) \tag{5.11a}$$

Now, since we can write

$$\Delta P_{RESP}(\%) = \Delta P_{RESP}(\text{pu})\cdot 100$$

and

$$\Delta f(\%) = \Delta f(\text{pu})\cdot 100$$

we can rewrite Eq. (5.11a) as

$$\frac{\Delta P_{RESP}(\%)}{100} = -k(\text{pu})\cdot\frac{\Delta f(\%)}{100}. \tag{5.11b}$$

Now if we set ΔP_{RESP} (%) = 100 to describe the machine output going from no load to full output, we have from Eq. (5.11b) that

$$1 = -k(\text{pu})\cdot\frac{\Delta f(\%)}{100}. \tag{5.11c}$$

Finally, rearranging Eq. (5.11c), we obtain

$$\Delta f(\%) = \frac{-100}{k(\text{pu})} \tag{5.11d}$$

for a machine load increase from no load to full output. Hence, recalling the definition of droop in percent as given in bold at the beginning of this paragraph, we see immediately that the droop in percent of the machine is given in terms of k as follows:

$$D(\%) = \frac{100}{k(\text{pu})} \tag{5.11e}$$

This can of course be written in the inverse form

$$k(\text{pu}) = \frac{100}{D(\%)} \tag{5.11f}$$

which gives k (*pu*) in terms of $D(\%)$.

5.4.7.3 Demand blocks

When demand blocks are tripped for frequency control purposes, the demand in question is lost *almost* instantaneously. As a "first approximation," this can be modeled as a unit step function u as follows:

$$\Delta P_{RESP} = -P_{TRIP}u(t - t_{TRIP}) \tag{5.12}$$

where $u(t - t_{TRIP})$ represents a unit step at time $t = t_{TRIP}$. This is modeling an "almost" instantaneous event as a fully instantaneous event.

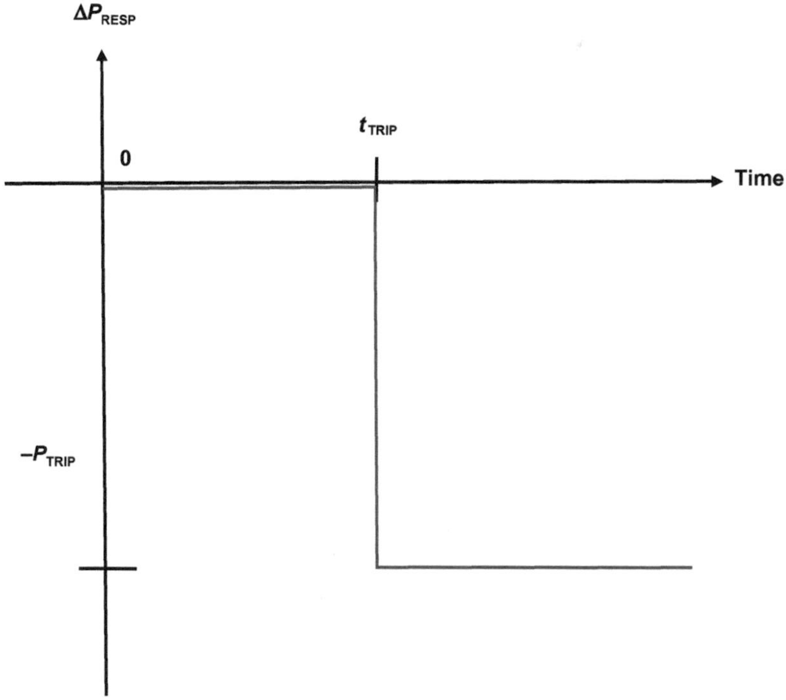

DIAGRAM 5.6 The instantaneous tripping of a block of demand.

In diagram form the earlier relation looks like Diagram 5.6.

Notice that we are modeling a very fast, but nevertheless noninstantaneous change by a fully instantaneous change. A possible consequence of making this approximation is that harmonics may be introduced into the solution where there may be none, or fewer, in the "real" system where nothing happens truly instantaneously.

5.4.7.4 The modeling of the profile of a general responsive device

We have discussed so far two kinds of responsive profile: "linear" (generator governor response) and "step change" (instantaneous demand tripping). In general responsive profiles are (or can be made artificially) more complicated than these.

Even the response characteristic (by which we mean power against frequency) of the speed governor of a synchronous machine is not linear for the whole of its range.

A general responsive device, that is one that has an irregular response characteristic, may in theory be modeled conveniently as a general

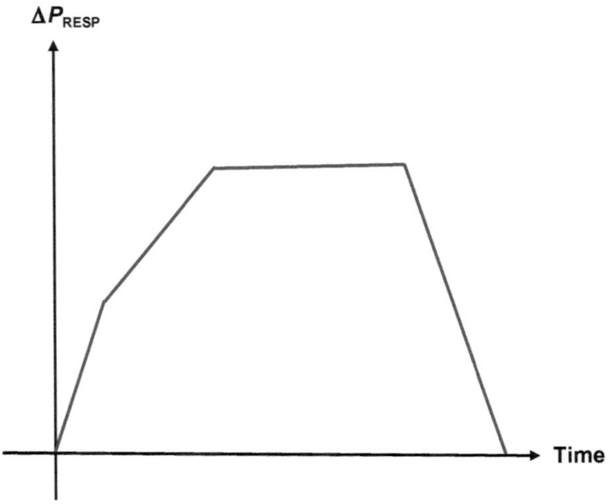

DIAGRAM 5.7 An example of a general response characteristic.

mathematical power series, with time as the independent variable. Hence, we may write a general definition:

$$\text{GRP} = a_0 + a_1 t + a_2 t^2 + a_3 t^3 + \cdots \tag{5.13}$$

where *GRP* is an abbreviation for "general responsive profile" and the coefficients a_i are constants.

As an example, in diagrammatic form, such a GRP could look, in general, something like Diagram 5.7.

Such a general curve can, of course, be modified to suit the representation of various kinds of responsive devices, such as wind turbines, solar photo-voltaic panels, and large batteries as required.

For each application, the unique set of coefficients a_i would need to be determined to define the model. In practice this information could be obtained from field measurements or from manufacturer's data.

In the next section, we piece together the various components of the "swing equation" that we have discussed earlier to produce a "general mathematical form" that will need to solve to obtain the evolution of the frequency with time and to find out the frequency response requirements for the system we are studying.

5.4.8 The general mathematical form of the "swing equation" to be solved

Taking Eqs. (5.2)–(5.10) inclusive together, we can deduce that the mathematical form of the "swing equation" to be solved in the general case is

$$\Sigma P_{\text{GEN(NR)}} + \Sigma P_{\text{GEN }n\text{(R)}}\{1 - k(f - f_n)\} - \Sigma P_{\text{LOAD }n}\{1 + \alpha(f - f_n)\} - P_{\text{TRIP}}u(t - t_{\text{TRIP}}) + \text{GRP} = 2 \cdot H \cdot \omega \cdot \frac{d\omega}{dt} \quad (5.14)$$

where "NR" indicates "nonresponsive," "R" indicates "responsive" and the abbreviation *GRP* denotes the sequence of interpolative functions representing a general response characteristic, as introduced in Eq. (5.13). As a reminder to the reader, we recall that *GRP* (in this book) is short for "GRP."

We note that not all the terms in Eq. (5.14) will necessarily be relevant at the same time.

As time progresses at the time of, and following, a system event, different ones become relevant at different times. For example, the value of H depends on how much real inertia is on the system at a given time. For example, if plant having inertia is lost, such as a synchronous generator, the inertia of the system changes is reduced.

As another example, the term "$-k\Delta f$" in Eq. (5.14) that models the generator speed governor characteristic is not present during the first second or so following a generation loss from the system because of the time delay before the output from the responsive generators begins to appear, as explained in Section 5.3.1.3 (1).

In the next section, we examine the frequency term $2 \times H \times \omega \times d\omega/dt$ on the right-hand side of Eq. (5.14) with a view to approximating it.

5.4.9 Approximating the frequency term $2 \times H \times \omega \times d\omega/dt$ on the right-hand side of Eq. (5.14)

It will be demonstrated in the next chapter that Eq. (5.14), in the form given, is very difficult to solve analytically. Therefore so that we may proceed any further, it is necessary to make some approximations to it.

Let us assume, unless otherwise indicated, that we are always working with "per unit" quantities. Then, since by definition $\omega_{pu} = f_{pu}$, let us first make Eq. (5.14) consistent so that we are dealing with the power frequency f throughout. Having done this, Eq. (5.14) becomes

$$\Sigma P_{\text{GEN(NR)}} + \Sigma P_{\text{GEN }n\text{(R)}}\{1 - k(f - f_n)\} - \Sigma P_{\text{LOAD }n}\{1 + \alpha(f - f_n)\} - P_{\text{TRIP}}u(t - t_{\text{TRIP}}) + \text{GRP} = 2 \cdot H \cdot f \cdot \frac{df}{dt} \quad (5.14a)$$

Now, we shall make an approximation that helps make Eq. (5.14a) much more tractable, by replacing the multiplying $f = f(t)$ term on the right-hand side of Eq. (5.14a) by $f(0) = f_0$, the frequency at time equals zero.

This is a reasonable approximation to make in terms of accuracy, provided $f(t)$ does not stray too far from $f(0)$ in the time period that we are considering. For practical purposes, less than about a 5% variation of the frequency from nominal should keep any errors to an acceptable level.

Having made this change to the frequency term on the right-hand side, Eq. (5.14a) becomes

$$\Sigma P_{\text{GEN (NR)}} + \Sigma P_{\text{GEN }n\text{(R)}}\{1 - k(f - f_n)\} - \Sigma P_{\text{LOAD }n}\{1 + \alpha(f - f_n)\}$$
$$- P_{\text{TRIP}}u(t - t_{\text{TRIP}}) + \text{GRP} = 2 \cdot H \cdot f_0 \cdot \frac{\text{d}f}{\text{d}t} \tag{5.15}$$

We now see that the right-hand side of Eq. (5.15) is now a term linear in f. This modification will enable us to use solution techniques that would not otherwise be applicable.

In the next section, we shall look at how we could simplify the "GRP" as well.

5.4.10 Approximating the "general responsive profile" term

The "GRP" term in Eq. (5.15) we recall is given by Eq. (5.13) and is an expression that is a general power series in the time "t." As it stands, this power series is highly nonlinear. Further, there are no guarantees that such a power series will converge quickly. A potentially slowly converging power series could would pose us an additional problem.

One helpful way to simplify Eq (6.13) is to limit the series to *two terms only*, thus conveniently linearizing the "GRP" series.

How is this approximation justified?

Well, providing that we treat any responsive profile as a *piecewise linear function of response with time*, we shall have simplified our task considerably by having made the whole of Eq. (5.15) linear in both f and t. This will be a great advantage to us.

Therefore let us write for one piecewise linear segment

$$\text{GRP(piecewise linear segment)} = a_0 + a_1 t \tag{5.16}$$

Now, by substituting the piecewise linear expression from Eq. (5.16) into Eq. (5.15), we obtain the following form of the "swing equation" after our second approximation:

$$\Sigma P_{\text{GEN(NR)}} + \text{SP}_{\text{GEN }n\text{(R)}}\{1 - k(f - f_n)\} - \Sigma P_{\text{LOAD }n}\{1 + \alpha(f - f_n)\}$$
$$- P_{\text{TRIP}} \cdot u(t - t_{\text{TRIP}}) + a_0 + a_1 t = 2 \cdot H \cdot f_0 \cdot \frac{\text{d}f}{\text{d}t} \tag{5.17}$$

We shall see in the next chapter that the "swing equation" in the form given by Eq. (5.17) may now be solved in a variety of different ways.

5.4.11 "A bit of tidying up"

Since we have already discussed this common feature in Sections 5.4.4 and 5.4.5, that the *system load (demand) varies with frequency* and the *response varies with frequency*, the former due to induction motors in the load and the latter due to the action of automatic speed governing installed on traditional synchronous machines.

The terms modeling these two similar ways that the system active power balance varies with system frequency in Eq. (5.17) are

$$-\Sigma P_{\mathrm{LOAD}\ n}\{\alpha\cdot(f-f_n)\} \tag{5.18}$$

and

$$-\Sigma P_{\mathrm{GEN}\ n(R)}\cdot k\cdot\Delta f \tag{5.19}$$

Here we note that the frequency difference Δf in Eq. (5.19) is given by

$$\Delta f = f - f_n \tag{5.20}$$

that is, Δf, being the difference between the *frequency* and the *nominal frequency*, is the same term as that occurring in Eq. (5.18).

Since the term "$f-f_n$" appears in both definitions, they can be combined to simplify further the mathematical form of the equation [Eq. (5.17)]. Therefore we may define a new constant k_1 by

$$k_1 = \Sigma P_{\mathrm{LOAD}\ n}\cdot\alpha\cdot + \Sigma P_{\mathrm{GEN}\ n(\mathrm{R})}\cdot k \tag{5.21}$$

so that the form of Eq. (5.17) now simplifies to

$$\begin{aligned}&\Sigma P_{\mathrm{GEN(NR)}} + \Sigma P_{\mathrm{GEN}\ n(\mathrm{R})} - \Sigma P_{\mathrm{LOAD}n} - k_1\cdot(f-f_n)\\&\quad - P_{\mathrm{TRIP}}\cdot u(t-t_{\mathrm{TRIP}}) + a_0 + a_1 t = 2\cdot H\cdot f_0\cdot\frac{\mathrm{d}f}{\mathrm{d}t}\end{aligned} \tag{5.22}$$

We can see that Eq. (5.22) is, mathematically, in a particularly simple form and therefore convenient to use for the purposes of our further analyses.

For the purposes of our analyses in Chapter 6, Analytical methods for solving the "swing equation", *and in subsequent chapters, we shall use Eq. (5.22) as our starting point.*

In the next section, we provide a summary of the solution techniques available to us that we shall be studying in much more detail in Chapters 6–8.

5.5 Available techniques for solving the "swing equation"

There are many different techniques available to us that we could use for the purposes of solving Eq. (5.22). Some of these techniques use analytical methods and some use numerical methods.

The main techniques will be described briefly here and then they will be treated in much greater depth in Chapters 6–8 which follow this chapter.

5.5.1 Analytical methods

Analytical methods for solving Eq. (5.22) will be looked at in Chapter 6, Analytical methods for solving the "swing equation".

We shall cover two analytical methods: direct solution of the differential equation and solution by Laplace transforms.

5.5.1.1 The "direct solution of the differential equation" method

It is possible, after the appropriate linearization has been made, to solve Eqs. (5.2)–(5.10) inclusive analytically, in other words, to solve Eq. (5.22) analytically using the known theory of differential equations. We shall look at this approach in detail in Section 6.2 of this book. We shall provide a short summary of this approach here.

1. *Step 1: Reduction of Eq. (5.22) to its simplest possible mathematical form*

   The first step required in the "direct solution of the differential equation" method is to reduce the mathematical form of the equation [Eq. (5.22)] to its simplest possible form by representing each coefficient in the equation by a single algebraic letter so that we obtain the form:

$$c_1 + c_2 f + c_3 t = \frac{\mathrm{d}f}{\mathrm{d}t} \tag{5.23}$$

   where

$$c_1 = \frac{\Sigma P_{\mathrm{GEN}} - \Sigma P_{\mathrm{LOAD}\ n} + k_1 \cdot f_n + a_0}{2 \cdot H \cdot f_0} \tag{5.24}$$

$$c_2 = \frac{-k_1}{2 \cdot H \cdot f_0} \tag{5.25}$$

   and

$$c_3 = \frac{a_1}{2 \cdot H \cdot f_0} \tag{5.26}$$

   We can see that Eq. (5.23) represents the simplest possible form mathematically of Eq. (5.22). This completes step one.

   Note that we have dropped the "step function" term $u(t - t_{\mathrm{TRIP}})$ which does not sit easily with this solution method. It can, however, still be dealt with by dividing the solution into time periods and applying boundary conditions at the boundaries of the time periods where there is a step change in power.

   We shall see in Chapter 6, Analytical methods for solving the "swing equation", that it is also possible to leave the step-function term intact in Eq. (5.22) and solve this equation directly using the Laplace transform solution method.

2. *Step 2: Solving the differential equation [Eq. (5.23)] directly analytically*

   The second step required in the "direct solution of the differential equation" method is to solve Eq. (5.23) directly by purely analytical means using the mathematical theory of differential equations.

This produces a solution to Eq. (5.23) of the form

$$f(t) = A_1 + B_1 \cdot t + C_1 \cdot \exp(D_1 \cdot t) \tag{5.27}$$

where A_1, B_1, C_1, and D_1 are constants.

3. *Step 3: Checking that the form of the solution is valid*

   The third step required in the "direct solution of the differential equation" method is to perform a basic check that the form of the solution given in Eq. (5.27) is correct. This can be done by substituting the solution for *f*(*t*) given by Eq. (5.27) directly back into Eq. (5.22) and verifying that a way exists to assign constants so that the equation can be satisfied.

4. *Step 4: Relating the new constants back to the original ones*

   The fourth step required in the "direct solution of the differential equation" method is to relate the constants A_1, B_1, C_1, and D_1 of Eq. (5.27) back to the original "physical" constants in Eq. (5.22). Having already substituted the solution for *f*(*t*) given by Eq. (5.27) back into Eq. (5.22) and comparing like terms so that A_1, B_1, C_1, and D_1 in Eq. (5.27) are expressed as functions of the original "physical" constants in Eq. (5.22).

In the next section, we shall summarize the "solution by Laplace transforms."

5.5.1.2 The "solution by Laplace transforms"

In our "Laplace transform approach" to the solution of the "swing equation" which we shall deal with in Section 6.3 of the following chapter, we are going to go back to Eq. (5.14) to see if is possible to deal successfully with the nonlinearities in the *GRP* term and with the nonlinear right-hand side of Eq. (5.14).

We know in advance that we shall be able to deal straightforwardly with the unit step function term $u(t - t_{TRIP})$ because there is a well-known and simple Laplace transform available for this term.

We shall now summarize the main steps we shall be covering in our Laplace transform analysis of the "swing equation" in the next chapter.

1. *Step 1: Processing the individual terms of the equation*

   We will look at the "swing equation" in the form given by Eq. (5.14).

   We shall examine each term with a view to its suitability for being Laplace transformed. We will also review the meaning and use of each term.

2. *Step 2: The need to make approximations in the equation*

   We shall look at the need to make approximations to Eq. (5.14), so that the problem is made soluble. The selected approximated version of Eq. (5.14) will be assembled. Unlike Eq. (5.22), the unit step function $u(t - t_{TRIP})$ will be retained as it is amenable to Laplace transforming.

3. *Step 3: Laplace transforming of the approximated equation*
   The approximated version will be Laplace transformed and the result simplified as far as possible before taking inverse transforms.
4. *Step 4: Taking inverse Laplace transforms*
   Inverse Laplace transforms will then be taken of the transformed equation and the resulting expression will be simplified as much as possible.
5. *Step 5: Checking the solution*
   The solution will be checked and compared against the solution obtained by the "direct solution of the differential equation" method.

In the next section, we preview the *numerical methods* covered in Chapter 7, Numerical methods for solving the "swing equation." There are many numerical methods available, and we shall be focusing our attention on a selection of them which we shall cover in some detail.

5.5.2 Numerical methods

The numerical methods available for solving the "swing equation" will be looked at in detail in Chapter 7, Numerical methods for solving the "swing equation," along with the "exact solution" or "direct solution" of the differential equation method.

5.5.2.1 The nature of numerical methods

As a rule, numerical methods for solving the "swing equation" all begin by converting the *differential equations* [Eqs. (5.2)–(5.10)] into the form of *difference equations*. Amongst other things, this method involves replacing the speed differential dω with the speed step change $\Delta\omega$, and the differential dt with the time-step Δt. Solutions are then iterated in time as successive time-steps are applied until adequate convergence is obtained.

5.5.2.2 The basic methods

In Section 7.1 we shall look at the basics of numerical methods and examine the different kinds of numerical methods available for solving our problem.

5.5.2.3 The choice of time-step

In Section 7.2 we shall examine how the choice of time-step affects the accuracy and speed of convergence of the results obtained from the numerical simulation.

5.5.2.4 The choice of simulation time

In Section 7.3 we shall look at the reasons for choosing different simulation times to suit different modeling situations.

5.5.2.5 Advantages and disadvantages of the methods

Finally, in Section 7.4 we shall compare the efficiencies of some of the different numerical methods available to us.

In the next section, we summarize the "control diagram" method of solving the "swing equation."

5.5.3 The "control diagram" method

This approach will be described in Chapter 8, The "control diagram" method, in a chapter of its own.

5.5.3.1 The nature of the "control diagram" method

The "control diagram" method uses a control-type feedback loop representation of the "swing equation" to model the load response and the system power balance. Elements of the control loop are assigned a set of starting numerical values. The loop is then iterated to generate successive outputs. This process continues until the level of convergence to the desired output has been achieved.

5.5.3.2 How the method is set up

In Section 8.2 we shall describe how the elements of the "control diagram method" are put together and how the different aspects of the problem are modeled.

5.5.3.3 An example of the "control diagram" method

In Section 8.3 we shall give an example of a possible use of the "control diagram" method.

5.5.3.4 Comparison of the "control diagram" method with some of the other methods

In Section 8.4 we shall compare the results of the "control diagram" method with the results of some of the other methods that we have looked at earlier in the book.

5.6 The next chapter: analytical methods for solving the "swing equation"

5.6.1 The "direct solution of the differential equation" and "Laplace transform" methods

In the next chapter, we shall consider two different but fundamentally related, analytical methods for solving the "swing equation": (1) the "direct solution of the differential equation" method and (2) the "Laplace transform" method.

5.6.2 Worked examples to be provided in MATLAB

In the next chapter, we shall also illustrate each of the two methods described earlier by a worked example in MATLAB.

References

Books

[1] P. Kundur, Electric Power Research Institute, Power System Engineering Series Power System Stability and Control, 22nd ed., McGraw Hill Education, 2017.

[2] B.M. Weedy, Electric Power Systems, second ed., John Wiley & Sons, London, New York, Sydney, and Toronto, 1972.

Chapter 6

Analytical methods for solving the "swing equation"

6.1 Introduction to the analytical solution methods

In this chapter, we shall be looking at two of the possible analytical methods available to us for solving the approximated per unit form of the "swing equation" that was given as Eq. (5.22) of Chapter 5, Techniques available for calculating frequency response requirements.

The first method we shall look at is the "direct solution of the differential equation" method. The second method we shall consider is the "solution by Laplace transforms."

6.1.1 Direct solution of the differential equation ("Method 1")

The first of the methods that we shall be considering is the "direct solution of the differential equation" method. We have already made some approximations to the structure of the equation to obtain a tractable solution (see Chapter 5: Techniques available for calculating frequency response requirements, for details). The mathematical theory we shall be using is contained in standard mathematical textbooks, such as Refs. [1,2].

6.1.2 Solution by Laplace transforms ("Method 2")

The second of the methods that we shall be considering is the *solution by Laplace transforms*. This time we shall take Laplace transforms of Eq. (6.1) to rearrange the resulting terms in s to find a closed expression for $F(s)$ and then take the inverse Laplace transform of $F(s)$ to find the desired solution $f(t)$.

6.1.3 The common objective in the case of the analytical methods

The objective in the case of all the analytical methods of solution is to find a closed expression for $f(t)$, since when we know $f(t)$, we will also know the time-evolution of the power system frequency following a system event such as the tripping of a generating set (generator).

Modern Aspects of Power System Frequency Stability and Control.
DOI: https://doi.org/10.1016/B978-0-12-816139-5.00006-0

6.1.4 Recalling the equation to be solved—the "swing equation"

For ease of references in this chapter we reproduce here the "swing equation" that was given by Eq. (5.22) at the end of the last chapter.

$$\Sigma P_{\mathrm{GEN(NR)}} + \Sigma P_{\mathrm{GEN}n\mathrm{(R)}} - \Sigma P_{\mathrm{LOAD}n} - k_1 \cdot (f - f_n) - P_{\mathrm{TRIP}} \cdot u\left(t - t_{\mathrm{trip}}\right) + a_0 + a_1 t = 2 \cdot H \cdot f_0 \cdot \frac{\mathrm{d}f}{\mathrm{d}t} \tag{6.1}$$

In the next section, we move on to the analysis of Eq. (6.1) using *Method 1:* the "direct solution of the differential equation."

6.2 Method 1: The direct solution of the differential equation

6.2.1 Solving the mathematical problem: what does Eq. (6.1) mean?

Let us start, as always in this chapter, with Eq. (6.1). Let us first remind ourselves of the meaning of each of the terms in this equation.

1. ΣP_{GEN}: This term represents the sum total of generation in pu at the beginning of the simulation, that is, unless stated otherwise, at $t = 0$.
2. $\Sigma P_{\mathrm{LOAD}n}$: This term represents the sum total of the system loads (demands) in pu at $t = 0$, the initial time, all measured at the nominal frequency.
3. k_1: This term, defined in Eq. (5.21) in the previous chapter, represents the combined effects of the variations of generation and load with frequency. It combines the model of the governor "droop" profile with the model of the "load damping" effect (the variation of load with frequency).
4. P_{TRIP}: This term represents the amount in pu of generation that is tripped at some time t_{trip} during the simulation, usually at $t = 0$ as a minimum. Multiple trips at different times may be modeled by using more than one term of this kind in the equation.
5. $a_0 + a_1\ t$: These terms represent the amount of response from "general sources" that is not represented by the "governor response" term encapsulated within k_1.
6. A general response profile may be built up by joining together multiple linear segments (piecewise linear form).
7. H: This term represents the equivalent inertia constant for the whole system (see Chapter 3: Per unit systems for frequency analysis, for an explanation of how this quantity may be calculated). It embodies all the individual inertias contained in the system into a single value.
8. $\mathrm{d}f/\mathrm{d}t$: This term is the rate of change of frequency in pu per second. The aim of all the analytical methods is to "remove" this term from the

equation and "replace" it with f, the frequency so that we obtain an expression of the form $f(t)$ describing the time-evolution of the frequency.

6.2.2 Some remarks on statements made in the preceding sections

1. *A note on the linearization of "$f \cdot \mathrm{d}f/\mathrm{d}t$"*

   We have already made the standard approximation that $f = f_0 = f(0)$ on the right-hand side of Eq. (6.1). In other words the term $f \cdot \mathrm{d}f/\mathrm{d}t$ has been replaced by $f(0) \cdot \mathrm{d}f/\mathrm{d}t$, where $f(0) = f_0$ is the system frequency at the beginning of the simulation, that is, at $t = 0$ (see Chapter 5: Techniques available for calculating frequency response requirements, for details).

   This approximation has achieved the linearization in f that we need to be able to make to apply the standard mathematical theory for the present chapter that is contained, for example, in Refs. [1,2]. It is a reasonable approximation in terms of accuracy, provided $f(t)$ does not depart too much from $f(0)$ in the time period that we are considering.
2. *The combined parameter k_1*

   As mentioned earlier, we have modeled both the *governor response* and the *variation of demand with frequency* under the single parameter k_1 (see Section 5.3.7 for more details on this). This is for mathematical convenience: we know that the two effects are different physically, but they have a common mathematical representation that helps to simplify the problem for us mathematically.
3. *Dropping of the $u(t - t_{\mathrm{trip}})$ term in the "direct solution of the differential equation" method*

   Further on in the present section, we shall disregard the $u(t - t_{\mathrm{trip}})$ term (which serves to model an instantaneous change in generation or demand) in favor of modeling all such changes in generation or load (or both) at any instant in time as piecewise linear changes with time. This is, in fact, more in keeping with reality, because, of course, nothing in nature changes truly instantaneously. In this way the term in $u(t - t_{\mathrm{trip}})$ will be removed from Eq. (6.1) and instead included in the term $c_0 + c_1 t$ in each time period.
4. *Modeling of all trips and response (except governor response) as piecewise linear segments in the "direct solution of the differential equation" method*

   In other words, Section 6.2 only (the "direct solution of the differential equation" method), we shall be modeling both trips and the general (nongovernor) response *GRP* (general responsive profile) as a sequence of piecewise linear segments, each of which will be described by a term such as $c_0 + c_1 t$.

5. *The equivalence to the modeling of changes in the system power balance*
   As we explained in Chapter 5, Techniques available for calculating frequency response requirements, these expressions, which are linear in t, are intended to describe a change in the power balance profile (in generation or demand) that is not attributable to the governors of synchronous machines.
   Governor response, as always, will be modeled as part of the parameter k_1.
6. *In the next section*
   In the next section, we shall begin to solve Eq. (6.1) with the additional simplification of modeling "step change" trips of generation or demand with piecewise linear segments. Once we have removed the step function model $u(t - t_{\text{trip}})$ from the equation, we shall then simplify the equation into as simple a form as possible from a mathematical point of view.

6.2.3 The differential equation [Eq. (6.1)] with the additional simplification noted earlier

Having now included the additional modification that we described earlier in Section 6.2.2 (3), we have obtained, finally, as the starting point for our "direct solution of the differential equation" method, the following equation:

$$\begin{aligned}\Sigma P_{\text{GEN(NR)}} + \Sigma P_{\text{GEN}n\text{(R)}} - \Sigma P_{\text{LOAD}n} - k_1 \cdot (f - f_n) \\ + a_0 + a_1 t = 2 \cdot H \cdot f_0 \cdot \frac{df}{dt}\end{aligned} \tag{6.2}$$

We notice that Eq. (6.2) now takes on the relatively simple mathematical form

$$c_1 + c_2 f + c_3 t = \frac{df}{dt} \tag{6.3}$$

where

$$c_1 = \frac{\Sigma P_{\text{GEN(NR)}} + \Sigma P_{\text{GEN}n\text{(R)}} - \Sigma P_{\text{LOAD}n} + k_1 \cdot f_n + a_0}{2 \cdot H \cdot f_0} \tag{6.4}$$

$$c_2 = \frac{-k_1}{2 \cdot H \cdot f_0} \tag{6.5}$$

$$c_3 = \frac{a_1}{2 \cdot H \cdot f_0} \tag{6.6}$$

The above results can, of course, be easily verified by the reader.
In the next section, we move on to the actual process of solving Eq. (6.3).

6.2.4 Solving the differential equation

Using the known theory for solving differential equations, available in, for example, Refs. [1,2], we shall now proceed to solve the differential equation given in Eq. (6.3).

As an initial step, to correspond to the theoretical form, it is useful to rearrange Eq. (6.3) in the following way:

$$\frac{df}{dt} - c_2 f = c_1 + c_3 t \tag{6.7}$$

It can be verified by the reader that Eq. (6.3) is now expressed in the standard form for a *first-order differential equation*, that is

$$\frac{df(t)}{dt} + b_1 \cdot f(t) = b_2 + b_3 \cdot t = g(t) \tag{6.8}$$

where $g(t)$ is, in our case, a *linear function of t*, which acts in the equation in the manner of a "driving function." The two sets of constants in Eqs. (6.7) and (6.8) are clearly related as follows:

$$b_1 = -c_2 = \frac{k_1}{2 \cdot H \cdot f_0} \tag{6.9}$$

$$b_2 = c_1 = \frac{\Sigma P_{\mathrm{GEN(NR)}} + \Sigma P_{\mathrm{GEN}n\mathrm{(R)}} - \Sigma P_{\mathrm{LOAD}n} + k_1 \cdot f_n + a_0}{2 \cdot H \cdot f_0} \tag{6.10}$$

$$b_3 = c_3 = \frac{a_1}{2 \cdot H \cdot f_0} \tag{6.11}$$

where in these results, we have also referred to Eqs. (6.4)–(6.6).

If we were to consult a standard text, such as Ref. [1], we would find that whether a first-order differential equation such as Eq. (6.8) may be solved analytically depends on the details of the right-hand side of the equation, that is, the details of the "driving function" part. In our case, we have a particularly simple form for the driving function, so we suspect that the chances of finding an analytical solution are quite high in our case. [It could have been much worse: $g(t)$ can in general be quite a complicated function in this kind of equation.]

In fact the solution [Eq. (6.8)] can be found relatively straightforwardly "from first principles," as it were, as follows:

The solution of a linear equation, such as Eq. (6.8), is always given by the sum of the "complementary function" and a "particular integral." What this means is that if we add the solution of the equation, with the "driving function" part equal to zero, to any solution of the differential equation (however simple), we then have the "general solution," that is, the complete solution of the equation.

However, in the case of the relatively simple, first-order differential equation given in Eq. (6.8), there is also available to us a more straightforward method (see, for example, Ref. [1], p. 13) that uses what is known as an "integrating factor." If we can find an integrating factor for the first-order differential equation, then finding the complete solution thereafter is relatively easy.

Proceeding from Eq. (6.8), we know from Ref. [1] that the integrating factor, with, of course, the variable y changed to f, and the variable x changed to t (to be suitable for our purposes here) is given by

$$\text{Integrating factor} = \int_0^t \exp(b_1 t)\mathrm{d}t = \exp(b_1 t) \tag{6.12}$$

Note that we have for simplicity taken the multiplying constant equal to one in the integrating factor on the right-hand side of Eq. (6.12).

Here, we notice that, in evaluating the integral expression to find the integrating factor, we have also taken the lower bound t_0 of the integral expression to be equal to *zero* for our case, since for our purposes we are assuming, for convenience, that time starts at *zero*.

Applying the *integrating factor* given in Eq. (6.12) to each term of our differential equation [Eq. (6.8)] in turn, we now obtain the following:

$$\frac{\mathrm{d}f}{\mathrm{d}t} \times \exp(b_1 t) + b_1 f \times \exp(b_1 t) = b_2 \cdot \exp(b_1 t) + b_3 \cdot t \cdot \exp(b_1 t) \tag{6.13}$$

which is the same thing as saying that

$$\frac{d}{\mathrm{d}t}\{f \cdot \exp(b_1 \cdot t)\} = b_2 \cdot \exp(b_1 t) + b_3 \cdot t \cdot \exp(b_1 t) \tag{6.14}$$

Upon integrating both sides of Eq. (6.14) between the limits 0 to t, we have

$$f \cdot \exp(b_1 \cdot t) - f_0 = b_2 \cdot \int_0^t \exp(b_1 \cdot t) \cdot \mathrm{d}t + b_3 \cdot \int_0^t (b_1 \cdot t) \cdot \mathrm{d}t \tag{6.15}$$

The first term on the right-hand side of Eq. (6.15) can be integrated straight away as it stands, and the second term on the right-hand side of Eq. (6.15) can be integrated by parts, so we now obtain

$$\begin{aligned} f \cdot \exp(b_1 t) - f_0 &= \left(\frac{b_2}{b_1}\right) \cdot \{\exp(b_1 \cdot t) - 1\} + \left(\frac{b_3}{b_1}\right) \cdot t \cdot \exp(b_1 t) \\ &\quad - \left(\frac{b_3}{b_1}\right) \cdot \int_0^t (b_1 \cdot t) \cdot dt. \end{aligned} \tag{6.16}$$

Then, from Eq. (6.16), we obtain

$$\begin{aligned} f\cdot\exp(b_1 t) - f_0 &= \left(\frac{b_2}{b_1}\right)\cdot\{\exp(b_1\cdot t) - 1\} + \left(\frac{b_3}{b_1}\right)\cdot t\cdot\exp(b_1\cdot t) \\ &- \left(\frac{b_3}{b_1^2}\right)\cdot\{\exp(b_1\cdot t) - 1\} \end{aligned} \tag{6.17}$$

A straightforward rearrangement of Eq. (6.17) then gives us the following expression for $f(t)$:

$$f(t) = \left(\frac{b_2}{b_1} - \frac{b_3}{b_1^2}\right) + \left(\frac{b_3}{b_1}\right)\cdot t + \left(f_0 - \frac{b_2}{b_1} + \frac{b_3}{b_1^2}\right)\cdot\exp(-b_1\cdot t) \tag{6.18}$$

We have thus achieved the goal of the analytical approach of obtaining f as a function of t, thus giving us the evolution of the power system frequency with time that we are trying to determine.

6.2.5 A check on the accuracy of the solution for $f(t)$ given by Eq. (6.18)

The form of the solution $f(t)$ given by Eq. (6.18) may be verified to be correct by substituting it and its derivative back into Eq. (6.8), as follows:

From Eq. (6.18) we can calculate the first derivative of f with respect to t, which is:

$$\frac{df}{dt} = \frac{b_3}{b_1} - \left(b_1\cdot f_0 - b_2 + \frac{b_3}{b_1}\right)\cdot\exp(-b_1\cdot t) \tag{6.19}$$

Substituting Eqs. (6.18) and (6.19) back into Eq. (6.8), we thereby obtain the following expression as a check on the accuracy of Eq. (6.8):

$$\begin{aligned} &\frac{b_3}{b_1} - \left(b_1\cdot f_0 - b_2 + \frac{b_3}{b_1}\right)\cdot\exp(-b_1\cdot t) \\ &+ b_1\cdot\left\{\left(\frac{b_2}{b_1} - \frac{b_3}{b_1^2}\right) + \left(\frac{b_3}{b_1}\right)\cdot t + \left(f_0 - \frac{b_2}{b_1} + \frac{b_3}{b_1^2}\right)\cdot\exp(-b_1\cdot t)\right\} \\ &= b_2 + b_3\cdot t \end{aligned} \tag{6.20}$$

It may be verified by the reader that, in Eq. (6.20), all terms cancel out with each other; and therefore, we can be confident that Eq. (6.18) is, formally at least, the correct solution of Eq. (6.8) for $f(t)$.

The solution given by Eq. (6.18) will also be checked again by a different method later in this chapter. In the next section, we look at the form of the solution with the intention of finding the coefficients in terms of the coefficients of the original equation.

6.2.6 Noting the form of the solution given by Eq. (6.18)

It can be seen readily by the reader upon inspection that the solution [Eq. (6.18)] for $f(t)$ is in fact of the basic form:

$$f(t) = A_1 + B_1 \cdot t + C_1 \cdot \exp(D_1 \cdot t) \tag{6.21}$$

in which A_1, B_1, C_1, and D_1 are constants, which, by comparing Eq. (6.21) with Eq. (6.18) can be found straightforwardly to be given by the following relationships:

$$A_1 = \left(\frac{b_2}{b_1} - \frac{b_3}{b_1^2}\right) \tag{6.22}$$

$$B_1 = \left(\frac{b_3}{b_1}\right) \tag{6.23}$$

$$C_1 = f_0 - b_2/b_1 + b_3/b_1^2 \tag{6.24}$$

$$D_1 = -b_1 \tag{6.25}$$

To summarize, we have obtained the relationships in Eqs. (6.22)–(6.25) by comparing each of the coefficients in Eq. (6.18) with the corresponding ones in Eq. (6.21).

In the next section, we shall finally relate the four constants A_1, B_1, C_1, and D_1 back to the four parameters H, k_1, a_0, and a_1 in the original equation [Eq. (6.2)] that was stated near the beginning of this chapter.

6.2.7 Relating the constants A_1, B_1, C_1, and D_1 of Eq. (6.21) back to H, k_1, a_0, and a_1 of Eq. (6.2)

Let us, for ease of comparison, repeat Eq. (6.2), the equation we are solving, and Eq. (6.21), and its solution is as follows:

$$\begin{aligned} &\Sigma P_{\mathrm{GEN(NR)}} + \Sigma P_{\mathrm{GEN}n\mathrm{(R)}} - \Sigma P_{\mathrm{LOAD}n} - k_1 \cdot (f - f_n) \\ &+ a_0 + a_1 t = 2 \cdot H \cdot f_0 \cdot \frac{\mathrm{d}f}{\mathrm{d}t} \end{aligned} \tag{6.26}$$

$$f(t) = A_1 + B_1 \cdot t + C_1 \cdot \exp(D_1 \cdot t) \tag{6.27}$$

To find the constants A_1, B_1, C_1, and D_1 of Eq. (6.27) in terms of the "original" parameters of Eq. (6.26), we can substitute for $f(t)$ from Eq. (6.27) back into Eq. (6.26).

We also need, of course, to calculate the *derivative of f with respect to t*, that is, df/dt, from Eq. (6.27) to substitute back into Eq. (6.26), as follows:

We note once more, by differentiating Eq. (6.26) with respect to t that

$$\frac{df}{dt} = B_1 + C_1 \cdot D_1 \cdot \exp(D_1 \cdot t) \tag{6.28}$$

Therefore by substituting Eqs. (6.27) and (6.28) into Eq. (6.26), we obtain the following expression:

$$\Sigma P_{\mathrm{GEN(NR)}} + \Sigma P_{\mathrm{GEN}n\mathrm{(R)}} - \Sigma P_{\mathrm{LOAD}n} - k_1 \cdot \{A_1 + B_1 \cdot t + C_1 \cdot \exp(D_1 \cdot t) - f_n\} + a_0 + a_1 t = 2 \cdot H \cdot f_0 \cdot \{B_1 + C_1 \cdot D_1 \cdot \exp(D_1 \cdot t)\} \tag{6.29}$$

By comparing the coefficients of the *constant*, *t* and *exponential terms* in Eq. (6.29) one by one, we now obtain, as may be verified easily by the reader, the following equations:

Constant terms:

$$\Sigma P_{\mathrm{GEN(NR)}} + \Sigma P_{\mathrm{GEN}n\mathrm{(R)}} - \Sigma P_{\mathrm{LOAD}n} - k_1 \cdot (A_1 - f_n) + a_0 = 2 \cdot H \cdot f_0 \cdot B_1 \tag{6.30}$$

Terms in t:

$$-k_1 \cdot B_1 + a_1 = 0 \tag{6.31}$$

Exponential terms:

$$-k_1 \cdot C_1 = 2 \cdot H \cdot f_0 \cdot C_1 \cdot D_1 \tag{6.32}$$

Initial value:

Since, to be able to completely solve a set of simultaneous equations, we need to have the same number of equations as unknowns, we also need to factor in to our calculation the value of $f(t)$ at $t = 0$, that is $f(0) = f_0$. We can, of course, easily do this by setting $t = 0$ in Eq. (6.27), and we obtain thereby the fourth equation that we need to be able to determine all four of the constants A_1, B_1, C_1, and D_1:

$$f(0) = f_0 = A_1 + C_1 \tag{6.33}$$

Considering together the four equations [Eqs. (6.30)–(6.33)], we can find straight away from Eq. (6.31) that

$$B_1 = \frac{a_1}{k_1} \tag{6.34}$$

Substituting for B_1 from Eq. (6.34) into Eq. (6.30) and rearranging the resulting expression, we can now also easily deduce that

$$A_1 = f_n + \left(a_0 + \Sigma P_{\mathrm{GEN(NR)}} + \Sigma P_{\mathrm{GEN}n\mathrm{(R)}} - \Sigma P_{\mathrm{LOAD}n}\right)/k_1 - 2 \cdot H \cdot f_0 \cdot a_1/k_1^2 \tag{6.35}$$

Given the result [Eq. (6.35)], by substituting for A_1 into Eq. (6.33), we can find out additionally that

$$\begin{aligned} C_1 &= f_0 - f_n - \left(a_0 + \Sigma P_{\mathrm{GEN(NR)}} + \Sigma P_{\mathrm{GEN}n\mathrm{(R)}} - \Sigma P_{\mathrm{LOAD}n}\right)/k_1 \\ &\quad + 2 \cdot H \cdot f_0 \cdot a_1/k_1^2 \end{aligned} \tag{6.36}$$

Finally, providing $C_1 \neq 0$, we can cancel the C_1's from both sides of Eq. (6.32) and rearrange the resulting expression to produce, finally, the expression for the constant D_1:

$$D_1 = \frac{-k_1}{2 \cdot H \cdot f_0} \tag{6.37}$$

So, to summarize, in Eqs. (6.35), (6.34), (6.36), and (6.37), respectively, we have now found all four of the parameters A_1, B_1, C_1, and D_1, which we have been able express in terms of our "original" four parameters H, k_1, a_0, and a_1 as required.

6.2.8 An alternative method for relating A_1, B_1, C_1, and D_1 back to H, k_1, a_0, and a_1

We can also find A_1, B_1, C_1, and D_1 in terms of H, k_1, a_0, and a_1 by the equally valid method of taking the definitions of b_1, b_2, and b_3 from Eqs. (6.9) to (6.11) inclusive and substituting directly for b_1, b_2, and b_3 back into Eqs. (6.22)–(6.25) inclusive.

For easy reference, here are Eqs. (6.9)–(6.11) again:

$$b_1 = -c_2 = \frac{k_1}{2 \cdot H \cdot f_0} \tag{6.38}$$

$$b_2 = c_1 = \frac{\Sigma P_{\mathrm{GEN(NR)}} + \Sigma P_{\mathrm{GEN}n\mathrm{(R)}} - \Sigma P_{\mathrm{LOAD}n} + k_1 \cdot f_n + a_0}{2 \cdot H \cdot f_0} \tag{6.39}$$

$$b_3 = c_3 = \frac{a_1}{2 \cdot H \cdot f_0} \tag{6.40}$$

and Eqs. (6.22)–(6.25) again:

$$A_1 = \left(\frac{b_2}{b_1} - \frac{b_3}{b_1^2}\right) \tag{6.41}$$

$$B_1 = \left(\frac{b_3}{b_1}\right) \tag{6.42}$$

$$C_1 = f_0 - \frac{b_2}{b_1} + \frac{b_3}{b_1^2} \tag{6.43}$$

$$D_1 = -b_1 \tag{6.44}$$

Thus, from the earlier seven equations, we have

$$
\begin{aligned}
A_1 &= \left(\frac{b_2}{b_1} - \frac{b_3}{b_1^2}\right) \\
&= \left[\left\{\Sigma P_{\mathrm{GEN(NR)}} + \Sigma P_{\mathrm{GEN}n\mathrm{(R)}} - \Sigma P_{\mathrm{LOAD}n} + k_1 \cdot f_n + a_0\right\}/2 \cdot H \cdot f_0\right] \\
&\quad /\left(k_1/2 \cdot H \cdot f_0\right) - \left(a_1/2 \cdot H \cdot f_0\right)/\left(k_1/2 \cdot H \cdot f_0\right)^2 \\
&= \left\{\Sigma P_{\mathrm{GEN(NR)}} + \Sigma P_{\mathrm{GEN}n\mathrm{(R)}} - \Sigma P_{\mathrm{LOAD}n} + k_1 \cdot f_n + a_0\right\} \\
&\quad /k_1 - 2 \cdot H \cdot f_0 a_1/k_1^2 \\
&=> \\
A_1 &= f_n + \left\{\Sigma P_{\mathrm{GEN(NR)}} + \Sigma P_{\mathrm{GEN}n\mathrm{(R)}} - \Sigma P_{\mathrm{LOAD}n} + a_0\right\}/k_1 \\
&\quad - 2 \cdot H \cdot f_0 a_1/k_1^2
\end{aligned}
\tag{6.45}
$$

It can be verified that the result in Eq. (6.45) matches that given in Eq. (6.35) of Section 6.2.7 as required.

Also

$$
\begin{aligned}
B_1 &= \frac{b_3}{b_1} \\
&= \frac{\left(a_1/2 \cdot H \cdot f_0\right)}{\left(k_1/2 \cdot H \cdot f_0\right)} \\
&=> \quad B_1 = \frac{a_1}{k_1}
\end{aligned}
\tag{6.46}
$$

The result in Eq. (6.46) matches that given in Eq. (6.34) of Section 6.2.7 as expected.

Further:

$$
\begin{aligned}
C_1 &= f_0 - \frac{b_2}{b_1} + \frac{b_3}{b_1^2} \\
&= f_0 - \left[\left\{\Sigma P_{\mathrm{GEN(NR)}} + \Sigma P_{\mathrm{GEN(NR)}} - \Sigma P_{\mathrm{LOAD}n} + k_1 \cdot f_n + a_0\right\}/2 \cdot H \cdot f_0\right] \\
&\quad /\left(k_1/2 \cdot H \cdot f_0\right) + \left(a_1/2 \cdot H \cdot f_0\right)/\left(k_1/2 \cdot H \cdot f_0\right)^2 \\
&=> \quad C_1 = f_0 - f_n - \left(\Sigma P_{\mathrm{GEN(NR)}} + \Sigma \mathrm{P}_{\mathrm{GEN}n\mathrm{(R)}} - \Sigma P_{\mathrm{LOAD}n} + a_0\right) \\
&\quad /\mathrm{k}_1 + 2 \cdot H \cdot f_0 \cdot a_1/k_1^2
\end{aligned}
\tag{6.47}
$$

We see that the result in Eq. (6.47) matches that given in Eq. (6.36) of Section 6.2.7 as required.

And finally, from Eqs. (6.38) and (6.44)

$$
\begin{aligned}
D_1 &= -b_1 \\
=> \quad D_1 &= \frac{-k_1}{2 \cdot H \cdot f_0}
\end{aligned}
\tag{6.48}
$$

And, it can be confirmed that the result in Eq. (6.48) matches that given in Eq. (6.37) of Section 6.2.7 as expected.

Thus we have derived again, using an alternative method from that followed in *Section 6.2.7*, the relationships between the constants A_1, B_1, C_1, and D_1 and the original constants H, k_1, a_0, and a_1. We have found that the results are the same. Hence, for the moment at least, we are confident of our results!

Let us now proceed briefly to remind ourselves, for the sake of clarity, what the group of parameters H, k_1, a_0, and a_1 serve to represent in the context of the power system as a whole.

6.2.9 The parameter *H*

We recall that the parameter "H" represents the total equivalent inertia constant of the power system we are considering, as discussed in Chapter 4, Initial analysis of the frequency control problem and a derivation of the swing equation. We will, in general, need to calculate the total equivalent inertia of a system from the inertia of the individual machines. Again, this process was described in detail in Chapter 4, Initial analysis of the frequency control problem and a derivation of the swing equation.

6.2.10 The parameter k_1

The parameter "k_1" represents the combination of the effects of the *variation of demand with frequency and the action of synchronous generator automatic speed governors.*

In Eq. (5.21) *of* Section 5.3.11, we saw that k_1 is given by the expression:

$$k_1 = \Sigma P_{\mathrm{LOAD}n} \cdot \alpha \cdot + \Sigma P_{\mathrm{GEN}n(\mathrm{R})} \cdot k \tag{6.49}$$

Note that here, as elsewhere, $P_{\mathrm{LOAD}n}$ represents the *active power load at nominal frequency* and $P_{\mathrm{GEN}n(\mathrm{R})}$ represents the *responsive generation measured at nominal frequency.*

6.2.11 The parameters a_0 and a_1

The parameters "a_0" and "a_1" are the remaining linear parts of the more general expression GRP that was introduced in Eq. (5.13). The reader will recall that we decided to reduce the general power series expression GRP to piecewise linear sections to enable the differential equation in Eq. (5.14) to be solved more easily. We also decided to subsume the "approximation" of the unit step function u [see *Section 6.2.2 (3)* and following, above] within these two parameters for reasons of tractability in the case of the "direct solution of the differential equation" method.

6.2.12 More about the variation of demand with frequency

The system demand (load) can vary with frequency if, for example, there are any induction machines (for example, induction motors) present as part of this demand. As a result, if the system frequency changes, so will the demand that is measured on the system by the "system operator."

From the macroscopic system point of view, this effect is relatively small, but it does have a significant effect on the process of the recovery of the frequency following the event of an unplanned generation or demand (load) loss from the power system. This effect, sometimes referred to as the "load damping" effect, is stabilizing for system frequency changes. Let us look a little more closely into this effect.

It can be seen from the foregoing analysis [see, for example, Eq. (5.14)] that the system load is known to vary with the system frequency according to the following approximately linear rule (see, for example, Ref. [3]):

$$\Sigma P_{\mathrm{LOAD}} = \Sigma P_{\mathrm{LOAD}n}\{1 + \alpha(f - f_n)\} \tag{6.50}$$

where α is a constant whose value is dependent on the nature and composition of the system demand (load).

From Eq. (6.50), we can see easily that if $f = f_n$ then $\Sigma P_{\mathrm{LOAD}} = \Sigma P_{\mathrm{LOAD}n}$ as we would expect as one confirmation of the accuracy of this model.

What is a typical value for α? The answer to this question depends, of course, on the frequency-dependent characteristics of the system we are studying. In one system, known to the author, whose nominal frequency is 50 Hz, the demand (load) is observed to decrease (alternatively, increase) by about 1% for each 0.5 Hz (1%) drop (alternatively rise) in the system frequency. If f and f_n in Eq. (6.50) are in Hz, then this implies that

$$\begin{aligned} & 0.01 = \alpha_{\mathrm{Hz}} \cdot 0.5 \\ => \quad & \alpha_{\mathrm{Hz}} = 0.02 \end{aligned} \tag{6.51}$$

However, since in this book we are writing all our equations in *per unit*, this is the same as saying that a 0.01 pu change in demand (load) is accompanied by a 0.01 pu change in frequency. Therefore *in per unit*, we have instead from Eq. (6.50):

$$\begin{aligned} & 0.01 = \alpha_{\mathrm{pu}} \cdot 0.01 \\ => \quad & \alpha_{\mathrm{pu}} = 1 \end{aligned} \tag{6.52}$$

Hence, we have demonstrated (not for the first time) that it is important to be certain, for the sake of accuracy of calculation, whether we are working in *physical units* or *per unit*!

In the next section, we shall use the solutions that we have obtained in the earlier sections to solve a worked example that we shall also program into *MATLAB* so that the parts of the calculation can be recorded, and a graph of frequency against time may be plotted out at the end.

6.3 MATLAB example simulation 1: using the "direct solution of the differential equation" method to solve the "swing equation"

6.3.1 Objective

Using the results of the analysis that we have obtained using the "direct solution of the differential equation" method detailed earlier, let us turn to an example. We shall be using MATLAB as a convenient programing tool to assist us.

6.3.2 Introduction to the first illustrative example

6.3.2.1 Basic description

We shall construct a practical example that uses the theory that we have been developing earlier in a concrete way. Let us start with the (imaginary) interconnected power system depicted in the single-line Diagram 6.1. Diagram 6.1 represents a small 50-Hz system whose peak demand is 3000 MW and consists of 10 identical generators of 300 MW-rated power output each. The inertia constant of each generator is $H = 5.0$ seconds. If all

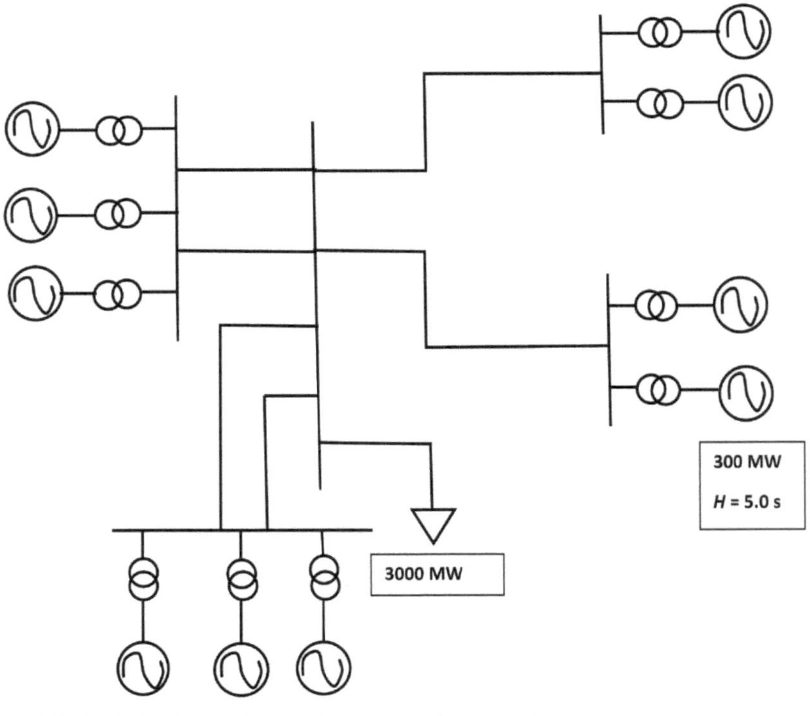

DIAGRAM 6.1 First example of an interconnected power system.

the generators are running at their rated outputs what will happen if one of the generators trips off the system, and how large will the effect be?

In this example, we are not considering the effects of response on the solution.

We shall consider the effects of response in MATLAB example simulation 2.

6.3.2.2 A note to the reader

We remind the reader that it is not necessary for the purposes of our frequency studies, using the analysis that we have derived earlier, to run any load-flow studies, fault current analyses, stability analyses, or any other of the many prepackaged computer analysis tools available to us as power system planning engineers. The problem can be solved without using a commercial package since we have been able to model the problem analytically.

6.3.2.3 The advantage of using a mathematical model

The advantage of performing a frequency analysis in this way using a mathematical model is that by removing the need to do the many repetitions of load flows that would be necessary if a commercial analysis package was used, the time needed to do the analysis is drastically reduced.

6.3.3 Steps to be taken in the illustrative example

In this illustrative MATLAB study, based on the "direct solution of the differential equation" method, we shall break the problem down into the following six steps:

STEP 1: "A statement of the problem"
STEP 2: "The general solution to the problem"
STEP 3: "The input data to the MATLAB code"
STEP 4: "The output data from the MATLAB code"
STEP 5: "The suggested MATLAB code"
STEP 6: "Instructions for the user"

6.3.4 A statement of the problem

With reference to the interconnected power system depicted in Diagram 6.1 earlier, we want to find out, using the analysis we have been developing, how the frequency would develop in time if one of the generators were to be tripped from the system. Let us say that GEN 1 is tripped from the system at $t = 0$. What will the trace of frequency against time look like, say in the first 10 seconds following the trip of GEN 1?

6.3.5 The general solution to the problem

6.3.5.1 *The quick answer*

To solve the problem posed in the previous section we need to solve the "swing equation" as given by Eq. (6.1) using the "direct" or "exact solution" of the form given in Eq. (6.21).

6.3.5.2 *The solution summarized in Appendix A*

The solution we have given in Eq. (6.21) is summarized in Appendix A where all the necessary details have been provided with it.

6.3.5.3 *No response provision in this example*

As mentioned earlier, in this introductory exercise, it is being assumed that there is no response provided from any of the generators in the system.

6.3.5.4 *In the next section*

In the next section, we shall build the input data that will be used by the MATLAB code that we shall be suggesting to solve this problem.

6.3.6 The input data to the MATLAB code

6.3.6.1 *Equations*

We first need to look at the elements of the "swing equation" given by Eq. (6.1) that we need to allocate values to. We reproduce Eq. (6.1) for convenience:

$$\begin{aligned}\Sigma P_{\text{GEN(NR)}} + \Sigma P_{\text{GEN}n\text{(R)}} - \Sigma P_{\text{LOAD}n} - k_1 \cdot (f - f_n) \\ - P_{\text{TRIP}} \cdot u\left(t - t_{\text{trip}}\right) + a_0 + a_1 t = 2 \cdot H \cdot f_0 \cdot \frac{\mathrm{d}f}{\mathrm{d}t}\end{aligned} \tag{6.1}$$

in which k_1 is given by Eq. (6.49) which we also repeat here:

$$k_1 = \Sigma P_{\text{LOAD}n} \cdot \alpha \cdot + \Sigma P_{\text{GEN}n(R)} \cdot k \tag{6.49}$$

As mentioned earlier, this mathematical problem and its solution are summarized in appendix, to which we refer.

6.3.6.2 *The input data in physical units*

Working from left to right in Eq. (6.1) we have the following data for our problem:

1. $\Sigma P_{\text{GEN(NR)}} = 3000$ MW
2. $\Sigma P_{\text{GEN}n\text{(R)}} = 0$ MW
3. $\Sigma P_{\text{LOAD}n} = 3000$ MW
4. $k_1 = 3000 \text{ MW} \times 0.02 \text{ Hz}^{-1} = 60$ MW/Hz

5. $f_n = 50$ Hz
6. $P_{TRIP} = 300$ MW
7. $a_0 = 0$
8. $a_1 = 0$
9. $H = 5.0$ seconds per machine
10. $f_0 = 50$ Hz

Note: in Eq. (6.49) we have assumed $\alpha = 0.02$ per Hz as suggested in Appendix A.

We have taken $k = 0$ since we are not considering any frequency response in this example.

6.3.6.3 *The input data in "per unit" quantities*

1. *Defining the base quantities*

   For input into the MATLAB code we need to convert the data listed earlier into "per unit" quantities, so first we define some base quantities for the problem.

   a. *System MVA base*

      Since the size of our system size is 3000 MW, we can create an MVA base from this figure and use a nominal machine power factor of 0.8 to give the MVA base = 3000/0.8 = 3750 MVA.

   b. *Frequency base*

      The obvious and convenient frequency base to take for our problem is the "nominal frequency" for our system, namely 50 Hz.

2. *Calculating the "per unit" quantities*

   The "per unit" counterparts of the quantities in physical units listed in Section 6.3.6.2 are therefore:

   a. $\Sigma P_{GEN(NR)} = 3000$ MW/3750 MVA = 0.8 pu
   b. $\Sigma P_{GENn(R)} = 0$ pu
   c. $\Sigma P_{LOADn} = 3000$ MW/3750 MVA = 0.8 pu
   d. $k_1 = 60$ MW/Hz × 50 Hz/3750 MVA = 0.8 pu
   e. $f_n = 50/50$ Hz = 1.0 pu
   f. $P_{TRIP} = 300$ MW/3750 MVA = 0.08 pu
   g. $a_0 = 0$
   h. $a_1 = 0$
   i. $H = 5$ seconds (the average H for the system)
   j. $f_0 = 50/50$ Hz = 1.0 pu

6.3.6.4 *In the next section*

In the next section, we shall specify the output data that we shall require from the MATLAB program. This will show us how the frequency behaves over time following the loss of a generating set from the system at time $t = 0$.

6.3.7 The output data from the MATLAB code

The output we require from the MATLAB program is the system frequency *f* in Hz plotted against time *t* in seconds for the first 10 seconds following the loss of one of the ten generating sets at time $t = 0$.

6.3.8 The suggested MATLAB code

A suitable MATLAB code that achieves our objective could be as follows:

```
% Inputting the data for the problem in "per unit" :
    PNR = input('Please enter the non-responsive generation in per unit')
    PR = input('Please enter the responsive generation in per unit')
    PL = input('Please enter the system load in per unit')
    K1 = input('Please enter the value of the constant k1in per unit')
    FN = input('Please enter the nominal frequency in per unit')
    PTRIP = input('Please enter the generation tripped in per unit')
    A_0 = input('Please enter the value of the constant a0in per unit')
    A_1 = input('Please enter the value of the constant a1in per unit')
    H = input('Please enter the value of the H-constant in seconds')
    F0 = input('Please enter the value of the initial frequency in per unit')
% Calculate the power balance at nominal frequency as a parameter:
    PBAL = PNR + PR − PL − PTRIP
% Calculate the value of the parameter k1:
    K1 = 1.00*PL
% Calculate the values of the constants A1, B1, C1 and D1:
    A1 = FN + (A_0 + PBAL)/K1
    B1 = A_1/K1
    C1 = F0 − (A_0 + PBAL)/K1 + 2*H*F0*A_1/(K1^2)
    D1 = −K1/(2*H*F0)
% Calculate the solution for the frequency f at a sequence of values
of t from 0 to 10
% seconds at intervals of 0.1 second:
    for k = 1,100
        t = k/10
        f(k) = A1 + B1*t + C1*EXP(D1*t)
    end
Note that the entries of the vector f are in "per unit" and so need to be
modified to
be in Hz (Quiz question: How do we do this?).
```

6.3.9 Instructions for the user

Run the MATLAB example with the input data from Section 6.3.6 and plot the graph of *frequency in Hz* against *time in seconds* for the first 10 seconds. What would happen if the *H* constant of each machine were only 2.5 seconds instead of 5.0 seconds? How low would the frequency fall in this case?

6.3.10 In the next section

In the next section, we shall move onto the second analytical method that we shall be considering in this chapter: the "solution by Laplace transforms."

6.4 Method 2: solution of the "swing equation" by Laplace transforms

6.4.1 Introduction

Looking at the "swing equation" represented by Eq. (5.14) of the previous chapter, can we see any possibility of being able to solve it using Laplace transforms?

Now since Eq. (5.14) is apparently "only" a first-order differential equation, it may be thought by the reader that it should be solvable by Laplace transforms in a relatively straightforward way. However, it must be recognized that there are three features of Eq. (5.14) which serve to complicate such a convenient initial assessment:

1. *The "$u(t - t_{\text{trip}})$" step-function term*

   This term, although apparently simple, is in fact nonlinear. However, we are in luck with a "Laplace transform" approach because this term has a straightforward, simple Laplace transform. Whether this fact enables us to solve the whole of Eq. (5.14) remains to be seen.
2. *The "GRP" term*

   The "GRP" term is a (nonlinear) power series in t. Although technically this does not make the *differential equation* nonlinear, it certainly complicates it. We shall see that there is a simple approximation to the series that may be made that considerably simplifies the solution process.
3. *The "$f \cdot \mathrm{d}f/\mathrm{d}t$" term*

   The "$f \cdot \mathrm{d}f/\mathrm{d}t$" term on the right-hand side of the equation, however, does technically make the *differential equation* [Eq. (5.14)] *nonlinear* and so poses a greater problem. However, again, there is a simple approximation that may be adopted that linearizes the equation.

In the next section, we look at each of the terms of Eq. (5.14) in turn to decide how they may be treated in a "Laplace transform" approach to the solution of the "swing equation."

6.4.2 Processing the individual terms

Let us state, once again, the full per unit form of the "swing equation" that was given as Eq. (5.14a) of the previous chapter:

$$\Sigma P_{\text{GEN(NR)}} + \Sigma P_{\text{GEN}n\text{(R)}}\left\{1 - k(f - f_n)\right\} - \Sigma P_{\text{LOAD}n}\left\{1 + \alpha(f - f_n)\right\} - P_{\text{TRIP}}u\left(t - t_{\text{trip}}\right) + \text{GRP} = 2 \cdot H \cdot f \cdot \frac{\mathrm{d}f}{\mathrm{d}t} \tag{6.53}$$

where, as before,

$$\mathrm{GRP} = a_0 + a_1 t + a_2 t^2 + a_3 t^3 + \cdots \tag{6.54}$$

is the power series representation for a "general responsive device", that is, one other than a traditional turbine-based generator that is equipped with a "speed governor" (which later we have included as well, but separately); and in the governor term:

$$\Delta f = f - f_n \tag{6.55}$$

where f_n represents the nominal frequency of the system.

Here are some remarks:

1. *A reminder of the meaning of the "GRP" term*

   As first introduced in Eq. (5.13), "GRP" is the abbreviation used in this book for "general responsive profile." This is intended to represent those sources of response other than the conventional kind of response from the governors of synchronous machines. The "general responsive" category may be appropriate for the modeling of such responsive devices as wind turbines, solar photovoltaic panels, large batteries, and demand-side providers. The suggested profile given by Eq. (5.13) is intended to be (mathematically) as general as possible, so that it can describe all possible response shapes (represented as graphs of active power response plotted against time).
2. *A note on the interchangeability of "f" and "ω" in "per unit"*

   Note that in Eq. (6.53) we have used for the sake of consistency the version of Eq. (5.14) with f on the right-hand side of the equation, rather than ω. Since our equation is expressed in *per unit*, and because $\omega_{\mathrm{pu}} = f_{\mathrm{pu}}$, no other changes to the equation are required as a result of this modification.
3. *A restatement of Eq. (6.53) in its simplest mathematical form*

   To simplify the task of solving the differential equation, let us now restate Eqs. (6.53) and (6.54) in a mathematically compact form, as follows:

$$c_1 + c_2 f + c_3 \cdot u\left(t - t_{\mathrm{trip}}\right) + c_4 t + c_5 t^2 + c_6 t^3 + \cdots = f \cdot \frac{\mathrm{d}f}{\mathrm{d}t} \tag{6.56}$$

   in which $u(t - t_{\mathrm{trip}})$ represents the unit step function describing a "trip" at some time t_{trip}, and the c_i, $i = 1,2,3,4,5,6$ are new constants related to the old ones by the relationships:

$$c_1 = \frac{\Sigma P_{\mathrm{GEN(NR)}} + \Sigma P_{\mathrm{GEN}n\mathrm{(R)}}(1 + k f_n) - \Sigma P_{\mathrm{LOAD}n}(1 - \alpha \cdot f_n) + a_0}{2H} \tag{6.57}$$

$$c_2 = \frac{-\Sigma P_{\mathrm{LOAD}n} \cdot \alpha + \Sigma P_{\mathrm{GEN}n\mathrm{(R)}} \cdot k}{2 \cdot H} \tag{6.58}$$

$$c_3 = \frac{-P_{TRIP}}{2 \cdot H} \tag{6.59}$$

$$c_4 = \frac{a_1}{2 \cdot H} \tag{6.60}$$

$$c_5 = \frac{a_2}{2 \cdot H} \tag{6.61}$$

and

$$c_6 = \frac{a_3}{2 \cdot H} \tag{6.62}$$

and so on.

4. *The equivalence of the equations*

   We notice that Eq. (6.56) has the same form as Eq. (6.53) except that all the constant terms in Eq. (6.53) have been "rolled up" into single constants wherever possible, to give the simplest possible mathematical form.

5. *The first main objective of the analysis*

   Since we have already replaced the angular speed ω by the rotational frequency f in Eq. (6.56), the problem presented to us now consists of finding $f(t)$, the desired development of frequency f with time t.

6. *In the next section*

   In the next section, we shall look at the need for making a further approximation to the "swing equation" in the "Laplace transform" method to make further progress.

6.4.3 The need to make a further approximation in the "Laplace transform" method

1. *Which terms of Eq. (6.56) are Laplace transformable?*

   In terms of attempting to solve Eq. (6.56) a good question to ask ourselves at this point is whether we can take Laplace transforms of all the terms in the equation without first making any modifications (making any approximations) to them?

2. *The critical answer to the question*

   The answer to the foregoing question depends critically, of course upon the complexity of the resulting individual terms in s taken together, and hence whether the resulting expression $F(s)$ can be inverted to find $f(t)$.

3. *The "GRP" expression*

   With the complexity of the equation [Eq. (6.56)] as it stands in mind, we need to consider whether, for example, it is appropriate to keep the full expression for the "GRP" term, as stated in Eq. (6.54). We recall that in the "direct solution of the differential equation" method of Section 6.2 this term was curtailed to the two linear terms $a_0 + a_1 t$.

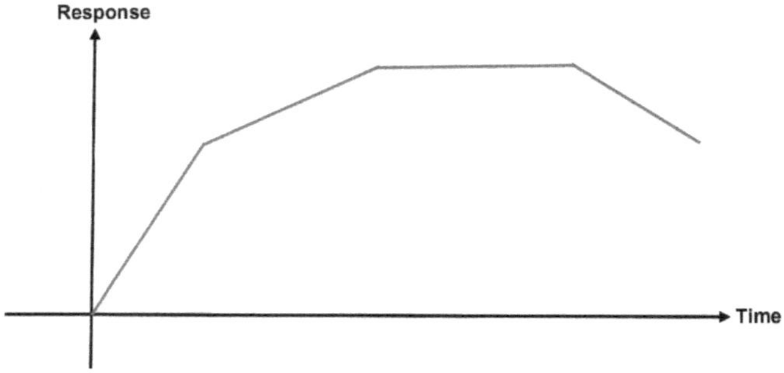

DIAGRAM 6.2 Piecewise linear representation.

4. *An argument based on reducing the complexity of the problem*
   Since the Laplace transforms of t^2, t^3, and higher powers of t make the solution unnecessarily complicated (such that we may indeed be unable to find the inverse transform of the resulting equation), this approximation, which was made in the "direct solution of the differential equation" method, will also be made here.
5. *Approximating the "GRP" expression to its first two terms*

As discussed earlier (see Section 6.2), restricting the "GRP" expression to its first two terms in practice may be accommodated easily by *expressing any* GRP *as a concatenation of piecewise linear segments* (please see Diagram 6.2 for an illustration of this).

Therefore we can write for a single segment, in the same manner as we did earlier:

$$\text{GRP(piecewise linear)} = a_0 + a_1 t \tag{6.63}$$

6. *The revised "swing equation" with the approximation in Eq. (6.63)*
   With the definition Eq. (6.62), Eq. (6.56) now becomes

$$c_1 + c_2 f + c_3 \cdot u(t - t_{\text{trip}}) + c_4 t = f \cdot \frac{\mathrm{d}f}{\mathrm{d}t} \tag{6.64}$$

where

$$c_1 = \frac{\Sigma P_{\text{GEN(NR)}} + \Sigma P_{\text{GEN}n\text{(R)}}(1 + kf_n) - \Sigma P_{\text{LOAD}n}(1 - \alpha \cdot f_n) + a_0}{2 \cdot H} \tag{6.65}$$

$$c_2 = \frac{-\Sigma P_{\text{LOAD}n} \cdot \alpha + \Sigma P_{\text{GEN}} \cdot k}{2 \cdot H} \tag{6.66}$$

$$c_3 = \frac{-P_{\text{TRIP}}}{2 \cdot H} \tag{6.67}$$

$$c_4 = \frac{a_1}{2 \cdot H} \tag{6.68}$$

7. *The revised basic form*

   If we now compare Eq. (6.64) with Eq. (6.3), which was used as a starting point for the "direct solution of the differential equation" method, which we have reproduced for ease of reference:

$$c_1 + c_2 f + c_3 t = \frac{df}{dt} \tag{6.69}$$

   we notice that the only difference in structure is the $u(t - t_{\text{trip}})$ term present in Eq. (6.64) but not present in Eq. (6.3). Indeed, we have left the unit trip term in the equation to be solved this time because a relatively simple Laplace transform exists for it that we can incorporate easily into our solution this time.

8. *In the next section*

   In the next section, we shall move forward to look at the Laplace transforms of the individual terms of Eq. (6.56).

6.4.4 Taking Laplace transforms of the individual terms of [Eq. (6.56)]

We shall now look at the different kinds of terms in Eq. (6.56) from the point of view of taking Laplace transforms of them.

If needed, for an explanation of the theory of the Laplace transform and a list of the more common transforms, the reader is directed at this point to Ref. [4]. For a more comprehensive list of Laplace transforms and a more in-depth treatment of the subject, the reader is directed to Ref. [5].

In the following text the Laplace transform of any quantity x will be denoted by $\mathcal{L}(x)$.

6.4.5 Definition of the Laplace transform

As perhaps (or not) a timely reminder to the reader, the formal definition of the *Laplace transform* $\mathcal{L}$ of a function $f(t)$ is given for reference. Here $\mathcal{L}$ is the shorthand form denoting the *Laplace operator*, also known in some texts as the "Laplacian" (see, for example, Ref. [4]):

$$\mathcal{L}\{f(t)\} \equiv F(s) = \int_0^{\infty} f(t) \cdot e^{-st} dt \tag{6.70}$$

The above formula transforms a function f of time t into a new function F of the new "Laplace" variable s.

We will now proceed to examine the Laplace transform of each of the terms of Eq. (6.53) in turn. When this has been done, we shall put all the terms together to produce the transformed function $F(s)$.

If we can do this, and if in addition we are able to take the *inverse Laplace transform* of $F(s)$, then we shall have obtained the sought-after solution $f(t)$ which represents the required *evolution of the power frequency with time*.

6.4.6 The Laplace transform of constant terms

Applying the definition of Eq. (6.70), the Laplace transform $\mathcal{L}$ of some constant c is

$$\mathcal{L}(c) = \frac{1}{s} \tag{6.71}$$

where s is the new Laplace operator-transformed variable that was defined by Eq. (6.70).

6.4.7 The Laplace transform of terms linear in *f*

As a reminder, we recall that f represents the unknown function $f(t)$, the *power system frequency*, which we are trying to discover by our subsequent calculations. Consequently, we shall leave the Laplace transform of $c \cdot f$, where c is some constant, as just $c \cdot \mathcal{L}(f)$ for the time being, until all the required terms have been collected together. So, for now, we simply write

$$\mathcal{L}(c \cdot f) = c \cdot \mathcal{L}(f) \tag{6.72}$$

6.4.8 The Laplace transform of the unit step function $u(t - t_{trip})$

The *unit step function* (also known as "Heaviside unit function"—see for instance Ref. [5], p. 8, for more details) represents a function that has the value *zero for t less than* t_{trip} and the value *1 for t greater than or equal to* t_{trip}.

The Laplace transform of the *unit step function* $c \cdot u(t - t_{trip})$ *is*

$$\mathcal{L}\{c \cdot u(t - t_{trip})\} = \left(\frac{c}{s}\right) \cdot \exp(-s \cdot t_{trip}) \tag{6.73}$$

where t_{trip} is the time of trip (either of generation or demand), and c is some constant.

6.4.9 The Laplace transform of terms linear in *t*

The Laplace transform of a term of the form $c \cdot t$, where c is some constant, and t represents time, is

$$\mathcal{L}(c \cdot t) = \frac{c}{s^2} \tag{6.74}$$

6.4.10 The Laplace transform of terms proportional to t^2

The Laplace transform of the "quadratic form" $c \cdot t^2$ is

$$\mathcal{L}\left(c \cdot t^2\right) = 2 \cdot \frac{c}{s^3} \tag{6.75}$$

where c is some constant.

6.4.11 The Laplace transform of terms of higher order in t

For the record the Laplace transform of terms in the nth *power of t* is given by the general formula:

$$\mathcal{L}(c \cdot t^n) = c \cdot n!/s^{n+1} \tag{6.76}$$

This formula applies in general where n is a *nonnegative integer.*

6.4.12 The Laplace transform of terms proportional to the first derivative of f with respect to t

The Laplace transform of terms proportional to the first derivative of f with respect to t is

$$\mathcal{L}\left(\mathrm{c} \cdot \frac{df}{dt}\right) = c \cdot \left\{s \cdot \mathcal{L}(f) - f(0)\right\} \tag{6.77}$$

where c (as always) is some constant, and $f(0)$ is the value of f at $t = 0$.

All or most of the results earlier should be familiar to any student who has attended an introductory course in Laplace transforms, but for extra clarity we have listed the results here.

6.4.13 Putting it all together

Let us now proceed to find our analytical solution $f(t)$ for Eq. (6.53), using the individual results [Eqs. (6.71)–(6.77)] that we have laid out earlier.

Applying the results from Eqs. (6.71) to (6.77) inclusive to Eq. (6.53) produces the following initial result:

$$c_1 + c_2 \cdot \mathcal{L}(f) + \left(\frac{c_3}{s}\right) \cdot \exp\left(-s \cdot t_{\text{trip}}\right) + \frac{c_4}{s^2} + 2 \cdot \frac{c_5}{s^3} + 6 \cdot \frac{c_6}{s^4} + \cdots = \mathcal{L}\left(f \cdot \frac{df}{dt}\right). \tag{6.78}$$

6.4.14 Dealing with the $f \cdot \mathrm{d}f/\mathrm{d}t$ term

We would now like to expand the Laplacian term on the right-hand side of Eq. (6.78), and by doing so convert Eq. (6.78) to a form containing the variable s *only* throughout, rather than the current mixture of s and t.

However, in its present form, converting the term $\mathcal{L}(f \cdot \mathrm{d}f/dt)$ to a function of s gives us a bit of a challenge. Although the Laplace transform of f is simply $F(s)$, and the Laplace transform of $\mathrm{d}f/\mathrm{d}t$ is $s \cdot F(s) - f(0)$ [see Eq. (6.76)], the Laplace transform of the product $f \cdot \mathrm{d}f/\mathrm{d}t$ requires some additional thought.

Essentially, what we need to do is to evaluate the Laplace transform of *a product of two* functions. Specifically, we need to find the Laplace transform $\mathcal{L}\{f(t) \cdot g(t)\}$, where in our case $f(t) = t$ and $g(t) = \mathrm{d}f/\mathrm{d}t$.

Now the closest general formula to the Laplace transform of a product of two functions is the *Convolution theorem* (see Ref. [5], Appendix A on p. 243 for more details). Could the convolution theorem help us in our case? Let's see

6.4.15 Trying to use the *convolution theorem* for Laplace transforms on the $f \cdot \mathrm{d}f/\mathrm{d}t$ term

The "convolution theorem" of Laplace transform theory can be stated in the following general way:

First, let $h(t)$ be the *convolution of $f(t)$ and $g(t)$* (see Ref. [5] p. 243), that is

$$h(t) = (f^* g)(t) \equiv \int_0^t f(u) \cdot g(t-u) \cdot \mathrm{d}u \tag{6.79}$$

With the definition given in Eq. (6.78) the convolution theorem states that

$$H(s) = F(s) \cdot G(s) \tag{6.80}$$

In which $F(s) = \mathcal{L}\{f(t)\}$, $G(s) = \mathcal{L}\{g(t)\}$, and $H(s) = \mathcal{L}\{h(t)\}$.

Can we use the above result to help us find the Laplace transform of the product $f(t) \cdot g(t)$ where $f(t) = t$, and $g(t) = \mathrm{d}f/\mathrm{d}t$?

Unfortunately, the *convolution theorem* cannot be applied in a straightforward manner to our case, because, in keeping with the definition of the convolution of $f(t)$ and $g(t)$ as given by Eq. (6.78), the subject of that integral is not the simple "$f(t) \cdot g(t)$" but instead "$f(u) \cdot g(t-u)$." This presents us with a difficulty.

In fact, Eqs. (6.78) and (6.79) are of much more practical use as an aid to the calculation of inverse Laplace transforms $F(s) \rightarrow f(t)$ than to Laplace transforms $f(t) \rightarrow F(s)$, so we shall abandon this method for the moment and try a different approach, described later.

6.4.16 Trying to use "integration by parts" on the $f \cdot df/dt$ term

As an alternative to trying to use the *convolution theorem*, could an expression for the Laplace transform of $f(t) \cdot g(t)$ instead be obtained using the method of *integration by parts*? Let us see in what follows.

Using the basic definition of the Laplace transform, stated earlier in Eq. (6.70) on the product $f \cdot df/dt$, we have

$$\mathcal{L}\left\{f(t) \cdot \frac{df}{dt}\right\} = \int_0^\infty f(t) \cdot \frac{df}{dt} \cdot e^{-st} dt \tag{6.81}$$

As a reminder to the reader, the mathematical procedure known as "integration by parts," in the form applicable to the integration of the product of two functions (usually called $u(t)$ and dv/dt) is usually expressed in the following way:

$$\int_a^b u(t) \cdot \frac{dv}{dt} \cdot dt = [u(t) \cdot v(t)]_{\text{eval.at } b,a} - \int_a^b v(t) \cdot \frac{du}{dt} \cdot dt \tag{6.82}$$

In our case, to attempt to evaluate $L\{f(t) \cdot df/dt\}$, we may select, for example, for our two variables:

$$u(t) \equiv e^{-st} \tag{6.83}$$

and

$$v(t) \equiv \frac{1}{2}\{f(t)\}^2 \tag{6.84}$$

With these two definitions for the variables, applying the definition of "integration by parts" given in Eqs. (6.80) and (6.81) produces the initial result:

$$\begin{aligned}\mathcal{L}\left\{f(t) \cdot \frac{df}{dt}\right\} &= \left[e^{-st} \cdot \frac{1}{2} \cdot \{f(t)\}^2\right] \infty, 0 - \left\{-s \cdot \int_0^\infty f(t)^2 \cdot e^{-st} \cdot dt\right\} \\ &= -\frac{1}{2}\{f(0)\}^2 + \left\{s \cdot \int_0^\infty f(t)^2 \cdot e^{-st} \cdot dt\right\}\end{aligned} \tag{6.85}$$

The author is not aware of any way to make any further progress analytically from this point on using the method of *integration by parts*. We must therefore look for some other way to solve Eq. (6.53). This will necessarily involve making an additional approximation, which we deal with in the next section.

6.4.17 The need to make an additional approximation

To summarize, we have attempted to find the Laplace transform of $f(t)\cdot \mathrm{d}f/\mathrm{d}t$ by a couple of methods, without success. To make any progress analytically with our Laplace transform approach, we therefore need to make some approximations to Eq. (6.53), in particular to the $f(t)\cdot \mathrm{d}f/\mathrm{d}t$ term on the right-hand side.

In the same manner as with the "exact solution of the differential equation," method of Section 6.2, we shall approximate the term $f(t)\cdot(\mathrm{d}f/\mathrm{d}t)$ with $f(0)\cdot \mathrm{d}f/\mathrm{d}t$ in order to make the problem tractable.

We shall also adopt the linear approximation to the *GRP* term, as previously. We shall not, however, approximate the $u(t-t_{\mathrm{trip}})$ term because it is not necessary to do so using the Laplace transform approach.

6.4.18 Approximating the $f(t)\cdot \mathrm{d}f/\mathrm{d}t$ term

We have seen from the earlier analysis how the Laplace transform of the product of terms $f(t)\cdot \mathrm{d}f/\mathrm{d}t$ as found by the method of *integration by parts* has introduced difficulties, and if we are to be able to find a practical solution, needs to be avoided. We recall also that we were unable to use the *convolution theorem* for *Laplace transforms* in this instance.

It is therefore necessary to make some approximation to the $f\cdot \mathrm{d}f/\mathrm{d}t$ term to simplify our task so that the problem may be solved.

As in the "direct solution of the differential equation" method described earlier in the book, in the expression "$f(t)\cdot \mathrm{d}f/\mathrm{d}t$" we shall also replace "$f(t)$" by "$f(0)$" in our "Laplace transform" method. This substitution makes things much easier for us.

What was previously a *nonlinear* term has been converted thereby to a *linear* term. We are therefore now able to make use of the standard result given in Eq. (6.53) to find the Laplace transform of $f(0)\cdot \mathrm{d}f/\mathrm{d}t$, which is as follows:

$$\mathcal{L}\left\{f(0)\cdot\frac{\mathrm{d}f}{\mathrm{d}t}\right\}=f(0)\cdot\{s\cdot\mathcal{L}(f)-f(0)\} \tag{6.86}$$

This approximation will be a good one, if $f(t)$ remains "not too far away from" $f(0)$ for all values of t.

What does the statement "not too far away from" really mean, in practical terms? The answer is that we are talking here of a "small amount," that is, a matter of *a few percentage points* of variation away from the initial value, no more. In round figures, if the variation away from the initial value was larger than about 2%, *say*, then we would not be happy to make this approximation, since the resulting error in our calculation would probably be too large.

6.4.19 Approximating the *GRP* term

For ease of reference, we repeat later the definition of "GRP" that was given near the beginning of the chapter as [Eq. (6.2)]:

$$\text{GRP} = a_0 + a_1 t + a_2 t^2 + a_3 t^3 + \cdots \tag{6.87}$$

As explained earlier, this term is intended to represent an "general frequency response profile," that is, it is intended to represent a frequency response profile in its most general form so that essentially any shape of response profile can be modeled.

As before, the problem for us with Eq. (6.87) is that the large number of terms needed in the expression to describe accurately the profile we want to represent, if we want to cover all values of time in the simulation and the corresponding complexity of the Laplace transforms. Is there a sound way round this, that is, a way of simplifying Eq. (6.87) to make our *Laplace transform* solution easier to calculate?

As touched upon earlier, one way to simplify the problem is to represent every response profile by a *sequence of contiguous linear segments*, thus making the profile "piecewise linear." By doing this we could find the solution in each of the linear segments and connect the solutions together at the boundaries.

As indicated earlier, for each individual section we need to only retain a linear representation, that is

$$\text{GRP(piecewise linear section)} = a_0 + a_1 t \tag{6.88}$$

This approximation simplifies the solution considerably for us. Possible drawbacks might be as follows:

1. Potential loss of accuracy in the solution;
2. The additional requirement to resolve the equation for each individual section, including the need to take multiple boundary conditions into consideration.

6.4.20 The $u(t - t_{\text{trip}})$ term

In this "Laplace transform solution method," we do not necessarily have to abandon the "unit step function" term (as we did in the "direct solution of the differential equation" method covered earlier), because there is a relatively simple standard Laplace transform available to us for the unit step function which we can easily use. *We are therefore going to retain the unit step term in the Laplace transform solution method.*

6.4.21 The final stages of the calculation to find $f(t)$ for $t \geq 0$

With the analysis that we have discussed so far earlier in mind, let us begin again with the original general form of the swing equation that was given in Eq. (6.53), which for convenience we repeat here:

$$\Sigma P_{\text{GEN(NR)}} + \Sigma P_{\text{GEN}n\text{(R)}}\{1 - k(f - f_n)\} - \Sigma P_{\text{LOAD}n}\{1 + \alpha(f - f_n)\} - P_{\text{TRIP}}u(t - t_{\text{trip}}) + \text{GRP} = 2 \cdot H \cdot f \cdot \frac{\text{d}f}{\text{d}t} \tag{6.89}$$

where

$$\text{GRP} = a_0 + a_1 t + a_2 t^2 + a_3 t^3 + \cdots \tag{6.90}$$

and

$$\Delta f = f - f_n. \tag{6.91}$$

In Eq. (6.89), let us now replace the term $f \cdot \text{d}f/\text{d}t$ by $f_0 \cdot \text{d}f/\text{d}t = f(0) \cdot \text{d}f/\text{d}t$, and in Eq. (6.90) restrict the series to the first two terms $a_0 + a_1 t$, making the expression *linear*.

The *GRP* term thus becomes

$$\text{GRP(piecewise linear section)} = a_0 + a_1 t \tag{6.92}$$

Then, taking this all together, we obtain the new general form for the *swing equation* which we shall solve by *Laplace transforms*:

$$c_1 + c_2 \cdot f(t) + c_3 \cdot u(t - t_{\text{trip}}) + c_4 t = f_0 \text{d}f/\text{d}t \tag{6.93}$$

We note here for the record that Eq. (6.93) contains four constants that are related to the parameters and constants in Eq. (6.89)–(6.91) in the following way:

$$c_1 = \frac{\Sigma P_{\text{GEN(NR)}} + \Sigma P_{\text{GEN}n\text{(R)}}(1 + kf_n) - \Sigma P_{\text{LOAD}n}(1 - \alpha \cdot f_n) + a_0}{2 \cdot H} \tag{6.94}$$

$$c_2 = \frac{-\Sigma P_{\text{LOAD}n} \cdot \alpha + \Sigma P_{\text{GEN}} \cdot k}{2 \cdot H} \tag{6.95}$$

$$c_3 = \frac{-P_{\text{TRIP}}}{2 \cdot H} \tag{6.96}$$

and

$$c_4 = \frac{a_1}{2 \cdot H} \tag{6.97}$$

We note in passing that Eq. (6.93) contains five different kinds of terms:

1. *Constants*;
2. Terms *linear in f*;
3. A term *linear in the unit step function*;
4. A term *linear in t*;
5. A term *linear in* d*f*/d*t*.

Having observed the mathematical form of Eq. (6.93), we can now proceed to use the Laplace transform results that we have recorded already in Eqs. (6.71)–(6.77) to find a solution for Eq. (6.93).

6.4.22 Laplace transforming the equation Eq. (6.93)

Applying the known Laplace transforms from the list given in Eqs. (6.71)–(6.77) (given earlier in the chapter) to Eq. (6.92), we obtain from it the following equation:

$$\frac{c_1}{s} + c_2 \cdot \mathcal{L}(f) + \left(\frac{c_3}{s}\right) \cdot \exp\left(-s \cdot t_{\text{trip}}\right) + \frac{c_4}{s^2} = f_0 \cdot \left\{s \cdot \mathcal{L}(f) - f_0\right\} \qquad (6.98)$$

It is clear that Eq. (6.98) can now be solved for $\mathcal{L}(f)$ in terms of s, as follows.

Upon rearranging Eq. (6.98), we find the following "closed expression" for $\mathcal{L}\{f(t)\}$:

$$\mathcal{L}\{f(t)\} = \frac{f_0}{s - b_2} + \frac{b_1}{s(s - b_2)} + \left[\frac{b_3}{s(s - b_2)}\right] \cdot \exp\left(-s \cdot t_{\text{trip}}\right) + \frac{b_4}{s^2(s - b_2)} \qquad (6.99)$$

where, to further simplify the equation we have defined

$$b_1 = \frac{c_1}{f_0}, \quad b_2 = \frac{c_2}{f_0}, \quad b_3 = \frac{c_3}{f_0}, \quad b_4 = \frac{c_4}{f_0}. \qquad (6.100)$$

Finally, to find our solution $f(t)$, we now need to find the *inverse Laplace transform* for each of the terms on the right-hand side of Eq. (6.99).

6.4.23 Taking inverse Laplace transforms

With the assistance (if needed) of a suitable textbook on *Laplace transforms* (for example Ref. [5]) and a bit of ingenuity, we can state or calculate the following inverses, to be applied subsequently to Eq. (6.99):

$$(1) \quad \mathcal{L}^{-1}\left\{\frac{1}{s - b_2}\right\} = \exp(b_2 t) \qquad (6.101)$$

$$(2)\quad \mathcal{L}^{-1}\left[\frac{1}{s(s-b_2)}\right] = \left(\frac{1}{b_2}\right)\cdot\mathcal{L}^{-1}\left\{\frac{1}{s-b_2}-\frac{1}{s}\right\} = \left(\frac{1}{b_2}\right)\cdot\{\exp(b_2\cdot t)-1\} \tag{6.102}$$

$$(3)\quad \mathcal{L}^{-1}\left[\frac{1}{s^2(s-b_2)}\right] = \left(\frac{1}{b_2^2}\right)\cdot\mathcal{L}^{-1}\left[\frac{1}{s-b_2}-\frac{s+b_2}{s^2}\right] = \left(\frac{1}{b_2^2}\right)\cdot\mathcal{L}^{-1}\left[1/(s-b_2)-1/s-b_2/s^2\right] = \left(\frac{1}{b_2^2}\right)\cdot\{\exp(b_2 t)-1-b_2.t\} \tag{6.103}$$

And finally:

$$(4)\quad \mathcal{L}^{-1}\left[\exp(-s\cdot t_{\text{trip}})/\{s^2(s-b_2)\}\right]$$
$$= \left(\frac{1}{b_2^2}\right)\cdot\mathcal{L}^{-1}\left[\{1/(s-b_2)-1/s-b_2/s^2\}\cdot\exp(-s\cdot t_{\text{trip}})\right]$$
$$= \left(\frac{1}{b_2^2}\right)\cdot\left[u(t-t_{\text{trip}})\cdot\exp\{b_2(t-t_{\text{trip}})\}-u(t-t_{\text{trip}}) - b_2\cdot\{(t-t_{\text{trip}})\cdot u(t-t_{\text{trip}})\}\right]$$
$$= \left(\frac{1}{b_2^2}\right)\cdot u(t-t_{\text{trip}})\cdot\left[\exp\{b_2(t-t_{\text{trip}})\}-1-\{b_2\cdot(t-t_{\text{trip}})\}\right]. \tag{6.104}$$

6.4.24 Putting the solution together

Having obtained the results in Eqs. (6.101)–(6.104) inclusive with which to make the relevant substitutions on the right-hand side of Eq. (6.99), taking the inverse Laplace transform of the whole equation [Eq. (6.99)] gives us the following expression for $f(t)$:

$$f(t)=f_0\cdot\exp(b_2 t)+\left(\frac{b_1}{b_2}\right)\cdot\{\exp(b_2\cdot t)-1\} + \left(\frac{b_4}{b_2^2}\right)\cdot\{\exp(b_2 t)-1-b_2\cdot t\} + \left(\frac{b_3}{b_2^2}\right)\cdot u(t-t_{\text{trip}})\cdot\left[\exp\{b_2(t-t_{\text{trip}})\}-1-\{\mathrm{b}_2\cdot(t-t_{\text{trip}})\}\right] \tag{6.105}$$

As a common sense check on the accuracy of Eq. (6.105), it can be seen readily that *by* putting $t=0$ gives $f=f_0$, the initial frequency, as expected.

Now, grouping together similar terms, we can rewrite Eq. (6.105) in the following form:

$$\begin{aligned} f(t) &= A_2 + B_2 \cdot t + C_2 \cdot \exp(D_2 \cdot t) \\ &+ u(t - t_{\text{trip}}) \cdot \{E_2 + F_2\} \cdot t + G_2 \cdot \exp(H_2 \cdot t) \end{aligned} \tag{6.106}$$

where

$$(1)\quad A_2 = -\frac{b_1}{b_2} - \frac{b_4}{b_2^2} \tag{6.107}$$

$$(2)\quad B_2 = -\frac{b_4}{b_2} \tag{6.108}$$

$$(3)\quad C_2 = f_0 + \frac{b_1}{b_2} + \frac{b_4}{b_2^2} \tag{6.109}$$

$$(4)\quad D_2 = b_2 \tag{6.110}$$

and

$$(5)\quad E_2 = -\frac{b_3}{b_2^2} + \frac{b_3}{b_2} \cdot t_{\text{trip}} \tag{6.111}$$

$$(6)\quad F_2 = -\frac{b_3}{b_2} \tag{6.112}$$

$$(7)\quad G_2 = -\frac{b_3}{b_2^2} \tag{6.113}$$

$$(8)\quad H_2 = b_2 \tag{6.114}$$

6.4.25 The form of the solution given in Eq. (6.105)

Inspecting Eq. (6.106), we see at once that there are the following kinds of terms in it, both multiplied and not multiplied by the step function $u(t - t_{\text{trip}})$:

1. *Constants*;
2. *Terms linear in* t;
3. *Exponentials in* $c \cdot t$, *where* c *is a constant.*

6.4.26 Some comments on the solution

We can see that the form of the solution given by Eq. (6.105) is essentially the same as that obtained in Eq. (6.21) of the "direct solution of the differential equation" method described in *Section 6.2, except for the addition of a set of four "unit trip function" terms.*

Since we did not consider trips to be instantaneous in nature in *Section 6.2*, but instead chose to include all trips as part of the linearized *GRP* term (a linear ramped term), the foregoing statement is not a surprise to us.

Since the solution given in Eq. (6.105) is more general than the solution given in Eq. (6.21), we should expect Eq. (6.21) to be the special case of Eq. (6.105) with $E_2 = F_2 = G_2 = H_2 = 0$.

Comparing the constants given in Eqs. (6.106)–(6.109) directly above with the similar constants given earlier in Eqs. (6.34)–(6.37), we should arrive at the "same answer" (within the terms of the definitions). In the next section, we shall see if we do.

6.4.27 A comparison of the Method 2 "Laplace transform" solution with the solution from Method 1 ("direct solution of the differential equation")

Let us now proceed to try to compare the results we have just obtained for "Method 2" with those we obtained earlier for "Method 1."

1. *A brief summary of Method 1*

   We recall that Eqs. (6.45), (6.46), (6.47), and (6.48), respectively, gave us the parameters A_1, B_1, C_1, and D_1 as functions of the terms in the simplified physical equation we started with, namely Eq. (6.2), which for ease of reference we repeat here:

$$\Sigma P_{\mathrm{GEN(NR)}} + \Sigma P_{\mathrm{GEN}n\mathrm{(R)}} - \Sigma P_{\mathrm{LOAD}n} - k_1 \cdot (f - f_n) + a_0 + a_1 t = 2 \cdot H \cdot f_0 \cdot \mathrm{d}f \qquad (6.115)$$

   We also recall that the parameter k_1 in Eq. (6.115) was defined in Eq. (6.49) as

$$k_1 = -\Sigma P_{\mathrm{LOAD}n} \cdot \alpha \cdot + \Sigma P_{\mathrm{GEN}n\mathrm{(R)}} \cdot k \qquad (6.116)$$

   in which, in turn, k was represented in the "original" equation [Eq. (5.14)] of the previous chapter.

   We recall that the parameter k_1 contains representations of both the *variation of load (demand) with frequency*, and *synchronous machine governor droop* effects.

2. *Back to Method 2*

Let us suppose, for the purposes of comparing the results of the two methods, that $E_2 = F_2 = G_2 = H_2 = 0$. In other words, we are not modeling any instantaneous trip events.

From here, to compare the results of the two solution methods, we need to find expressions for the "Method 2" parameters A_2, B_2, C_2, and D_2 in terms of the parameters in the approximated version "physical" equation [Eq. (6.53)] we are starting from in *Method 2* for this comparison, namely:

$$\Sigma P_{\mathrm{GEN(NR)}} + \Sigma P_{\mathrm{GEN}\ n\mathrm{(R)}} \cdot [1 - k \cdot (f - f_n)] - \Sigma P_{\mathrm{LOAD}\ n}[1 + \alpha(f - f_n)] - P_{\mathrm{TRIP}} \cdot u\left(t - t_{\mathrm{trip}}\right) + a_0 + a_1 \cdot t = 2 \cdot H \cdot f_0 \cdot \frac{\mathrm{d}f}{\mathrm{d}t} \tag{6.117}$$

in which, as explained earlier, for the purposes of comparison, we shall be assuming in addition that $P_{\mathrm{TRIP}} = 0$. With this extra assumption, and "rolling up" α and k into a single constant k_1 as defined in Eq. (6.116), Eq. (6.117) becomes

$$\Sigma P_{\mathrm{GEN(NR)}} + \Sigma P_{\mathrm{GEN}n\mathrm{(R)}} - \Sigma P_{\mathrm{LOAD}n} - k_1 \cdot (f - f_n) + a_0 + a_1 t = 2 \cdot H \cdot f_0 \cdot \frac{\mathrm{d}f}{\mathrm{d}t} \tag{6.118}$$

As a brief check, we can see that Eq. (6.118) is identical to Eq. (6.2), which we note was the starting point in the "Method 1" section.

3. *Comparing Method 1 with Method 2*

Let us, then, now examine the definitions of A_2, B_2, C_2, and D_2, as given by our *Method 2* solution represented by Eqs. (6.107), (6.108), (6.109), and (6.110), respectively.

We can see that each of these expressions is a function of the three parameters b_1, b_2, and b_4, which in turn are defined in Eq. (6.100) in terms of the parameters c_1, c_2, and c_4. This latter set of parameters is in turn defined by Eqs. (6.94), (6.95), and (6.97).

Using this information, we can relate the three parameters b_1, b_2, and b_4 back to the "original" parameters k_1, H, a_0, and a_1 contained in Eq. (6.118), in which we recall that Eq. (6.117) has been expressed in the form:

$$b_1 + b_2 f(t) + b_3 \cdot u\left(t - t_{\mathrm{trip}}\right) + b_4 \cdot t = \frac{\mathrm{d}f}{\mathrm{d}t} \tag{6.119}$$

where we assume, for the purposes of this comparison, that $b_3 = 0$. Hence, by comparing Eq. (6.119) directly with Eq. (6.118), we obtain for the remaining three parameters b_1, b_2, and b_3:

$$b_1 = \frac{\Sigma P_{\mathrm{GEN(NR)}} + \Sigma P_{\mathrm{GEN}n\mathrm{(R)}} - \Sigma P_{\mathrm{LOAD}n} + k_1 \cdot f_n + a_0}{2 \cdot H \cdot f_0} \tag{6.120}$$

$$b_2 = -\frac{k_1}{2 \cdot H \cdot f_0} \tag{6.121}$$

and

$$b_4 = \frac{a_1}{2 \cdot H \cdot f_0} \tag{6.122}$$

where we have also used the definition of k_1 as given in Eq. (6.116).

Substituting for b_1, b_2, and b_4, as given by Eqs. (6.120)–(6.122), into Eqs. (6.107)–(6.110), we obtain thereby the following four results for A_2, B_2, C_2, and D_2 in terms of k_1, H, a_0, and a_1:

$$\begin{aligned}
A_2 &= -\frac{b_1}{b_2} - \frac{b_4}{b_2^2} \\
=> \quad A_2 &= \frac{\Sigma P_{\text{GEN(NR)}} + \Sigma P_{\text{GEN}n\text{(R)}} - \Sigma P_{\text{LOAD}n} + k_1 \cdot f_n + a_0}{k_1} \\
&\quad - 2 \cdot H \cdot f_0 \cdot \frac{a_1}{k_1^2} \\
=> \quad A_2 &= f_n + \frac{\Sigma P_{\text{GEN(NR)}} + \Sigma P_{\text{GEN}n\text{(R)}} - \Sigma P_{\text{LOAD}n} + a_0}{k_1} \\
&\quad - 2 \cdot H \cdot f_0 \cdot \frac{a_1}{k_1^2}
\end{aligned} \tag{6.123}$$

We note that Eq. (6.123) matches the corresponding Method 1 result given in Eq. (6.45).

Further, we have

$$\begin{aligned}
B_2 &= -\frac{b_4}{b_2} \\
=> \quad B_2 &= \frac{a_1}{k_1}
\end{aligned} \tag{6.124}$$

We note that Eq. (6.124) matches the corresponding Method 1 result given in Eq. (6.46).

Continuing

$$\begin{aligned}
C_2 &= f_0 + \frac{b_1}{b_2} + \frac{b_4}{b_2^2} \\
&= f_0 - \frac{\Sigma P_{\text{GEN(NR)}} + \Sigma P_{\text{GEN}n\text{(R)}} - \Sigma P_{\text{LOAD}n} - k_1 \cdot f_n + a_0}{k_1} + 2 \cdot H \cdot f_0 \cdot \frac{a_1}{k_1^2} \\
=> \quad C_2 &= f_0 - f_n - \frac{\Sigma P_{\text{GEN}} - \Sigma P_{\text{LOAD}n} + a_0}{k_1} + 2 \cdot H \cdot f_0 \cdot \frac{a_1}{k_1^2}
\end{aligned} \tag{6.125}$$

We note that Eq. (6.124) matches the corresponding Method 1 result given in Eq. (6.47).

And finally, we have

$$D_2 = b_2$$
$$=> \quad D_2 = -\frac{k_1}{2 \cdot H \cdot f_0} \tag{6.126}$$

We note that Eq. (6.125) also matches the corresponding Method 1 result given in Eq. (6.48).

From the earlier results, we see that the results from *Method 2, the solution by Laplace transforms, are the same as those from Method 1, the direct solution of the differential equations, for the same problem.*

We remind the reader that the comparison above does not include the $u(t - t_{trip})$ term, which was not included in *Method 1*.

6.5 MATLAB example simulation 2: using the "Laplace transform" method to solve the "swing equation"

6.5.1 Objective

Using, this time, the results from the "Laplace transform" method given in Eq. (6.106), here is another illustrative example using MATLAB as our programing tool. We shall use the fact that the "Laplace transform method" is equivalent mathematically to the "direct solution of the differential equation" method as proved earlier.

6.5.2 Introduction to the second illustrative example

We shall again construct a practical example that uses in a concrete way the theory that we have developed earlier. Let us start with the interconnected power system depicted by the single-line Diagram 6.3.

We remind the reader once more that is not necessary, for the purposes of our frequency studies using the analysis that we have derived earlier, to run any load-flow studies, fault current analyses, stability analyses, or any other of the many prepackaged computer analysis tools available to us as power system planning engineers.

The reason for this is that the intended advantage of doing the analysis this way is to cut out the need to do the many repetitions of load flows that would be needed with a commercial analysis package, thus reducing greatly the time needed to do the analysis.

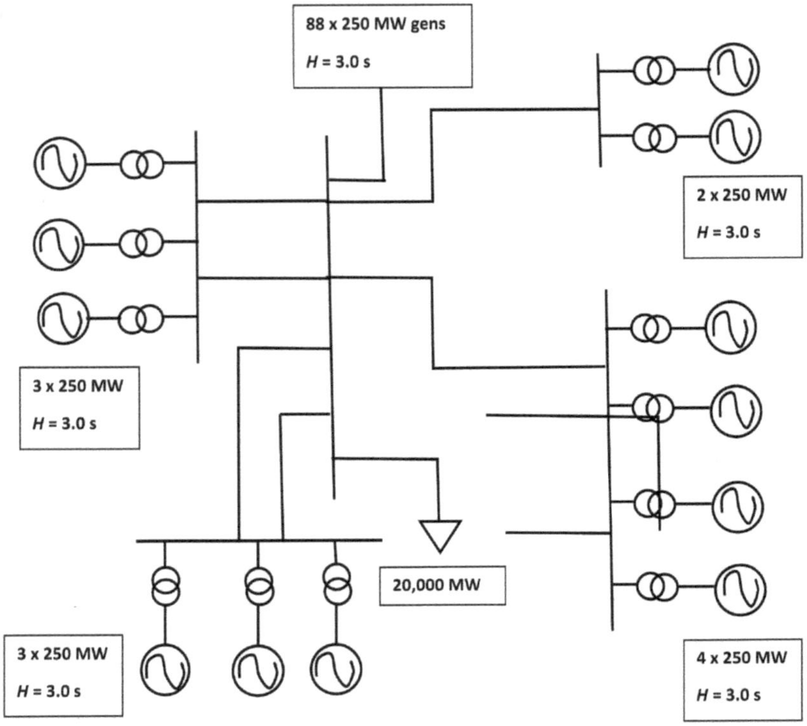

DIAGRAM 6.3 Second example of an interconnected power system.

6.5.3 Steps to be taken in the second illustrative example

In this illustrative *MATLAB* study, based on the "Laplace transform/direct solution of the differential equation" solution method covered earlier, we shall again break the problem down into the following six steps:

STEP 1: "A statement of the problem"
STEP 2: "The general solution to the problem"
STEP 3: "The input data to the MATLAB code"
STEP 4: "The output data from the MATLAB code"
STEP 5: "The suggested MATLAB code"
STEP 6: "Instructions for the user"

6.5.4 A statement of the problem

6.5.4.1 The network

Diagram 6.3 shows a (yet again imaginary) interconnected 50 Hz system in which this time there are 100 identical generators *available*, each of rated output 250 MW and inertia constant $H = 3.0$ seconds. The total system load (demand) is 20,000 MW.

6.5.4.2 The challenges

1. *Generation loss with no response available*
   a. *Details of the generators running*
   In this first case, we are assuming that 80 generators are running at rated output, and no other generators are running. This meets the 20,000 MW load (demand) exactly, but of course there is no provision for response if any generators should be lost from the system.
   b. *Details of the system event*
   At time $t=0$, there is a loss of *three* of the 250 MW generators simultaneously. Assume the losses are instantaneous.
   c. *Details of the challenge*
   The challenge is to calculate the frequency evolution for the first 10 seconds. Also, what is most noticeable about the frequency during this period?
2. *Generation loss with response available*
   a. *Details of the generators running*
   In this second case, assume that 53 generators are running at their full output of 250 MW each, 30 generators are running de-loaded at 90% output = 225 MW each, and that 17 generators are not running at all, thus meeting the total load of $53 \times 250 + 30 \times 0.9 \times 250 = 20{,}000$ MW.
   b. *Details of the system event*
   At time $t=0$, there is a loss of *three* of the generators simultaneously.
   Assume the losses are instantaneous. Assume that the three generators lost are fully loaded ones.
   c. *Details of the challenge*
   Assume that the de-loaded generators are operating on a "droop" profile of 4%. How does the evolution of the frequency in the first 10 seconds compare with that in part (1)? Is the amount of response provision adequate?

6.5.5 The general solution to the challenges

6.5.5.1 Generation loss with no response available

In this case, we may use the solution method summarized in Appendix A since we know already from our analysis earlier in this chapter that the "Laplace transform" solution is equivalent to the "exact solution" if we do not model a step function as such but model the system just after $t=0$ when the three generators have just been disconnected from the system. Therefore the procedure to find the solution is the same as that used for MATLAB example 1.

6.5.5.2 Generation loss with response available

In this case, we are faced with a new challenge: how can the solution summarized in Appendix A be used to model de-loaded responsive generators?

The answer is that we must first modify the Appendix A equations slightly, as follows, before we can represent the system behavior properly.

1. *Set points and slopes of the governor response characteristic*

In Diagram 6.4 two of the many possible set point/slope ("droop") combinations for a synchronous generator speed governor are shown.

One of the straight lines (the upper one) passes through P_{rated} on the power output axis and $f_{\text{nom}} + D(\text{pu}) \cdot f_{\text{nom}}$ on the frequency axis. D(pu) here is the "droop" of the characteristic in pu $= D(\%)/100$. The upper line could be termed the "default" characteristic. The equations given in Appendix A correspond to this "default" characteristic, which assumes that the responsive generator is generating at full output at the rated frequency f_n.

The second straight line (the lower one) passes through $p \cdot P_{\text{rated}}$ on the power output axis, where p is some number between 0 and 1, and otherwise remains parallel to the first straight line. The second line intersects the frequency axis at a smaller value of frequency than the first line, in fact at a frequency value of $p \cdot f_{\text{nom}}\ (1 + D)$, a result obtained by comparing similar triangles.

The lower line characteristic is the one used if a generator is de-loaded prefault to allow it to provide extra power if the frequency falls due to a loss of generation from the system.

2. *How the equations in Appendix A should be modified to account for de-loading of responsive generators at nominal frequency*

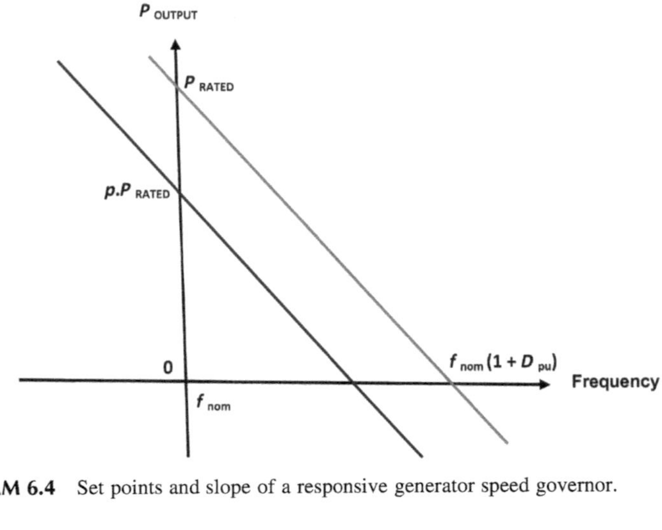

DIAGRAM 6.4 Set points and slope of a responsive generator speed governor.

To correctly account for the fact that, in the case of a set de-loaded for response purposes, its power output at nominal frequency is less than its rated power output, the equations in Appendix A need to be modified.

Specifically, the second term of Eq. (A.1) and the definition of the parameter k in Eq. (A.10) require attention.

From Diagram 6.4, we note immediately that the upper and the lower lines have the same gradient (slope) which is $\Delta P/\Delta f = P_{RATED}/f_{nom} \cdot D_{pu}$. Therefore the equation of the "default" (upper) characteristic is

$$P = P_{RATED} + \left(P_{RATED}/f_{nom} \cdot D_{pu}\right) \times (f - f_{nom}) \tag{6.127}$$

whereas the equation of the lower line characteristic is

$$P = p \cdot P_{RATED} + \left(P_{RATED}/f_{nom} \cdot D_{pu}\right) \times (f - f_{nom}) \tag{6.128}$$

The latter version of the "exact solution" thus modified with the factor p is given for the purposes of reference in Appendix B.

6.5.6 The main input data to the MATLAB code

1. *Generation loss with no response available*

   In this case the main input data to the MATLAB code consists of

   a. The total system generation;
   b. The total system load (demand);
   c. The total amount of generation tripped at $t = 0$
   d. The system inertia constant.

2. *Generation loss with response available*

   In this case the main input data to the MATLAB code in this case consists of

   a. The total system nonresponsive generation;
   b. The total system *responsive* generation at nominal frequency;
   c. The total system load (demand);
   d. The total amount of generation tripped at $t = 0$
   e. The system inertia constant;
   f. The generator governor "droop";
   g. The initial responsive generation de-loading proportion p.

6.5.7 The output data from the MATLAB code

In both cases the output data that we require from the MATLAB code is a set of system frequency values for times between 0 and 10 seconds so that an evolution curve for the system frequency can be printed out and commented on.

6.5.8 The suggested MATLAB code

We give suggested MATLAB code for the two cases in turn. But first we shall summarize the solution method that is applicable to each case and the relevant input data.

1. *Generation loss with no response available*
   a. *Solution method*
      In this first case, since the "solution by Laplace transforms" has been proven to be equivalent to the "direct solution of the differential equation," we may use the form of solution provided in Appendix A.
   b. *A summary of the input data*
      From Diagram 6.3 and from Sections 6.5.4.1 and 6.5.4.2 (1) we can summarize the input data for *Case* (1) here as follows:
      i. $f_n = 50$ Hz
      ii. P_{rated} *(1 set)* = 250 MW
      iii. *H (1 set)* = 3.0 seconds
      iv. $P_{\text{LOAD}} = 20,000$ MW
      v. *No. of sets running* = 80
      vi. $P_{\text{TRIP}} = 3 \times 250$ MW
      vii. $f_0 = 50$ Hz (assumed)
   c. *The input data for Case (1) according to Eq. (6.1) in physical units*
      Working from left to right in Eq. (6.1) we therefore have the following data for Case (1) in physical units:
      i. $\Sigma P_{\text{GEN(NR)}} = 20{,}000$ MW
      ii. $\Sigma P_{\text{GENn(R)}} = 0$ MW
      iii. $\Sigma P_{\text{LOADn}} = 20{,}000$ MW
      iv. $k_1 = 20{,}000 \times 0.02\ \text{Hz}^{-1} = 400$ MW/Hz
      v. $f_{\text{n}} = 50$ Hz
      vi. $P_{\text{TRIP}} = 750$ MW
      vii. $a_0 = 0$
      viii. $a_1 = 0$
      ix. $H = 3.0$ seconds per machine
      x. $f_0 = 50$ Hz
   d. *The input data according to Eq. (6.1) in "per unit"*
      Taking an *MVA base* of 20,000/0.8 = 25,000 and a frequency base of 50 Hz, the data in the preceding paragraph become, in "per unit":
      i. $\Sigma P_{\text{GEN(NR)}} = 20{,}000$ MW/25,000 MVA = 0.8 pu
      ii. $\Sigma P_{\text{GEN}n\text{(R)}} = 0$ pu
      iii. $\Sigma P_{\text{LOAD}n} = 20{,}000$ MW/25,000 MVA = 0.8 pu
      iv. $k_1 = (400$ MW/Hz)/25,000 MVA = 0.016 Hz^{-1}
      v. $f_n = 50/50$ Hz = 1.0 pu
      vi. $P_{\text{TRIP}} = 750$ MW/25,000 MVA = 0.03 pu
      vii. $a_0 = 0$

viii. $a_1 = 0$
ix. $H = 3.0$ seconds
x. $f_0 = 50/50\ \text{Hz} = 1.0$ pu

e. *The suggested MATLAB code for Case (1)*

A suitable MATLAB code that achieves our objective is essentially the same as that in MATLAB example simulation 1 since we have no response operating in this example:

```
% Inputting the data for the problem in "per unit" :
  PNR = input('Please enter the non-responsive generation in per unit')
  PR = input('Please enter the responsive generation in per unit')
  PL = input('Please enter the system load in per unit')
  K1 = input('Please enter the value of the constant k1 in per unit')
  FN = input('Please enter the nominal frequency in per unit')
  PTRIP = input('Please enter the generation tripped in per unit')
  A_0 = input('Please enter the value of the constant a0 in per unit')
  A_1 = input('Please enter the value of the constant a1 in per unit')
  H = input('Please enter the value of the H-constant in seconds')
  F0 = input('Please enter the value of the initial frequency in
       per unit')
% Calculate the power balance at nominal frequency as a parameter:
  PBAL = PNR + PR - PL - PTRIP
% Calculate the value of the parameter k1:
  K1 = 1.00*PL
% Calculate the values of the constants A1, B1, C1 and D1:
  A1 = FN + (A_0 + PBAL)/K1
  B1 = A_1/K1
  C1 = F0 - (A_0 + PBAL)/K1 + 2*H*F0*A_1/(K1^2)
  D1 = -K1/(2*H*F0)
% Calculate the solution for the frequency f at a sequence of values of t
% from 0 to 10 seconds at intervals of 0.1 second:
  for k = 1,100
      t = k/10
      f(k) = A1 + B1*t + C1*EXP(D1*t)
  end
% Note again that the entries of the output vector f are in "per unit"
and
% so need to be modified to be in Hz by multiplying by the nominal
% frequency for the system in Hz (in this example 50 Hz).
```

2. *Generation loss with response available*

a. *Solution method*

In this second case, we are de-loading sets for response purposes, so we need to use the form of solution given in Appendix B.

b. *A summary of the input data*

From Diagram 6.3 and from Sections 6.5.4.1 and 6.5.4.2 (2) we summarize the input data for Case (2) here as follows:

- **i.** $f_n = 50$ Hz
- **ii.** P_{rated} *(1 set)* = 250 MW
- **iii.** *H (1 set)* = 3.0 seconds
- **iv.** $P_{\text{LOAD}} = 20,\ 000$ MW
- **v.** *No. of nonresponsive sets running* = 53 × 250 MW
- **vi.** *No. of responsive sets running* = 30 × 225 MW
- **vii.** $P_{\text{TRIP}} = 3 \times 250$ MW
- **viii.** $f_0 = 50$ Hz *(assumed)*

c. *The input data for Case (2) according to Eq. (6.1) in physical units*

Working from left to right in Eq. (6.1) we therefore have the following data for Case (2) in physical units:

1. $\Sigma P_{\text{GEN(NR)}} = 53 \times 250 = 13{,}250$ MW
2. $\Sigma P_{\text{GENn(R)}} = 30 \times 250 = 7500$ MW
3. $\Sigma P_{\text{LOADn}} = 20{,}000$ MW
4. $k_1 = 20{,}000 \times 0.02 + [100/(4 \times 50)] \times 7500 = 4150$ MW/Hz
5. $f_n = 50$ Hz
6. $P_{\text{TRIP}} = 750$ MW
7. $a_0 = 0$
8. $a_1 = 0$
9. $H = 3.0$ seconds *(per machine)*
10. $f_0 = 50$ Hz

d. *The input data for Case (2) according to Eq. (6.1) in "per unit"*

Taking an *MVA base* of 20,000/0.8 = 25,000 and a *frequency base* of 50 Hz, the data in the preceding paragraph become, in "per unit":

1. $\Sigma P_{\text{GEN(NR)}} = 13{,}250$ MW/25,000 MVA = 0.53 pu
2. $\Sigma P_{\text{GEN}n\text{(R)}} = 7500$ MW/25,000 MVA = 0.30 pu
3. $\Sigma P_{\text{LOAD}n} = 20{,}000$ MW/25,000 MVA = 0.8 pu
4. $k_1 = (4150$ MW/Hz)/25,000 MVA = 0.166 seconds
5. $f_n = 50$ Hz/50 Hz = 1 pu
6. $P_{\text{TRIP}} = 750$ MW/25,000 MVA = 0.03 pu
7. $a_0 = 0$
8. $a_1 = 0$
9. $H = 3.0$ seconds × (13,250 + 7500)/20,000 = 3.1125 seconds *(system average)*
10. $f_0 = 50/50$ Hz = 1.0 pu

e. *The suggested MATLAB code for Case (2)*

This time we need to include the loading factor $p = 0.9$ for the responsive machines. We shall also assume a value of droop = 4% for the responsive machines. We are following the solution as given in Appendix B because the responsive machines are de-loaded prefault. Therefore a suitable MATLAB code could be as follows:

```
% Inputting the data for the problem in "per unit" :
  PNR = input('Please enter the rated non-responsive generation in
        per unit')
  PR = input('Please enter the total rated responsive generation in
       per unit')
  P = input('Please enter the loading factor p for the responsive
      machines in per unit')
  PL = input('Please enter the system load in per unit')
  FN = input('Please enter the nominal frequency in per unit')
  PTRIP = input('Please enter the generation tripped in per unit')
  A_0 = input('Please enter the value of the constant a0in per unit')
  A_1 = input('Please enter the value of the constant a1in per unit')
  H = input('Please enter the value of the H-constant in seconds')
  F0 = input('Please enter the value of the initial frequency in per
       unit')
  DPC = input('Please enter the value of the "droop" of the responsive
        generators in per cent')
%
% Calculate the power balance at nominal frequency as a parameter:
  PBAL = PNR + P*PR - PL - PTRIP
% Calculate the value of the parameter k1:
  K1 = 1.00*PL + (100/DPC)
%
% Calculate the values of the constants A1, B1, C1and D1:
  A1 = FN + (A_0 + PBAL)/K1
  B1 = A_1/K1
  C1 = F0 - (A_0 + PBAL)/K1 + 2*H*F0*A_1/(K1^2)
  D1 = -K1/(2*H*F0)
% Calculate the solution for the frequency f at a sequence of values of t
% from 0 to 10 seconds at intervals of 0.1 second:
  for k = 1,100
      t = k/10
      f(k) = A1 + B1*t + C1*EXP(D1*t)
  end
% Note again that the entries of the output vector f are in "per unit" and
% so need to be modified to be in Hz by multiplying by the nominal
% frequency for the system in Hz (in this example 50 Hz).
```

6.5.9 Instructions for the user

In each case enter the required input data when prompted and process the output data in a suitable way in MATLAB so that the relevant graphs can be plotted out.

6.5.10 In the next section

In the next section of this chapter, we shall examine the advantages and disadvantages of the "direct solution of the differential equations" and "Laplace transform solution" methods.

6.6 Advantages and disadvantages of the "direct solution" and "Laplace transform" methods

We are now going to compare critically the two very different, although fundamentally closely related, approaches that we have described in this chapter.

6.6.1 Direct solution of the differential equation

6.6.1.1 Advantages

1. *Straightforwardness*
   One advantage of the "direct solution of the differential equation" method is its straightforwardness. With a minimum of assumptions, the solution is obtained in a direct, logical fashion.
2. *Universal applicability*
   Another advantage of this method is that once the solution has been found, it is there to be used for any value of t that we wish: there are no simulations to be rerun, only reevaluations.
3. *Choice of parameters*
   Different parameters may be chosen to suit our requirements. Again, no simulation needs to be rerun, the solution just needs to be reevaluated for a different set of parameters.
4. *Accuracy*
   Apart from the single approximation of $f = f_0$ to make the "swing equation" linear, we can be confident that the solutions is accurate, no additional approximations needing to be made.

6.6.1.2 Disadvantages

The "direct solution" method does not cope easily with the unit step function, which is complicated to deal with. Presumably other "special functions" are also difficult to handle.

6.6.2 Laplace transform solution

6.6.2.1 Advantages

1. *Familiarity*
   One advantage of using the "Laplace transform" method is that it is a familiar method in engineering. So, if you happen to be an engineer, it is a particularly easy method to follow.

2. *Works well with the unit step function*

The method, involving as it does an integral in its basic definition, copes well with discontinuous functions such as the unit step function.

6.6.2.2 Disadvantages

Notwithstanding the earlier two advantages, in the end, except for step function terms, the solution is equivalent to that for the "direct solution" method so is not really an independent method.

6.7 The next chapter: numerical methods for solving the "swing equation"

6.7.1 Introduction

In the next chapter, we look at numerical solution methods for solving the "swing equation." Numerical methods may be useful in cases where the complexity of the changes on the system that are being modeled are relatively high, thus rendering an analytical solution difficult or perhaps even impossible.

6.7.2 Summary

The numerical methods that we shall examine in the next chapter may be conveniently divided into three broad categories:

1. Extrapolation methods
2. Runge–Kutta type methods
3. Predictor–Corrector methods

We shall look at each of the earlier methods in turn and assess them according to their suitability for solving the "swing equation."

References

Books

[1] D. Greenspan, Theory and Solution of Ordinary Differential Equations, The MacMillan Company, New York, 1960.

[2] G. Birkhoff, G.-C. Rota, Ordinary Differential Equations, third ed., John Wiley and Sons, New York, Chichester, Brisbane, Toronto and Singapore, 1978.

[3] B.M. Weedy, Electric Power Systems, second ed., John Wiley and Sons, London, New York, Sydney and Toronto, 1972.

[4] R.D. Strum, J.R. Ward, Laplace Transform Solution of Differential Equations, Prentice-Hall, 1968.

[5] M.R. Spiegel, Laplace Transforms (Schaum Outline Series), McGraw-Hill Education, 1965.

Chapter 7

Numerical methods for solving the "swing equation"

7.1 An introduction to "numerical methods" for solving the "swing equation"

7.1.1 A resume of what we covered in the previous chapter

7.1.1.1 Two analytical methods

In the previous chapter, we looked at some of the "analytical methods" for solving the "swing equation." We found that we were able to find a solution of the "swing equation" using two main analytical techniques, the "direct solution of the differential equation" and the "Laplace transform" methods, although in both cases, we were obliged to accept some degree of approximation before being able to proceed analytically.

7.1.1.2 The main approximation to the equation

Most importantly, we found that we had to approximate the frequency f by the initial frequency f_0 on the right-hand side of (Eq. 5.14a) to be able to proceed satisfactorily analytically. We shall therefore always start instead with Eq. (5.22) as our basic starting point for further analysis.

7.1.2 What we will cover in this chapter

1. *Various numerical methods*

   In this chapter, we shall be examining a variety of "numerical methods," any one of which could, in principle, be used to address our given problem of solving the "swing equation" for use in practical applications. We shall compare the "numerical methods," not only with each other but also with the "analytical methods" covered in the previous chapter.
2. *A list of the methods to be covered*

   The numerical methods that we shall be covering in this chapter are as follows: (1) "Euler's method"; (2) the "modified Euler's method"; (3) the "Runge–Kutta Method"; (4) the "Milne–Hamming method"; and finally (5) the "Adams–Bashforth–Moulton method." We shall apply all of these to the solution of the "swing equation" given in Eq. (5.22).

Modern Aspects of Power System Frequency Stability and Control.
DOI: https://doi.org/10.1016/B978-0-12-816139-5.00007-2

3. *The next section*

In the next section, we move on to discuss from a mathematical point of view what kind of problem the "swing equation" is.

7.1.3 What kind of mathematical problem do we have?

1. *The starting point: the "swing equation"*

For ease of reference, we repeat again here the basic starting point of the simplified "swing equation" from Chapter 5, Techniques available for calculating frequency response requirements, techniques available for calculating frequency response requirements, techniques available for calculating frequency response requirements, that is, Eq. (5.22):

$$\Sigma P_{\mathrm{GEN(NR)}} + \Sigma P_{\mathrm{GEN}n\mathrm{(R)}} - \Sigma P_{\mathrm{LOAD}n} - k_1 \cdot (f - f_n) - P_{\mathrm{TRIP}} \cdot u(t - t_{\mathrm{TRIP}}) + a_0 + a_1 t = 2 \cdot H \cdot f_0 \cdot \frac{\mathrm{d}f}{\mathrm{d}t} \tag{7.1}$$

where

$$k_1 = \Sigma P_{\mathrm{LOAD}n} \cdot \alpha \cdot + \Sigma P_{\mathrm{GEN}n\mathrm{(R)}} \cdot k \tag{7.1a}$$

is a constant.

We recall in Eq. (7.1a) that α is the "load damping" coefficient given by the equation:

$$\Sigma P_{\mathrm{LOAD}} = \Sigma P_{\mathrm{LOAD}n}\{1 + \alpha \cdot (f - f_n)\} \tag{7.1b}$$

and that k is the constant of proportionality in the equation that describes the speed governor response. It is related to the governor "droop," expressed in percentage, by the relationship:

$$D(\%) = \frac{100}{k}(\mathrm{pu}) \tag{7.1c}$$

We notice that in Eq. (7.1), we have approximated the general responsive profile expression by its first two terms $a_0 + a_1 \times t$ and approximated $f \times \mathrm{d}f/\mathrm{d}t$ to $f_0 \times \mathrm{d}f/\mathrm{d}t$ but have retained the step function term $P_{\mathrm{TRIP}} \times u(t - t_{\mathrm{TRIP}})$ intact.

2. *The mathematical structure of Eq. (7.1)*

We can see from the structure of Eq. (7.1) that it is an "initial value problem (IVP)," also referred to as an "IVP" for short. As such, it models a system that has some defined starting value [for our application, this is $f(t_0)$, the value of the frequency f at $t = 0$], and there is some subsequent evolution of the dependent variable [for us, $f(t)$] with the independent variable, which, as for us, is usually meant the time t.

3. *Specification of the initial condition*

   Since the problem being addressed is an "IVP" then we must always include in our solution method the statement:

$$f(t_0) = f_0 \tag{7.2}$$

   which describes the "initial condition." The formal requirement expressed in Eq. (7.2) applies to all the numerical methods that we shall be considering later for solving the "swing equation."

4. *Order and linearity of Eq. (7.1)*

   We can also see that Eq. (7.1) is of "first order" and is also "linear" in f and t. Formally, Eq. (7.1) is a subset of the following general type of equation:

$$\frac{\mathrm{d}y}{\mathrm{d}x} = y'(x) = \phi(x, y) \tag{7.3}$$

   with the "initial condition":

$$y(x_0) = y_0 \tag{7.4}$$

   (see Ref. [1], p. 1 for more details). For us, $y \equiv f$ and $x \equiv t$.

5. *"Dependent" and "independent" variables*

   We note that in Eq. (7.3), x is the "independent variable," y is the "dependent variable" and the symbol ϕ denotes some arbitrary (i.e., yet to be specified) "mathematical function."

6. *Some preliminary definitions*

   Before we proceed to describe the main "numerical methods" available to us for solving Eq. (7.1), as a preamble, we shall now review, briefly, some important definitions.

7. *In the next section*

   We begin in the next section with the underlying structural feature of most "numerical methods," the "Taylor series expansion."

7.1.4 The definition of the "Taylor series expansion"

As a reminder to the reader, the "Taylor series expansion" of a function $f(t)$ about some starting point $f(t_0) = f_0$ is defined as follows:

$$f(t_0 + h) = f(t_0) + h \cdot f'(t_0) + \left(\frac{h^2}{2!}\right) \cdot f''(t_0) + \cdots + \left(h^n \frac{f^{(n)}}{n!}\right) + \cdots \tag{7.5}$$

where $h = t_{n+1} - t_n$ is the numerical step length, the "time step" in our case because our mathematical problem is posed in terms of *frequency (f)* as a function of *time (t)*.

The "Taylor series expansion" is of fundamental importance to all numerical integration methods. At this point the reader may wish to refer to Ref. [1], Section 1.2, p. 3, for more details.

Note that, when using a "Taylor series expansion," we usually assume (if we do not know already) that the series expressed by Eq. (7.5) is "convergent." In modeling a real process, we do not always know this is true for sure at the start: often we must find out afterward!

In the case of using the "Taylor series expansion" as the basis of a numerical integration technique, usually only the first few terms of the expansion are used in practice.

In the next section, we introduce the concepts of "forward differences" and "backward differences" in numerical methods.

7.1.5 The use of the "forward difference" and the "backward difference" in numerical methods

Numerical techniques that use "differences" that are defined between one iteration of the independent variable x and the next can be expressed in terms of "forward differences" or "backward differences."

7.1.6 The definition of "forward difference methods"

In "forward difference methods" the "difference" here is defined by subtracting the value of the dependent variable for the "nth iteration" of the independent variable from the value of the dependent variable for the "$(n+1)$th iteration" of the independent variable x. The "forward difference" is therefore equal to $y_{n+1} - y_n$.

7.1.7 The definition of "backward difference methods"

In "backward difference methods" the "difference" here is defined by subtracting the value of the dependent variable for the "$(n-1)$"th iteration of the independent variable from the value of the dependent variable for the "n"th iteration of the independent variable. The "backward difference" is therefore equal to $y_n - y_{n-1}$.

7.1.8 The definition of the "forward difference operator"

The "forward difference operator" Δ is a shorthand symbol that has been defined for use in conjunction with the "forward difference methods" to describe the "forward difference" and extensions of it to higher order differences.

It is defined as follows:

$$\Delta y_n = y_{n+1} - y_n \tag{7.6a}$$

Similarly,

$$\Delta^2 y_n = \Delta y_{n+1} - \Delta y_n \tag{7.6b}$$

$$\Delta^3 y_n = \Delta^2 y_{n+1} - \Delta^2 y_n \tag{7.6c}$$

and so on.

See Ref. [1], p. 5, for more details on this part of the subject.

7.1.9 The definition of the "backward difference operator"

The "backward difference operator" ∇, by contrast, is a shorthand symbol that has been defined for use in conjunction with the "backward difference" methods to describe the "backward difference" and definitions of higher order differences.

It is defined as follows:

$$\nabla y_n = y_n - y_{n-1} \tag{7.7a}$$

Similarly,

$$\nabla^2 y_n = \nabla y_n - \nabla y_{n-1} \tag{7.7b}$$

$$\nabla^3 y_n = \nabla^2 y_n - \nabla^2 y_{n-1} \tag{7.7c}$$

and so on.

Again, see Ref. [1], p. 5, for more details on this.

7.1.10 Defining the time step for the numerical solution of the "swing equation"

For us in the case of the "swing equation" the independent variable x is the time t. Therefore the "difference" relevant to our case is a "time step," either a "forward time step" Δt or a "backward time step" ∇t.

7.2 Some important practical considerations

7.2.1 "Single-step" and "multistep" integration methods

7.2.1.1 An introduction to "single-step" and "multistep" integration methods

Some of the numerical methods available for integrating differential equations are termed "single-step integration methods," whereas others are termed "multistep integration methods," but what do these names mean precisely? In the next two subsections, we look at these two common definitions in more detail.

7.2.1.2 Single-step integration methods

1. *The definition of a "single-step integration method"*

   A "single-step integration method" is defined as one in which the successive values (y_{n+1}, x_{n+1}) are obtained through knowing only the current values (y_n, x_n) (see Ref. [1], p. 39, for more details on this). As such, the accuracy of the solution naturally tends to decline quite fast as more and more iterations are performed. However, for example, with "Runge–Kutta" type methods, it is possible to increase the general level of accuracy by using multiple evaluations within a single step (see Ref. [1], p. 39).

   In "single-step integration methods," knowledge of the single starting value y_n at the step x_n provides the knowledge of the value y_{n+1} that corresponds to the subsequent step x_{n+1}.
2. *Which methods are classified as "single-step integration methods"?*

   Methods that are classified as "single-step methods" include (1) "Euler's" method; (2) the "modified Euler's" method; and (3) the *single-step* "Runge–Kutta" methods.

   The "modified Euler's" method is termed, in addition, an "implicit method," because y_{n+1} is given as a function of both y_n and y_{n+1}.

7.2.1.3 Multistep integration methods

1. *The definition of a "multistep integration method"*

   A "multistep integration method" is defined as one which uses values at more than two points to define each iteration. For example, the "Adams–Bashforth–Moulton Formula" uses four points. Therefore in these methods, more than the single point value y_n is needed to start the iteration process.
2. *Which methods are considered "multistep integration methods"?*

   Examples of "multistep integration methods" are (1) the "Milne–Hamming" and (2) the "Adams–Bashforth–Moulton" methods.

   In addition, the foregoing methods are called "predictor–corrector methods" because they employ *two stages*: the "predictor" stage and the "corrector" stage.
3. *The next section*

   In the next section, we look at the general question of how the *size of the* "time step" affects (1) the "accuracy" and (2) the "stability" of the numerical solution.

7.2.2 The effect of the size of the time step on the accuracy and stability of the numerical solution

7.2.2.1 An introduction to the "time step"

The size of the *time step* $\Delta t = t_{n+1} - t_n$ (the *forward difference*) or $\nabla t = t_n - t_{n-1}$ (the *backward difference*) has a critical bearing on both the *accuracy* and *stability* of all numerical integration methods.

In the next two subsections, we discuss these two effects.

7.2.2.2 The effect of the size of the "time step" on the "accuracy" of the solution

In general, *reducing* the size of the "time step" would be expected to *increase* the *accuracy* of the solution. The drawback to reducing the time step to very small values is that the shorter the time step, the more iterations of the numerical scheme will be required and hence if the scheme is automated then the *longer* the *run time* of the program will be and consequently the more expensive the computation will be.

Therefore in a practical environment, it is necessary when using a numerical algorithm to balance the need to *maximize* the *accuracy* of the solution against the need to *minimize* the run time of the program.

7.2.2.3 The effect of the size of the "time step" on the stability of the solution

The size of the time step also affects the *stability* of the solution, that is, whether it *converges* in a numerical sense or not. In general the *smaller* the *time step*, the more likely the numerical algorithm will *converge* satisfactorily. Again, with reference to the previous subparagraph, a balance needs to be struck. However, *convergence* is the primary requirement because without *convergence* there can be no acceptable solution.

7.2.3 The effect of the length of the simulation time on the numerical solution

Does the *length of the simulation time* affect either the *accuracy* or the *stability* of the numerical solution? This might seem like a silly question, but there is a serious point to be made here, and that is, if the simulation time is too small, one may be duped into thinking that a simulation has converged satisfactorily, but this may be found not to be the case if a longer simulation time were to be chosen. In other words, one must choose the simulation time to be long enough to be able to see any possible instabilities that may occur later in the simulation to ensure that the solution is acceptable for an extended length of time. How long this "extended length of time" needs to be in practice will ultimately depend on the requirements of the study and the specific application.

7.2.4 The "truncation errors" associated with numerical integration schemes

7.2.4.1 Introduction

All numerical integration schemes are *approximate* in some way, and therefore have *errors* associated with them. In general, provided a numerical scheme is *convergent*, the error associated with a scheme *decreases* as the number of terms used in the *Taylor series* that was used to create it *increases*. Thus multistep methods are in general more accurate than single-step methods.

7.2.4.2 Newton's finite interpolation formulas

1. *Introduction*

   Many, if not all, numerical methods use "interpolation" between two points on the x-axis ("single-step" methods) or between three or more points on the x-axis ("multistep" methods) as the basis of their schemes. *Sir Isaac Newton* (1643–1727) devised two forms of a "multipoint interpolation formula": one expressed in terms of "forward differences" as defined in Eqs. (7.6a)–(7.6c), the other expressed in terms of "backward differences" according to the definition given in Eqs. (7.7a)–(7.7c).

   Both these formulas are of great practical use in defining a wide range of possible numerical integration schemes.

2. *Newton's forward difference formula*

   This version of the formula uses *forward differences* (see Eqs. 7.6a–7.6c for definitions):

$$\begin{aligned} y_\alpha &= y_0 + \alpha\cdot\Delta y_0 + \frac{\alpha\cdot(\alpha-1)}{2!}\cdot\Delta^2 y_0 + \cdots \\ &+ \frac{\alpha\cdot(\alpha-1)\cdots(\alpha-n+1)}{n!}\cdot\Delta^n y_0 \end{aligned} \tag{7.8a}$$

   where

$$\alpha = \frac{x_\alpha - x_0}{h} \tag{7.8b}$$

   is a positive integer (see Ref. [1], p. 5, for more details on this topic).

   We note that Eq. (7.8a) is a "finite interpolative polynomial" that terminates at the nth term.

3. *Newton's backward difference formula*

   This version of the formula by contrast uses *backward differences* (see Eqs. 7.7a–7.7c for definitions).

$$\begin{aligned} y_\alpha &= y_n + \alpha\cdot\Delta y_n + \frac{\alpha\cdot(\alpha+1)}{2!}\cdot\Delta^2 y_0 + \cdots \\ &+ \frac{\alpha\cdot(\alpha+1)\cdots(\alpha+n-1)}{n!}\cdot\Delta^n y_0 \end{aligned} \tag{7.9a}$$

   where

$$\alpha = \frac{x_\alpha - x_n}{h} \tag{7.9b}$$

   is a negative integer (see Ref. [1], p. 5, for more details on this topic).

   Again, we note that Eq. (7.9a) is a "finite interpolative polynomial" that terminates at the nth term.

7.2.5 "Truncation error" formulas for numerical integration methods

7.2.5.1 *The magnitude of the "truncation error"*

An expression for the magnitude of the "truncation error" T_α of a given numerical scheme is usually described in terms of the *first neglected term* in the Δ or ∇ sequence, evaluated at some point $x = \zeta$, where $x_n < \zeta < x_{n+\alpha h}$, the range across which the interpolation is being applied. Here α denotes a positive integer and h is the step length.

7.2.5.2 *An example of a "truncation error" formula*

For example, the "truncation error" for a scheme terminated after the Δ^2 term is given by

$$T_\alpha = -\left(\frac{h^5}{90}\right) \cdot y^{[4]}(\zeta) \tag{7.10}$$

where [4] denotes the fourth derivative with respect to x (see Ref. [1], p. 8 ff, for more details on this topic).

7.2.5.3 *The "order" of the "truncation error"*

In the literature, you may find a "truncation error" defined in a form like that in Eq. (7.10) referred to in an alternative way as

$$T_\alpha = O\left(h^5\right) \tag{7.11}$$

That is, the "truncation error" is "of order h^5" (see Ref. [1], p. 11).

7.3 The numerical techniques available for solving first-order ordinary differential equations

As noted earlier, the "swing equation" is a "first-order ordinary differential equation."

Numerical integration methods available for solving "first-order ordinary differential equations" include (1) "Euler's" method; (2) the "modified Euler's" method; (3) the "Runge–Kutta" methods; (4) the "Milne–Hamming" method; and (5) the "Adams–Bashforth–Moulton" methods (see, for example, Ref. [3], Chapters 26 and 27, for further details about these methods).

In the next section, we begin by looking at "Euler's" method, the simplest of all numerical integration schemes.

7.4 "Euler's method"—the simplest method

7.4.1 Introduction to "Euler's" method

"Euler's" method is the simplest of the numerical methods available to us for solving first-order ordinary differential equations. It is a *single-step method*, and each iteration of the method is defined as follows (see, for example, Ref. [3], pp. 236–237):

$$y_{n+1} = y_n + h \cdot y'_n \tag{7.12}$$

where h is the "step length" (step size) and $y_n' \equiv dy_n/dx$. Beginning with the "initial value" x_0, the "step size" h is the interval between successive values of x_n for $n > 0$.

7.4.2 The "truncation error" associated with "Euler's" method

1. *The definition of the "truncation error"*

   The "truncation error" is basically defined as the difference between the "approximate solution" y_n and the "exact solution" $y(x_n)$ at *step* "*n*."

   The "local truncation error" is defined as the error introduced in performing one step of the numerical method. The "global truncation error" is defined as the absolute error at a step of the numerical scheme.
2. *A formula for the "local truncation error" of "Euler's" method*

   The formula for the "local truncation error" T_α associated with "Euler's" method is given by

$$T_\alpha = \frac{1}{2} \cdot h^2 \cdot y^{[2]}(\zeta) \tag{7.13}$$

   where the superscript [2] denotes the second derivative of y with respect to x and $(x_n < \zeta < x_{n+1})$, that is, ζ is some value within the single integration step (see Ref. [1], p. 11).
3. *The effect on the solution of varying the "step size" h*

   The "step size" h in Eq. (7.13) may be chosen at will. We can see from Eq. (7.13) that the smaller the "step size" is, the smaller the "truncation error" will be.
4. *The "order" of "Euler's" method*

   We can also see from Eq. (7.13) that the "truncation error" associated with "Euler's" method is of $O(h^2)$. This means that "Euler's" method itself is a *first-order method* (see Ref. [1] p. 11, for further discussion on this topic).

7.4.3 The "numerical stability" of "Euler's" method

1. *The definition of "numerical instability"*

   If the "step size" h in Eq. (7.13) is chosen to be too large, the difference between the "approximate solution" and the "exact solution" may oscillate wildly and/or grow larger with increasing n. If either of these things should happen, then the "approximate solution" is said to be "numerically unstable" (see Ref. [3], p. 237, for more details on this topic).

2. *How "numerical instability" is avoided in "Euler's" method*

   "Numerical instability" in "Euler's" method is typically avoided by *halving* the "step size" h after each individual iteration (see Ref. [3], p. 237, for more details on this topic).

7.4.4 Formulating "Euler's" method for solving the "swing equation"

1. *Applying "Euler's" method to the "swing equation"*

   In the following section, we shall be programming "Euler's" method for solving the "swing equation." To do this, we need to go back to a version of the "swing equation" that is expressed in a suitable format.

2. *Reviewing the "standard form" of the "swing equation"*

   Let us first review the "standard form" of the "swing equation" given by Eq. (7.1) which was stated at the beginning of this chapter. For convenience, we repeat this equation below:

$$\Sigma P_{\mathrm{GEN(NR)}} + \Sigma P_{\mathrm{GEN}n\mathrm{(R)}} - \Sigma P_{\mathrm{LOAD}n} - k_1 \cdot (f - f_n) - P_{\mathrm{TRIP}} \cdot u(t - t_{\mathrm{TRIP}}) + a_0 + a_1 t = 2 \cdot H \cdot f_0 \cdot \frac{\mathrm{d}f}{\mathrm{d}t} \tag{7.14}$$

   where

$$k_1 = \Sigma P_{\mathrm{LOAD}n} \cdot \alpha \cdot + \Sigma P_{\mathrm{GEN}n\mathrm{(R)}} \cdot k \tag{7.14a}$$

   as described earlier in this chapter.

   Our main objective will be to formulate Eq. (7.14) to match the standard mathematical form of Eqs. (7.3) and (7.4). However, before we try to do this, it is useful to decide upon a "particular form" of Eq. (7.14) with which to work soon in an example in *MATLAB* using "Euler's" method.

3. *Choosing a "particular form" of Eq. (7.14) to analyze using "Euler's" method*

   a. *Modeling of the governor response*

   A convenient form of Eq. (7.14) to work with in all our subsequent examples is to put $k = 0$ and model the governor response plus any other response by the general "$a_0 + a_1 \times t$" term. In addition, without loss of generality, we may take the time of the trip t_{TRIP} to be zero.

   b. *The rationale behind specific modeling*

   By modeling the response in this way, we simplify the mathematics for the purposes of studying our examples. It is easier to see what is going on when we model a ramped response than by representing the response by a governor model with $k \neq 0$.

c. *Structure of the examples generally*

In some examples, we shall not be modeling response, whereas in others, there will be two cases: (1) without response and (2) with response.

d. *Structure of the "Euler's" method example*

For this the Euler's method we shall provide an example with no response being provided, thus, in this case, $a_0 = a_1 = 0$. As such, we shall therefore see the frequency in "free fall" following the loss of an amount of generation from the system.

4. *Putting Eq. (7.14) into standard mathematical form*

To match the "standard mathematical form" of Eqs. (7.3) and (7.4) with our simplifying assumption of $k = 0$ and $t_{\mathrm{TRIP}} = 0$, we proceed to rewrite Eq. (7.14) as follows:

$$\frac{\mathrm{d}f}{\mathrm{d}t} = \frac{1}{2 \cdot H \cdot f_0}[-k_1 \cdot f + a_1 \cdot t + k_2] \tag{7.15}$$

where

$$k_1 = \Sigma P_{\mathrm{LOAD}n} \cdot \alpha \quad \text{only} \tag{7.15a}$$

and

$$k_2 = \Sigma P_{\mathrm{GEN(NR)}} + \Sigma P_{\mathrm{GEN}n\mathrm{(R)}} - \Sigma P_{\mathrm{LOAD}n} + k_1 \cdot f_n - P_{\mathrm{TRIP}} \cdot u(t) + a_0 \tag{7.15b}$$

5. *How to use "Euler's" method to solve Eq. (7.15)*

a. *"Euler's" formula for our case*

"Euler's" formula in Eq. (7.12) translates in our case to

$$f_{n+1} = f_n + h \cdot f'_n \tag{7.16}$$

where f represents the power frequency in per unit (pu).

b. *The "initial condition"*

On the "zeroth" iteration (the "initial condition"), we take $t = 0^+$, that is, at a time just after the loss of P_{TRIP} occurs. This is of course the same as saying that $u(t) = 1$ in Eq. (7.15). We also set $n = 0$ in Eq. (7.16) to give

$$f_1 = f_0 + h \cdot f'_0 \tag{7.17}$$

If we assume that we know a value for the initial frequency, this value (or an estimate of it if we do not know exactly what the initial frequency is) can be entered as the first term on the right-hand side of Eq. (7.17).

c. *The "final piece of the jigsaw"*

The "final piece in the jigsaw" of the right-hand side of Eq. (7.16) is the derivative $f_0{}'$. This may be evaluated using Eq. (7.15) with $f = f_0$ and $t = 0^+$.

d. *The first new value of "f"*

Taking together the preceding results, we are now able to evaluate f_1, our first new value in the predictive sequence given by "Euler's" method.

e. *Further iterations*

We then proceed to calculate f_2, f_3, and so on in a similar way, knowing the current value of f and calculating $f' = df/dt$ using Eq. (7.15).

f. *When to stop iterating*

We stop iterating when a sufficiently large value for the *time (t)* has been reached. This value will be determined by the timescale of interest for the simulation, say 0–10 seconds or something of a similar magnitude.

g. *The next section*

In the next section, we look at a practical example of using "Euler's" method to solve the "swing equation." We shall again use MATLAB as our tool. For general example problems in MATLAB please see Ref. [4].

7.5 MATLAB example simulation 3: using "Euler's" method to solve the "swing equation"

7.5.1 Introduction

7.5.1.1 Objective

In this example, we shall use "Euler's" method as described above to generate a solution of the "swing equation" as represented by Eqs. (7.15), (7.15a), and (7.15b). This means that we shall be calculating the solution for time steps starting at $t = 0$ seconds when a generator is tripped and progressing in time to the solution at $t = 10$ seconds. In this problem, we shall choose a value of the "time step" of $h = 0.05$ seconds.

7.5.1.2 Following example 1

For the network data, we shall be following *MATLAB example simulation 1* in Section 6.3.

7.5.2 A statement of the problem

1. *The network*

To summarize, we have constructed an imaginary 50 Hz system in which there is a total demand of 3000 MW that is being met by 10 identical generators, all of which are operating at their "rated output" of

300 MW each to meet the demand exactly. There is no response provision in this example. The "inertia constant" of each generator is $H = 5.0$ seconds.

2. *The challenge*

   The problem is to find the time evolution of the system frequency using "Euler's" *method if we* assume that the initial frequency is $f_0 = 50$ Hz and one generator is lost from the system at $t = 0$ seconds. We shall extend the simulation time to $t = 10$ seconds.

7.5.3 The general solution of the problem

From Section 7.4.1 with $y \equiv f$ and $x \equiv t$, we know that each step of "Euler's" method is described by the simple formula:

$$f_{n+1} = f_n + h \cdot f'_n \qquad (7.16a)$$

as in Eq. (7.16). The initial step is obtained by putting $n = 0$ in this formula, giving

$$f_1 = f_0 + h \cdot f'_0 \qquad (7.17a)$$

as in Eq. (7.17).

In both these equations, h is the "time step," which should be small enough to maintain reasonable accuracy in the simulation but not so small that a very large number of iterations must be performed thus taking a very long time to run the simulation.

7.5.4 The input data to MATLAB

The input data to MATLAB for this problem is given in physical units in Section 6.3.6.2 and in per unit (pu) terms in Section 6.3.6.3(2).

7.5.5 The output data from MATLAB

The output data from *MATLAB* is a sequence of values for the system frequency plotted against the corresponding values of time t, thus giving the evolution of frequency with time for the *first 10 seconds*.

7.5.6 The suggested MATLAB code

A suitable MATLAB code to solve our *IVP* by "Euler's" *method* could be as follows:

```
% Input the parameter values applicable to the "Swing Equation" as represented
% by (Eq 7.15), (Eq 7.15a) and (Eq 7.15b) in "per unit":
      PNR = input('Please enter the non-responsive generation in per unit')
      PR = input('Please enter the responsive generation in per unit')
      PL = input('Please enter the system load in per unit')
```

```
        K1 = input('Please enter the value of the constant k₁in per unit')
        FN = input('Please enter the nominal frequency in per unit')
        PTRIP = input('Please enter the generation tripped in per unit')
        A_0 = input('Please enter the value of the constant a₀in per unit')
        A_1 = input('Please enter the value of the constant a₁in per unit')
        H = input('Please enter the value of the H-constant in seconds')
        F0 = input('Please enter the value of the initial frequency in per unit')
%   Input the size of the time step for "Euler's" method:
        TSTEP = input('Please enter the value of the time step h in seconds')
%   Calculate the power balance at time = 0⁺seconds (that is, just following the
    loss of generation) at nominal frequency as a parameter:
                    PBAL = PNR + PR - PL - PTRIP
%   Calculate the value of the parameter k₁in "per unit":
                    K1 = 1.0 * PL
%   Calculate the parameter k₂in "per unit":
                    K2 = PBAL + K1*FN + A_0
%   Now calculate the solution for the frequency f at a sequence of values of t
    from 0
%   to 10 seconds at intervals of h = 0.05 seconds using "Euler's" method:
%
%   Zeroth iteration, following (Eq 7.17):
%
%   First, calculate df/dt in "per unit" using (Eq 7.15):
                    DFDT_0 = (- K1*F0 + A_1*0 + K2)/(2*H*F0)
%   Second, calculate f₁using (Eq 7.17):
                    F(1) = F0 + HSTEP*DFDT_0
%   Third, calculate the rest of the values of f in the sequence up to t = 10
    seconds
%   in steps of 0.05 seconds:
                    for k = 1,199
                        T(k) = TSTEP*k
                        DFDT(k) = (- K1*F(k) + A_1*T(k) + K2)
                        F(k + 1) = F(K) + TSTEP*DFDT(k)
                    end
                        T(200) = 10.00
%   The vector F contains the required values of frequency in "per unit".
%
%   These values can now be printed out as a graph of frequency against time
%   after converting the F values to Hz from "per unit".
```

7.5.7 Instructions for the user

When prompted, enter the input data as given in Section 7.5.2. Then run the program and print out the results.

7.5.8 In the next section

In the section, we shall move on to a description of the "modified Euler's" method, which is a variant of "Euler's" method.

7.6 The "modified Euler's" method

7.6.1 An introduction to the "modified Euler's" method

The "modified Euler's" method is an improved version of the "Euler's" method that was given in Eq. (7.12). In the "modified Euler's" method the approximate solution at each step is calculated in *two stages*, a "predictor" and a "corrector," which are defined as follows:

1. *Stage 1: the "predictor"*
   In the first "stage" of the "modified Euler's" method the basic "Euler's" method is first used to obtain an *estimated value* for the dependent variable y_{n+1} at the next step.
2. *Stage 2: the "corrector"*
   In the second "stage" of the "modified Euler's" method, a revised formula for the dependent variable y_{n+1} at the next step based on the average of the derivatives of the dependent variable y at the nth and $(n+1)$th steps is applied as a correction.

7.6.2 Formulation of the "modified Euler's" method

1. *The defining equations for the "modified Euler's" method*
   The "modified Euler's" method is defined formally as follows:

$$py_{n+1} = y_n + h \cdot y'_n \tag{7.18a}$$

   for the "predictor stage" at each iteration, and

$$y_{n+1} = y_n + \left(\frac{h}{2}\right) \cdot \left(py'_{n+1} + y'_n\right) \tag{7.18b}$$

   for the "corrector stage" at each iteration (see, for example, Ref. [3], p. 254, for more details about this formulism). The symbol "p" here indicates a "predicted" value of the variable concerned.
2. *Completing the formulation*
   To complete the formulation of the "modified Euler's" method, we still need to find an expression for py_{n+1}' to substitute into the right-hand side of Eq. (7.18b). We explain how this expression may be found in the next.
3. *Finding an expression for py_{n+1}' to substitute into Eq. (7.18b)*
   If the original equation that is to be integrated can be expressed in the form:

$$y'_n = \phi(x_n, y_n) \tag{7.18c}$$

   where $\phi(x,y)$ is a *known function*, then, by this definition, we must therefore have by substituting the predicted value of y_{n+1}' in Eq. (7.18c):

$$py'_{n+1} = \phi(x_{n+1}, py_{n+1}) \tag{7.18d}$$

Eq. (7.18d) is hence the correct formula for the "predicted derivative" at the $(n+1)$th *step* for substitution into Eq. (7.18b).

This completes the formulation of the "modified Euler's" method.

4. *A note on the "predicted value" notation*

We note again that the letter "p" in Eqs. (7.18a), (7.18b), and (7.18d) denotes the "predicted" value of the quantity that follows it. See Ref. [3], p. 254, for further details on this topic, but we provide a short summary in the next paragraph.

5. *A general remark about "predictor–corrector" methods*

a. *The "modified Euler's" method*

The "modified Euler's" method is a simple example of a "predictor–corrector" method. In these methods an initial estimate, the "predictor" stage, is followed by a modification, the "corrector" stage. The "modified Euler's" method is a *single-step* "predictor–corrector" method.

b. *Further methods*

Later in this book, in Sections 7.10 and 7.12, we shall look in detail at two further *multistep* "predictor–corrector" methods, the "Milne–Hamming" method, and the "Adams–Bashforth–Moulton" method, respectively.

7.6.3 The "truncation error" associated with the "modified Euler's" method

1. *The formula for the "truncation error" associated with the "modified Euler's" method*

The formula for the "truncation error" associated with the "modified Euler's" method is

$$T_\alpha = -\left(\frac{h^3}{12}\right)\cdot y^{[3]}(\zeta) \tag{7.19}$$

where T_α is the standard notation for the "truncation error," and the superscript [3] denotes the *third derivative* with respect to x. The quantity ζ is some value of x satisfying the inequality:

$$x_n < \zeta < x_{n+1} \tag{7.20}$$

(see Ref. [1], p. 12, for further details on this topic).

From the formula for the truncation error, we see that the "modified Euler's" method is a *second-order method* since the power of h is 3.

2. *A comparison of the accuracy of the "modified Euler's" method with that of "Euler's" method*

It can be seen readily from Eq. (7.19) that the "truncation error" associated with the "modified Euler's" method is of $O(h^3)$ compared with the "truncation error" for the basic "Euler's" method which is only of $O(h^2)$ (see Eq. 7.12 for the definition of this). Therefore the "modified Euler's" method is a *second-order method* and consequently, in general, we can say that it is *an order of magnitude more accurate* than the basic "Euler's" method.

7.6.4 Formulating the "modified Euler's" method for solving the "swing equation"

1. *Applying the "modified Euler's" method to the "swing equation"*

   In the following section, we shall be programming into MATLAB the "modified Euler's method" for solving the "swing equation." To achieve this goal, we need to go back to a version of the "swing equation" that is expressed in a suitable format for the purposes of programming.

2. *Reviewing the "standard version" of the "swing equation"*

   Let us first review again the "standard version" of the "swing equation" given by Eq. (7.1) given at the beginning of this chapter. For convenience, we repeat this equation here:

$$\Sigma P_{\text{GEN(NR)}} + \Sigma P_{\text{GEN}n\text{(R)}} - \Sigma P_{\text{LOAD}n} - k_1 \cdot (f - f_n) - P_{\text{TRIP}} \cdot u(t - t_{\text{TRIP}}) + a_0 + a_1 t = 2 \cdot H \cdot f_0 \cdot \frac{df}{dt} \tag{7.21}$$

   where

$$k_1 = \Sigma P_{\text{LOAD}n} \cdot \alpha + \Sigma P_{\text{GEN}n(R)} \cdot k \tag{7.21a}$$

   as previously.

   We should like, as with the simple "Euler's" method, to simplify Eq. (7.21) a little for the purposes of programming a straightforward example that demonstrates the method into MATLAB.

3. *Choosing a "particular form" of Eq. (7.21) to analyze with the "modified Euler's" method*

   We shall follow the same route that we took with the simple "Euler's" method and simplify Eq. (7.21) by setting $k = 0$ and $t_{\text{TRIP}=}0$. The first of these choices means that we are not modeling the speed governor "slope" (assuming we are dealing with a traditional "synchronous generator") directly but are modeling it alongside all the other forms of response in the "$a_0 + a_1 \times t$" term. The second assumption simply means that the "system event," that is, the unplanned loss of generation or demand, occurs at $t = 0$. This is an arbitrary assumption that does not alter the physics but simplifies the mathematics for us.

4. *Putting Eq. (7.21) into the "standard mathematical form"*

Maintaining consistency, we shall adopt Eqs. (7.15), (7.15a), and (7.15b) as before in the simple "Euler's" method. We repeat these equations here for convenience:

$$\frac{df}{dt} = \frac{1}{2 \cdot H \cdot f_0}[-k_1 \cdot f + a_1 \cdot t + k_2] \tag{7.22}$$

where

$$k_1 = \Sigma P_{LOADn} \cdot \alpha \quad \text{only} \tag{7.22a}$$

and

$$k_2 = \Sigma P_{GEN(NR)} + \Sigma P_{GENn(R)} - \Sigma P_{LOADn} - P_{TRIP} \cdot u(t) + k_1 \cdot f_n + a_0 \tag{7.22b}$$

These equations are the ones we shall use as the basis for our subsequent programming.

5. *How to use the "modified Euler's" method to solve the "swing equation"*

a. *The "modified Euler's" formulas for our case*

The "modified Euler's" formulas of Eqs. (7.18a) and (7.18b) translate to our case, by replacing y with f and x with t as follows:

$$pf_{n+1} = f_n + h \cdot f'_n \tag{7.23a}$$

as the "predictor stage" at each iteration, and

$$f_{n+1} = f_n + \left(\frac{h}{2}\right) \cdot (pf'_{n+1} + f'_n) \tag{7.23b}$$

as the "corrector stage" at each iteration.

b. *The "initial condition"*

The "initial condition" corresponds to $n = 0$. For this value of n the "predictor" formula (Eq. 7.23a) becomes

$$pf_1 = f_0 + h \cdot f'_0 \tag{7.24a}$$

and the "corrector" formula Eq. (7.23b) becomes

$$f_1 = f_0 + \left(\frac{h}{2}\right) \cdot (pf'_1 + f'_0) \tag{7.24b}$$

c. *The "initial condition" for the "predictor"*

It may be verified that we can evaluate all the terms on the right-hand side of Eq. (7.24a) using the selected values for the *initial frequency* f_0 and the *time step* h, with f_0' evaluated using Eq. (7.22) directly using $f = f_0$ and $t = 0^+$.

d. *The "initial condition" for the "corrector"*

We can then, having already a predicted value for f_1 obtained from part (c), follow the same general procedure as in that part to obtain the "corrected" value for f_1 using Eqs. (7.22) and (7.24b) as before.

e. *Further iterations*

To calculate values for f_2, f_3, and f at subsequent iterations, we continue in the same general way as detailed above for f_1.

f. *In the next section*

In the next section, we shall go through an example calculation using the "modified Euler's" method in MATLAB as formulated above. At the end of the calculation, we shall compare the results of the "modified Euler's" method with those of the simple "Euler's" method and with the "direct solution of the differential equation" method as well.

7.7 MATLAB example simulation 4: using the "modified Euler's" method to solve the "swing equation"

7.7.1 Introduction

In this example, we shall use the "modified Euler's" method as described above to generate a solution of the "swing equation" as represented by Eqs. (7.22), (7.22a), and (7.22b). This means that we shall be calculating the solution for time steps starting at $t = 0$ when some generation is tripped and progressing the solution in time to $t = 10$ seconds.

7.7.2 A statement of the problem

1. *Objective*

   The objective of the exercise is to generate a plot of frequency against time for the first 10 seconds following the generation loss event at $t = 0$.

2. *Following example 2*

   a. *The network*

   This time, for the network data, we shall be following *MATLAB example simulation* 2 in Section 6.5.

   b. *Details of the generation*

   Thus we have 53 generators running at their full output of 250 MW each, 30 generators running deloaded at 90% of full output = 225 MW each and 17 generators not running, making a total of 100 generators in all. This meets a total load of $53 \times 250 + 30 \times 0.9 \times 250 = 20{,}000$ MW. The inertia constant of each generator is 3.0 seconds.

c. *The system event*

At $t = 0$ second, 3×250 MW units are tripped, thereby unbalancing the system. We need to calculate the subsequent evolution of the frequency.

d. *Response provision*

We shall assume in this case that no response is available to check the frequency drop that occurs after the 750 MW of generation is lost.

7.7.3 The general solution to the problem

From Section 7.6.2, we know that, with $y \equiv f$ and $x \equiv t$,

$$pf_{n+1} = f_n + h \cdot f'_n \tag{7.25a}$$

is the "predictor stage" at each iteration and

$$f_{n+1} = f_n + \left(\frac{h}{2}\right) \cdot (pf'_{n+1} + f'_n) \tag{7.25b}$$

is the "corrector stage" at each iteration.

The term f_n' on the right-hand side of Eq. (7.25a) is evaluated using the "swing equation" evaluated at the nth time step. For this, we can use Eq. (7.22) to give us:

$$f'_n = \frac{1}{2 \cdot H \cdot f_0}[-k_1 \cdot f_n + a_1 \cdot t_n + k_2] \tag{7.25c}$$

in terms of f_n and t_n, where k_1 and k_2 are given by Eqs. (7.22a) and (7.22b).

The term pf_{n+1}' on the right-hand side of Eq. (7.25b) is, in this instance, evaluated using the "swing equation" evaluated at the $(n + 1)$th time step. This means that we use Eq. (7.22) at time step $n + 1$ to evaluate the derivative. Thus we have

$$pf'_{n+1} = \frac{1}{2 \cdot H \cdot f_0}[-k_1 \cdot f_{n+1} + a_1 \cdot t_{n+1} + k_2] \tag{7.25d}$$

where k_1 and k_2 are given by Eqs. (7.22a) and (7.22b), respectively, as previously.

We notice that the right-hand side of Eq. (7.25d) contains the term f_{n+1} which we evaluate using the predicted value pf_{n+1} given by Eq. (7.25a). This, then, completes a single step of the "modified Euler's" method, from f_n to f_{n+1}.

This process is repeated until $t = 10$ seconds is reached, when we stop iterating.

We note, as with "Euler's" method, that for the "first" (or "zeroth") iteration, we must put $n = 0$ in Eqs. (7.25a)–(7.25d), and this gives the following expressions:

$$pf_1 = f_0 + h \cdot f'_0 \tag{7.25e}$$

$$f_1 = f_0 + \left(\frac{h}{2}\right) \cdot (pf_1' + f_0') \tag{7.25f}$$

$$f_0' = \frac{1}{2 \cdot H \cdot f_0}[-k_1 \cdot f_0 + a_1 \cdot t_0 + k_2] \tag{7.25g}$$

$$pf_1' = \frac{1}{2 \cdot H \cdot f_0}[-k_1 \cdot pf_1 + a_1 \cdot t_1 + k_2] \tag{7.25h}$$

Note also that in Eq. (7.25g), for our case, we have $t_0 = 0$.

The order of evaluation of the above four equations is first to calculate f_0' using Eq. (7.25g).

The result is then substituted on the right-hand side of Eq. (7.25e) to find pf_1. Then pf_1 is substituted into the right-hand side of Eq. (7.25h) to enable pf_1' to be calculated. Finally, f_1 can be evaluated from Eq. (7.25f) using the now known values for pf_1' and f_0'.

7.7.4 The input data

We first need to convert the data provided in Section 7.7.2(2) from *physical units* to per unit (pu) for use in the calculation. This may be done as follows:

By using a "standard machine power factor" of 0.8, we shall adopt an *MVA base* equal to 20,000 MW/0.8 = 25,000 MVA. For the frequency base, we shall choose 50 Hz, the nominal frequency for the system.

Then the variables and parameters in Eqs. (7.22), (7.23a) and (7.23b), and (7.24a) and (7.24b) become in per unit (pu) terms

$$P_{\text{TRIP}} = \frac{3 \times 250}{(25,000)} \tag{7.26a}$$

$$k_1 = \frac{(20,000)}{(25,000)} \cdot (\alpha \times 50) \tag{7.26b}$$

$$k_2 = \frac{(53 \times 250) + (30 \times 250) - 20,000 - (3 \times 250)}{(25,000)} + k_1 + a_0 \tag{7.26c}$$

where we shall assign values of zero to the "ramp" parameters a_0 and a_1 because we are not considering generator response to be operating in this example.

The parameters a_0 and a_1 could have modeled a responsive profile if we had chosen to include one in the simulation. It would have been necessary in that case to calculate a_0 and a_1 in "per unit" to represent whatever was the desired ramped response to be applied. We would have needed a "starting value" and a "ramp rate" to do this.

7.7.5 The output data

The output data will consist of a sequence of values of the frequency *f* for each of the time values separated by the chosen time step of 0.05 seconds.

7.7.6 The MATLAB code

A suggested MATLAB code could be as follows:

```
% Input the parameter values applicable to the "Swing Equation" as represented
% by (Eq 7.15), (Eq 7.15a) and (Eq 7.15b) in "per unit":
      PNR = input('Please enter the non-responsive generation in per unit')
      PR = input('Please enter the rated responsive generation in per unit')
      PL = input('Please enter the system load in per unit')
      FN = input('Please enter the nominal frequency in per unit')
      PTRIP = input('Please enter the generation tripped in per unit')
      A_0 = input('Please enter the value of the constant a0in per unit')
      A_1 = input('Please enter the value of the constant a1in per unit')
      H = input('Please enter the value of the H-constant in seconds')
      F0 = input('Please enter the value of the initial frequency in per unit')
% Input the size of the time step for the "Modified Euler's" method:
      TSTEP = input('Please enter the value of the time step h in seconds')
% Calculate the power balance at time = 0+seconds (that is, just follow-
ing the loss of generation) at nominal frequency as a parameter:
                  PBAL = PNR+ PR - PL - PTRIP
% Calculate the value of the parameter k1in "per unit":
                  K1 = 0.02 *50* PL
% Calculate the parameter k2in "per unit":
                  K2 = PBAL+ K1*FN+ A_0
% Now calculate the solution for the frequency f at a sequence of values of t
  from 0
% to 10 seconds at intervals of h = 0.05 seconds using the "Modified Euler's"
% method:
%
% Calculate the "zeroth iteration" following (Eq 7.25e) - (Eq 7.25h):
%
                  DFDT_0 = (- K1*F0+ K2)/(2*H*F0)
                  PF1 = F0+ TSTEP*DFDT_0
                  PDFDT_1 = (- K1*PF1+ K2)
                  F(1) = F0+ (TSTEP/2)*(PDFDT_1+ DFDT_0)
%
% Calculate all subsequent values of the frequency F:
%
                  for k = 1,199
                    DFDT(k) = (- K1*F(k)+ K2)/(2*H*F0)
                    PF(k+1) = F(k)+ TSTEP*DFDT(k)
                    PDFDT(k+1) = (- K1*F(k+1)+ K2)/(2*H*F0)
                    F(k+1) = F(k)+ (TSTEP/2)*(PDFDT(k+1)+ DFDT(k))
                  end
%
```

```
% The vector F contains the required values of frequency in "per unit".
%
% These values can now be printed out as a graph of frequency against time
% after converting the F values to Hz from "per unit".
%
% We have not included the activation of any response in this example.
%
```

7.7.7 Instructions for the user

Run the MATLAB program, entering the data when prompted. The results may then be plotted out as a graph of frequency against time.

7.7.8 In the next section

In the next section, we shall examine the "Runge–Kutta" type methods and their suitability for solving the "swing equation."

7.8 The "Runge–Kutta" type methods

7.8.1 An introduction to the "Runge–Kutta" type methods

7.8.1.1 Characteristics of the "Runge–Kutta" type methods

The "Runge–Kutta" and related methods are "single-step" methods that are characterized by performing multiple evaluations of $\phi(x,y)$ (see Eq. 7.18c for the definition of ϕ) *within the interval itself* rather than spanning several intervals (see Ref. [1], p. 39, for further details on this point).

7.8.1.2 The "order" of a "Runge–Kutta" method

"Runge–Kutta" methods are available in different *orders* (see Ref. [1], p. 46). The *order* of a "Runge–Kutta" method essentially depends upon *how many subintervals* each interval is subdivided into (see Ref. [1], p. 40, for more details on this). As an example, in this book, we shall be looking at a *fourth-order* "Runge–Kutta" algorithm, whose "truncation error" is therefore, by definition, of $O(h^5)$.

7.8.1.3 The general mathematical form of the "Runge–Kutta" methods

The general mathematical form of all the "Runge–Kutta" methods is

$$y_{n+1} = y_n + \Sigma_{i=1,m} w_i k_i \tag{7.27}$$

The w_i are called the weighting factors, and the k_i follow a specific sequence such that the overall approximation agrees with the "Taylor series" for y_{n+1} expanded in powers of h up to the selected number of terms (again, see Ref. [1], p. 40, for more details on this).

7.8.2 The first-order "Runge–Kutta" method (="Euler's" method)

If we choose, for example, $m = 1$ in Eq. (7.27), then we obtain the form:

$$y_{n+1} = y_n + 1 \cdot h \cdot \phi(x_n, y_n) \tag{7.28}$$

since $w_1 = 1$ and $k_1 = h \times \phi(x_n, y_n) \equiv h \times y_n{}'$ by Eq. (7.16c). Thus it can be seen that Eq. (7.28) is simply the formula for "Euler's" method as given by Eq. (7.12).

7.8.3 Higher order "Runge–Kutta" methods

In general, the *accuracy* of the "Runge–Kutta" method solution *increases* by *increasing* the "order" of the method, at the expense of having to expend more computational effort.

For example, here are the formulas corresponding to the "classic fourth-order Runge–Kutta" algorithm:

$$y_{n+1} = y_n + \left(\frac{1}{6}\right)(k_1 + 2 \cdot k_2 + 2 \cdot k_3 + k_4) \tag{7.29a}$$

where

$$k_1 = h \cdot \phi(x_n, y_n) \tag{7.29b}$$

$$k_2 = h \cdot \phi\left(x_n + \frac{1}{2}h, y_n + \frac{1}{2}k_1\right) \tag{7.29c}$$

$$k_3 = h \cdot \phi\left(x_n + \frac{1}{2}h, y_n + \frac{1}{2}k_2\right) \tag{7.29d}$$

and

$$k_4 = h \cdot \phi(x_n + h, y_n + k_3) \tag{7.29e}$$

(see Ref. [1], p. 42, or Ref. [3], p. 255, for further details.)

In the next section, we shall look briefly at the "truncation errors" associated with the various "Runge–Kutta" methods referred to above.

7.8.4 "Truncation errors" associated with the various "Runge–Kutta" methods

1. *Calculating the "local truncation error" of "Runge–Kutta" methods*

   Calculating the "local truncation errors," that is, the "truncation error" for a *single step* of a "Runge–Kutta" method may be accomplished, although this can be very complicated. For example, the reader may be interested to see the exact formula for the local truncation error $T(x,h)$ in the case of the *fourth-order Runge–Kutta* algorithm.

This formula is not given here but is given in Ref. [3], Section 2.3.4 on p. 49. It is derived from a formulation with 11 equations and 13 unknowns. The two "free parameters" are called c_2 and c_3, whose values can be chosen in different ways to give different versions of the *fourth-order Runge–Kutta* method. For example, the "classic Runge–Kutta method" has $c_2 = c_3 = h/2$ (see Ref. [1], p. 49, for more details).

2. *Estimating the "local truncation error" of "Runge–Kutta" methods*

One method of (indirectly) measuring the "local truncation error" $T(x,h)$ associated with a single step of the *fourth-order Runge–Kutta* method by specifying the "correct" magnitude of the *step h* is known as the "Collatz Rule of Thumb" (see Ref. [1], p. 69) and suggests that

$$\left|\frac{(k_3 - k_2)}{(k_2 - k_1)}\right| = 1.0 \tag{7.30}$$

when the *step h* is chosen "correctly" to minimize the "truncation error" (see Ref. [1], p. 70).

In the words of *Lapidus and Seinfeld* (see Ref. [1], p. 70): "If the ratio is much greater than 1, the local truncation error is too large in this step, and h must be decreased. If the ratio is much smaller than 1, the h must be increased. The use of Eq. (7.30) to specify h is, of course, subject to considerable interpretation and should be used only with care."

Again, according to *Lapidus and Seinfeld*: "Warten (see Ref. [5]) has, however, used the idea with some success" (see Ref. [1], p. 70). In the view of this author, the "Collatz Rule of Thumb" is simple to use and therefore worth trying out in practice as a quick method.

The reader is referred to Ref. [1], pp. 68–74 and 100–102 for more extensive information on this topic.

3. *The "error bounds" of a finite "Taylor series" expansion of $\phi(x_n, y_n)$*

If y_{n+1} is expanded as a "Taylor series" of the function $\phi(x_n, y_n)$ up to the term in h^p then the theoretical error $|\varepsilon_n|$ is bounded according to the formula:

$$|\varepsilon_n| \le \frac{h^p}{(p+1)!} \cdot x_n \cdot \exp(x_n) \tag{7.31}$$

(see Ref. [1], Eq. 2.13-8, p. 80).

a. *The dependency of the absolute error $|\varepsilon_n|$ on h*

It can be seen from Eq. (7.31) that for any fixed value of p the "absolute error" $|\varepsilon_n|$ decreases with decreasing h, where h is the "step size."

b. *The dependency of the absolute error $|\varepsilon_n|$ on p*

It can also be seen from Eq. (7.31) that for a fixed value of h the "absolute error" $|\varepsilon_n|$ decreases with increasing p, where p is the *number of terms in the* "Taylor series" *minus two.*

4. *In the next section*

   In the next section, we look briefly at the "stability" of the various "Runge–Kutta" methods referred to above.

7.8.5 The "stability" of the various "Runge–Kutta" methods

1. *The dual requirements of "accuracy" and "stability"*

   In general, as well as the "accuracy" (alternatively, the size of the "truncation error," as considered above) of a numerical method, we also need to consider its "stability."
2. *The definition of the "stability" of a numerical method*

   The "stability" of a numerical method is defined in terms of the *difference* between the "approximate solution" $y_n(x_n)$ and the "exact solution" $y(x_n)$ at *step n* of the algorithm. More specifically, the "stability" of a numerical method depends on the behavior of the difference $\varepsilon = y_n(x_n) - y(x_n)$ as n increases to large values (see Ref. [2], p. 109, for more details on this).
3. *A statement regarding the "stability" of all explicit "Runge–Kutta" methods*

   According to *Lapidus and Seinfeld*: "All explicit 'Runge–Kutta' methods have a finite stability boundary and will be stable for a small enough h. In general, as the order [of the method] increases, the stability bounds [also] increase" (see Ref. [1], p. 131).

7.8.6 Formulating the *classic explicit fourth-order* "Runge–Kutta" *method* for solving the "swing equation"

1. *Applying the classic explicit fourth-order "Runge–Kutta" method to the "swing equation"*

   In the following section, we shall be programming into *MATLAB* the *classic explicit fourth-order* "Runge–Kutta" *method* to solve the "swing equation." To do this, we need to go back to a version of the "swing equation" that is expressed in a suitable format.
2. *Reviewing the "standard version" of the "swing equation"*

   Let us again first review the "standard version" of the "swing equation" given by Eq. (7.1) given at the beginning of this chapter. For convenience, we repeat this equation below:

$$\Sigma P_{\text{GEN(NR)}} + \Sigma P_{\text{GEN}n\text{(R)}} - \Sigma P_{\text{LOAD}n} - k_1 \cdot (f - f_n) - P_{\text{TRIP}} \cdot u(t - t_{\text{TRIP}}) + a_0 + a_1 t = 2 \cdot H \cdot f_0 \cdot \frac{df}{dt} \tag{7.32}$$

where

$$k_1 = \Sigma P_{\text{LOAD}n} \cdot \alpha + \Sigma P_{\text{GEN}n\text{(R)}} \cdot k \tag{7.32a}$$

3. *Choosing a "particular form" of Eqs. (7.24a) and (7.24b) to analyze with the classic explicit fourth-order "Runge–Kutta" method*

   In keeping with our choice for the previous two numerical solution methods that we are considering in this book, we shall simplify Eqs. (7.24a) and (7.24b) by assuming (without loss of generality) that $t_{TRIP} = 0$ and by setting $k = 0$. Any response requirements, if selected, may be obtained by choosing suitable values for a_0 and a_1.

4. *Putting Eqs. (7.24a) and (7.24b) into the "standard mathematical form"*

   Further maintaining consistency with previous practice in this book, we shall adopt Eqs. (7.15), (7.15a), and (7.15b) as before, and we repeat these equations here for convenience:

$$\frac{df}{dt} = \frac{1}{2 \cdot H \cdot f_0}[-k\_1 \cdot f + a_1 \cdot t + k\_2] = \phi(f, t) \tag{7.33}$$

   where

$$k\_1 = \Sigma P_{LOADn} \cdot \alpha \quad \text{only} \tag{7.33a}$$

   and

$$k\_2 = \Sigma P_{GEN(NR)} + \Sigma P_{GENn(R)} - \Sigma P_{LOADn} - P_{TRIP} \cdot u(t) + k\_1 \cdot f_n + a_0 \tag{7.33b}$$

   In the above three equations, we have used k_1 instead of k_1 and k_2 instead of k_2 to avoid confusion with the k parameters used in the "classic fourth-order Runge–Kutta" algorithm used in the next.

5. *How to use the "classic fourth-order Runge–Kutta" algorithm to solve the "swing equation"*

   a. *The "classic fourth-order Runge–Kutta" formulas for our case*

      The "classic fourth-order Runge–Kutta" formulas given in Eqs. (7.29a)–(7.29e) inclusive translate to our case by setting $y = f$ and $x = t$. They are then

$$f_{n+1} = f_n + \left(\frac{1}{6}\right)(k_1 + 2 \cdot k_2 + 2 \cdot k_3 + k_4) \tag{7.34a}$$

      where

$$k_1 = h \cdot \phi(t_n, f_n) \tag{7.34b}$$

$$k_2 = h \cdot \phi\left(t_n + \frac{1}{2}h, f_n + \frac{1}{2}k_1\right) \tag{7.34c}$$

$$k_3 = h \cdot \phi\left(t_n + \frac{1}{2}h, f_n + \frac{1}{2}k_2\right) \tag{7.34d}$$

and

$$k_4 = h\cdot\phi(t_n + h, f_n + k_3) \tag{7.34e}$$

As mentioned earlier, there are many other ways of formulating these equations by altering the coefficients to different values, but the above formulation is the standard one that we shall use in this book for the purposes of illustrating the use of the "Runge–Kutta" algorithm to solve the "swing equation."

b. *The "initial condition"*

The "initial condition," as introduced earlier, corresponds to having $n = 0$ in the above set of equations. Putting $n = 0$ in these produces

$$f_1 = f_0 + \left(\frac{1}{6}\right)(k_1 + 2\cdot k_2 + 2\cdot k_3 + k_4) \tag{7.35a}$$

where

$$k_1 = h\cdot\phi(t_0, f_0) \tag{7.35b}$$

$$k_2 = h\cdot\phi\left(t_0 + \frac{1}{2}h, f_0 + \frac{1}{2}k_1\right) \tag{7.35c}$$

$$k_3 = h\cdot\phi\left(t_0 + \frac{1}{2}h, f_0 + \frac{1}{2}k_2\right) \tag{7.35d}$$

and

$$k_4 = h\cdot\phi(t_0 + h, f_0 + k_3) \tag{7.35e}$$

which provides us with f_1, the first updated value for the system frequency f.

c. *Further iterations*

To calculate f for subsequent iterations, we simply repeat the process with $n = 1$, $n = 2$, $n = 3$, and so on in Eqs. (7.34a)–(7.34e) inclusive.

d. *The next section*

In the next section, we shall set up an example of the *classic explicit fourth-order "Runge–Kutta" method* for running in *MATLAB*. We shall then compare the results with those of the "Euler" method and with the "direct solution of the differential equations" method.

7.9 MATLAB example simulation 5: using the classic explicit fourth-order "Runge–Kutta" method to solve the "swing equation"

7.9.1 Introduction

7.9.1.1 The popularity of the "Runge–Kutta" method

The "Runge–Kutta" method is one of the most popular methods used by numerical analysts, and it is easy to see why: it has relatively high accuracy ($O(h^5)$ for the fourth-order method) compared with, say the "modified Euler's" method, whose "local truncation error" is $O(h^3)$ and yet is a "single-step," "explicit" method which makes it very easy to use.

7.9.1.2 The data to be used

In this example, we shall use the same data that we used for both the "Euler's" method in Section 7.4 and for the "direct solution of the differential equation" method that is the 3000 MW system example that is depicted in Diagram 6.1 toward the beginning of Chapter 6, Analytical solutions.

7.9.2 A statement of the problem

7.9.2.1 The data

We shall restate here for convenience a summary of the data for the example we are going to use. We have a small 50 Hz system with a demand of 3000 MW. It consists of 10 identical generators, each of which has an output of 300 MW and an "inertia constant" $H = 5.0$ seconds. At time $t = 0$ second, one of these 300 MW generators trips off the system.

7.9.2.2 The challenge

Our task is to calculate the subsequent evolution of the system frequency with time using the fourth-order "Runge–Kutta" method given above. We shall plot a graph of frequency against time for the first 10 seconds of the evolution.

7.9.3 The general solution of the problem

To solve the problem, we use Eqs. (7.34a)–(7.34e) inclusive for the general iteration and Eqs. (7.35a)–(7.35e) for the first iteration only. We note that throughout the function, ϕ is identical with df/dt as given by Eq. (7.33) in keeping with the fundamental definition of "ϕ" that was given in Eq. (7.3) near the beginning of this chapter.

7.9.4 The input data

A summary of the input data is given in Section 7.9.2.1 but these (and the rest of the data for the "swing equation") need to be converted to per unit (pu) values since all our calculations are to be done in per unit (pu) values rather than in *physical units*. A suitable MVA base is 3000/0.8 = 3750 MVA, and a suitable frequency base is 50 Hz. A suitable "time step" is 0.05 second. We give below all the relevant data for the problem, first in "physical units" and then in "per unit."

7.9.4.1 The input data for the problem in "physical units"

1. $\Sigma P_{\text{GEN(NR)}} = 3000$ MW
2. $\Sigma P_{\text{GEN}n\text{(R)}} = 0$ MW
3. $\Sigma P_{\text{LOAD}n} = 3000$ MW
4. $f_n = 50$ Hz
5. $P_{\text{TRIP}} = 300$ MW
6. $s_0 = 0$
7. $a_1 = 0$
8. $H = 5.0$ seconds per machine
9. $f_0 = 50$ Hz

7.9.4.2 The input data for the problem in "per unit"

1. $\Sigma P_{\text{GEN(NR)}} = 3000\ \text{MW}/3750\ \text{MVA} = 0.8$ pu
2. $\Sigma P_{\text{GEN}n\text{(R)}} = 0$ pu
3. $\Sigma P_{\text{LOAD}n} = 3000\ \text{MW}/3750\ \text{MVA} = 0.8$ pu
4. $f_n = 50\ \text{Hz}/50\ \text{Hz} = 1.0$ pu
5. $P_{\text{TRIP}} = 300\ \text{MW}/3750\ \text{MVA} = 0.08$ pu
6. $a_0 = 0$
7. $a_1 = 0$
8. $H = 5$ seconds (the average H for the system)
9. $f_0 = 50\ \text{Hz}/50\ \text{Hz} = 1.0$ pus

7.9.5 The output data

The output data should consist of a value for the system frequency f for each *time step*. From these data a graph should be plotted to be able to see the time evolution of the system frequency.

7.9.6 The MATLAB code

A suggested MATLAB code to program the "classic fourth-order Runge–Kutta" algorithm for the same input data as the "direct solution of the differential equation" example might be as follows:

```
% "Classic Fourth-Order Runge-Kutta" algorithm to solve the "Swing Equation"
%
% Inputting the data for the problem in "per unit":
%
% Input the parameter values applicable to the "Swing Equation" as represented
% by (Eq 7.15), (Eq 7.15a) and (Eq 7.15b) in "per unit":
    PNR = input('Please enter the non-responsive generation in per unit')
    PR = input('Please enter the responsive generation in per unit')
    PL = input('Please enter the system load in per unit')
    FN = input('Please enter the nominal frequency in per unit')
    PTRIP = input('Please enter the generation tripped in per unit')
    A_0 = input('Please enter the value of the constant a0in per unit')
    A_1 = input('Please enter the value of the constant a1in per unit')
    H = input('Please enter the value of the H-constant in seconds')
    F0 = input('Please enter the value of the initial frequency in per unit')
% Input the size of the time step for the "Classic Fourth-Order Runge Kutta"
  method:
    TSTEP = input('Please enter the value of the time step h in seconds')
% Calculate the power balance at time = 0+seconds (that is, just follow-
ing the loss of generation) at nominal frequency as a parameter:
        PBAL = PNR + PR - PL - PTRIP
% Calculate the value of the parameter k1 in "per unit":
        K_1 = 0.02*50*PL
% Calculate the parameter k2in "per unit":
        K_2 = PBAL + K_1 + A_0
% Calculate the parameters for the "initial condition":
        K1 = TSTEP*(- K_1*F0 + K_2)/(2*H*F0)
        K2 = TSTEP*(- K_1*(F0 + 0.5*K1) + K_2)/(2*H*F0)
        K3 = TSTEP*(- K_1*(F0 + 0.5*K2) + K_2)/(2*H*F0)
        K4 = TSTEP*(- K_1*(F0 + 0.5*K3) + K_2)/(2*H*F0)
% Calculate the frequency F(1) for the first step:
        F(1) = F0 + (K1 + 2*K2 + 2*K3 + K4)/6
% Calculate the frequency for all subsequent time-steps in the interval 0 - 10
% seconds:
        for k = 1,199
            % Calculate the coefficients for this time step:
            K1 = TSTEP*(- K_1*F(k) + K_2)/(2*H*F0)
            K2 = TSTEP*(- K_1*(F(k) + 0.5*K1) + K_2)/(2*H*F0)
            K3 = TSTEP*(- K_1*(F(k) + 0.5*K2) + K_2)/(2*H*F0)
            K4 = TSTEP*(- K_1*(F(k) + 0.5*K3) + K_2)/(2*H*F0)
            % Calculate the frequency at the next time-step:
            F(k + 1) = F(k) + (K1 + 2*K2 + 2*K3 + K4)/6
        end
% The vector F contains the solution frequencies at time intervals of TSTEP = 0.05
% seconds
% Using the calculated data a graph of frequency F against time t may be plotted.
```

7.9.7 Instructions for the user

Run the *MATLAB* code and input the per unit data when prompted.

7.9.8 In the next sections

1. *Two "multistep predictor–corrector" methods*
   We finish off the chapter with two "multistep predictor–corrector" methods: the "Milne–Hamming multistep predictor–corrector" method and the "Adams–Bashforth–Moulton multistep predictor–corrector" method.
2. *"Multistep predictor–corrector" methods in general*
   Before either of the above methods is considered in detail, we give a general introduction to "multistep predictor–corrector" methods in general.

7.10 The "Milne–Hamming multistep predictor–corrector" method

7.10.1 A general introduction to "multistep predictor–corrector" methods

1. *"Single-step predictor–corrector" methods*
   In Section 7.6, we described an example of a "single-step predictor–corrector" method, the "modified Euler's" method. In "single-step predictor–corrector" methods, we recall from earlier, although there are two stages at each iteration, the "predictor" and the "corrector" stages; each iteration is carried out across a *single step*, that is from x_n to x_{n+1}. By contrast, in this and the next section, we shall be looking at some "multistep predictor–corrector" methods, for which each iteration spans *several steps*.
2. *Several starting values are needed for "multistep methods"*
   We note here that "multistep predictor–corrector" methods require more than one initial value to be provided by some other, preferably "single-step" method such as the "modified Euler" method or the "Runge–Kutta" method to provide enough starting values for the first few iterations before the "multistep" method can begin to be applied. The "time step" for the first few steps should be small enough that a good set of starting values is obtained. Clearly, there is little point in using a sophisticated method such as a "multistep predictor–corrector" method if the starting values are inaccurate.
3. *"Multistep predictor–corrector" methods*
   In the "multistep predictor–corrector" methods, we still have the separate "predictor" and "corrector" stages as in the "modified Euler's"

method above, but in this case, each of these stages involves *evaluating functions and their first derivatives* simultaneously at *multiple steps*.

For example, in the "Milne–Hamming" method that we shall be programming in Section 7.11, we shall be selecting a "predictor" having *4 steps* ($k = 4$) and a "corrector" having *3 steps* ($k = 3$).

We note here that it is not a critical requirement of the "predictor–corrector" combination that the number of *steps* in the "corrector" stage be exactly equal to the number of *steps* in the "predictor" stage but it helps if the "predictor" and "corrector" stages are of comparable accuracy.

4. *Matching the "order" of the "corrector" stage to that of the "predictor" stage*

   Although it is not critical that the number of "steps" in the "corrector" stage be equal to the number of *steps* in the "predictor" stage, it is, however, normal practice to ensure that the *order* of the scheme selected for the "corrector" stage be *equal* to the *order* of the "predictor" stage to keep the truncation errors compatible with each other and the overall accuracy of the algorithm as high as possible.

   We recall from earlier in the chapter that the "order" of a method refers to the power of the "time step" h in the formula for the "truncation error" associated with the method *minus one*. Thus from Eq. (7.19), for example, we see that the "order" of the "modified Euler's" method is $3 - 1 = 2$. The "modified Euler's" method is therefore a "second-order" method.

5. *In the next subsection*

   In the next subsection, we shall be looking in detail at the "Milne–Hamming" *multistep predictor–corrector* method.

7.10.2 A brief background to the "Milne–Hamming" method

1. *"Milne's" method*

   "Milne's" method was one of the earliest numerical integration methods to have a significant degree of "accuracy"—it has a "truncation error" of $O(h^5)$—but it suffered from very poor "stability" performance. It was found by subsequent analysis that the "region of stability" is in fact very small for this method compared to that for comparable methods, such as the *"classic fourth-order Runge–Kutta"* method (see Ref. [1], p. 170, to see a graph of how just bad the stability performance of the original version of "Milne's" method was). "Milne's" method was a basic "predictor–corrector" method.

2. *Hamming's "corrector replacement" for "Milne's" method*

   Consequently, to improve the "stability" of Milne's original method, *Hamming* proposed a *replacement* for the existing "corrector" of

"Milne's" method, while keeping the "predictor" of "Milne's" method unchanged (again, see Ref. [1] for more details on this).

3. *The "Milne–Hamming" predictor–corrector combination*

   The net effect of the "corrector replacement" for "Milne's" method as proposed by *Hamming* was to increase substantially the "stability region" of Milne's original method to far more acceptable proportions (see Ref. [1], p. 179, to see a graph of the huge improvement achieved by *Hamming's* modification).

7.10.3 The definition of the "Milne–Hamming" method

The "Milne–Hamming" method, which is a *fourth-order* method (i.e., the "truncation error" is of $O(h^5)$—see below), is defined as follows:

1. *The "predictor" for the "Milne–Hamming" method*

   The "predictor" for the "Milne–Hamming" method is

$$py_{n+1} = y_{n-3} + \left(\frac{4h}{3}\right)\cdot\left[2\cdot y'_n - y'_{n-1} + 2\cdot y'_{n-2}\right] \tag{7.36}$$

   where "p" denotes a "predicted value." We note in passing that Eqs. (7.24a) and (7.24b) is a *four-step* equation because "y" must be evaluated at four different points for "x."

2. *The "corrector" for the "Milne–Hamming" method*

   The "corrector" for the "Milne–Hamming" method, provided by Hamming, is

$$y_{n+1} = \left(\frac{1}{8}\right)\cdot[9\cdot y_n - y_{n-2}] + \left(\frac{3\cdot h}{8}\right)\cdot\left[py'_{n+1} + 2\cdot y'_n - y'_{n-1}\right] \tag{7.37}$$

   where again "p" denotes a "predicted value."

   We note that Eq. (7.37) defines a *three-step* method because "y" must be evaluated at three different points for "x" (see Ref. [1], p. 174, for further details).

7.10.4 The "truncation errors" associated with the "Milne–Hamming" method

1. *The "truncation error" associated with the "predictor" stage*

   The "truncation error" $T(x,h)$ associated with the "predictor" stage of the "Milne–Hamming" method is given by

$$T(x,h) = \left(\frac{112\cdot h^5}{360}\right)\cdot y^{[5]}(\zeta) \tag{7.38}$$

where the superscript [5] indicates the *fifth derivative* and $x_{n-3} < \zeta < x_{n+1}$ (four steps). The order of h in Eqs. (7.26a)–(7.26c) indicates that it is a *fourth-order* method.

2. *The "truncation error" associated with the "corrector" stage*

   The "truncation error" $T(x,h)$ associated with the "corrector" stage of the "Milne–Hamming" method is given by

$$T(x,h) = -\left(\frac{9 \cdot h^5}{360}\right) \cdot y^{[5]}(\zeta) \tag{7.39}$$

   where again the superscript [5] indicates the *fifth derivative* but here we have $x_{n-2} < \zeta < x_{n+1}$ (three steps).

   See Ref. [1], p. 174, for further details on the subject of "truncation errors" in the "Milne–Hamming" method.

7.10.5 A comparison of the "stability" of the "Milne–Hamming" method with that of other methods

The "stability region" of the "Milne–Hamming" method is somewhat less than that of the corresponding *fourth-order* "Adams–Bashforth–Moulton" predictor–corrector combination (in fact, around *half* the size). However, it is still a great improvement over the "stability region" of the original "Milne's" method (in fact, about *five times* the size) (see Ref. [1], pp. 170, 179, and 187, for the relevant comparisons in the $h\lambda$ plane).

7.10.6 Formulating the "Milne–Hamming method" for solving the "swing equation"

1. *Applying the "Milne–Hamming method" to the "swing equation"*

   In the following section, we shall be programming into *MATLAB* the "Milne–Hamming method" to solve the "swing equation." To do this, we need to go back to a version of the "swing equation" that is expressed in a suitable format.

2. *Reviewing the "standard version" of the "swing equation"*

   Let us first review the "standard version" of the "swing equation" given by Eq. (7.1) given at the beginning of this chapter. For convenience, we repeat this equation below:

$$\Sigma P_{\mathrm{GEN(NR)}} + \Sigma P_{\mathrm{GEN}n\mathrm{(R)}} - \Sigma P_{\mathrm{LOAD}n} - k_1 \cdot (f - f_n) - P_{\mathrm{TRIP}} \cdot u(t - t_{\mathrm{TRIP}}) + a_0 + a_1 t = 2 \cdot H \cdot f_0 \cdot \frac{\mathrm{d}f}{\mathrm{d}t} \tag{7.40}$$

   where

$$k_1 = \Sigma P_{\mathrm{LOAD}n} \cdot \alpha + \Sigma P_{\mathrm{GEN}n\mathrm{(R)}} \cdot k \tag{7.40a}$$

3. *Choosing a "particular form" of Eq. (7.28) to analyze the "swing equation" using the "Milne–Hamming" method*

   In keeping with our choice for the previous three numerical solution methods that we are considering in this book, we shall simplify Eqs. (7.40) and (7.40a) by assuming (without loss of generality) that $t_{\mathrm{TRIP}}=0$.

   We shall model two cases: (1) No governor response is available and (2) response is made available. However, we shall this time model governor action with a *droop of 4%* in the second case for which response is made available.

4. *Putting Eq. (7.28) into "standard mathematical form"*

   Further maintaining consistency, we shall adopt Eqs. (7.15), (7.15a), and (7.15b) as before, which we repeat here for convenience:

$$\frac{\mathrm{d}f}{\mathrm{d}t}=\frac{1}{2\cdot H\cdot f_0}[-k_1\cdot f+a_1\cdot t+k_2] \tag{7.41}$$

where

$$k_1=\Sigma P_{\mathrm{LOAD}n}\cdot\alpha+k\cdot\Sigma P_{\mathrm{GEN}n(\mathrm{R})} \tag{7.41a}$$

and

$$k_2=\Sigma P_{\mathrm{GEN(NR)}}+\Sigma P_{\mathrm{GEN}n(\mathrm{R})}-\Sigma P_{\mathrm{LOAD}n}-P_{\mathrm{TRIP}}\cdot u(t)+k_1\cdot f_n+a_0 \tag{7.41b}$$

5. *Using the "Milne–Hamming" algorithm to solve the "swing equation"*

   a. *The "Milne–Hamming" method for our case*

   In our case, we have $y=f$ and $x=t$ so that Eqs. (7.24a) and (7.24b) and (7.25a)–(7.25h) become for us:

$$pf_{n+1}=f_{n-3}+\left(\frac{4h}{3}\right)\cdot\left[2\cdot f'_n-f'_{n-1}+2\cdot f'_{n-2}\right] \tag{7.42}$$

   for the "predictor" and

$$f_{n+1}=\left(\frac{1}{8}\right)\cdot[9\cdot f_n-f_{n-2}]+\left(\frac{3\cdot h}{8}\right)\cdot\left[pf'_{n+1}+2\cdot f'_n-f'_{n-1}\right] \tag{7.43}$$

   for the "corrector."

   b. *The "initial conditions"*

   As explained in Section 7.10.1(2), we need more than just a value for the initial frequency f_0 to be able to start off the "Milne–Hamming" method. If we look at Eq. (7.30), we see that we need to have values for f_0, f_1, f_2, and f_3 to be able to evaluate the right-hand side of the equation, that is, we need to have values for the *first four points* in the sequence of *t-values*.

c. *How to assign values to the first four points in the "Milne–Hamming" method*

We note first that the first value is the "initial frequency" f_0, and this is a known quantity since it is specified as part of the input data.

To find values for the three remaining points, f_1, f_2, and f_3, of the first four points in the sequence in the "Milne–Hamming" method, we may use a "single-step method" three times over.

Suitable "single-step methods" include the "modified Euler's" method and the "Runge–Kutta" method. Although the "standard fourth-order Runge–Kutta" method as formulated in this book is more accurate than the (only) *second-order* "modified Euler's" method, if the time step h is chosen small enough which is used should not matter too much. On the other hand the *Runge–Kutta* method is more straightforward to use because it is an "explicit method," whereas the "modified Euler's" method is an "implicit method."

d. *Further iterations*

To calculate f_4 and the estimated frequency at higher order time steps, the formulas (7.30) and (7.31) may now be used since the values of the frequency f at the required previous time steps are now known quantities.

Eqs. (7.29a)–(7.29e) evaluated at the appropriate time step should be used to evaluate the derivative terms on the right-hand sides of Eqs. (7.30) and (7.31).

This completes the description of the "Milne–Hamming" method.

e. *The next section*

In the next section, we shall go through an example calculation in *MATLAB* using the "Milne–Hamming" four-step, fourth-order predictor–corrector method. Having completed this, we shall compare the results with those of the "Euler's" method, the "modified Euler's" method, the fourth-order "Runge–Kutta" method, and the "exact solution of the differential equation" method.

7.11 MATLAB example simulation 6: using the "Milne–Hamming" method to solve the "swing equation"

7.11.1 Introduction

1. *Starting off the "Milne–Hamming" algorithm*

As mentioned above, the main difference between the "Milne–Hamming" method and the other methods that we have considered up to now in this book is that more than one starting point is required. Therefore the "Milne–Hamming" method needs to be started off by some other method before it can be used.

2. *Using the "Runge–Kutta" algorithm*

   In what follows we shall use the "Runge–Kutta" method to generate the first four points in time t for the "Milne–Hamming" method before we start using the equations for the "Milne–Hamming" method itself.

7.11.2 A statement of the problem

1. *General description*

   The problem that we shall address is that given for the "Laplace transform" method and the "modified Euler's" methods, for which we have a 20,000 MW system in which there are both "fully loaded" and "partially loaded" generators supplying the system. Thus we have some "unresponsive" sets and have available some "responsive" sets in the generation mix. We shall analyze the cases where (1) no response provision is made and (2) when response is activated.
2. *Numerical values*

   Following *MATLAB example simulation 2* and *MATLAB example simulation 4*, we have a 50 Hz system in which there are 100 generating sets, each of 250 MW, available 53 of these are running at their full output of 250 MW, 30 generators are running deloaded for response at 90% output of 225 MW each, and 17 generators have been selected not to run. Altogether, the generators are meeting a load of $(53 \times 250) + (30 \times 0.9 \times 250) = 20{,}000$ MW.
3. *The full data in "physical units"*

   The full data for the "swing equation" in "physical units" is

   a. *No response is provided*
      i. $\Sigma P_{\mathrm{GEN(NR)}} = 20{,}000\ \mathrm{MW}$
      ii. $\Sigma P_{\mathrm{GEN}n\mathrm{(R)}} = 0\ \mathrm{MW}$
      iii. $\Sigma P_{\mathrm{LOAD}n} = 20{,}000\ \mathrm{MW}$
      iv. $k_1 = 20{,}000\ \mathrm{MW} \times 0.02\ \mathrm{Hz}^{-1} = 400\ \mathrm{MW/Hz}$
      v. $f_n = 50\ \mathrm{Hz}$
      vi. $P_{\mathrm{TRIP}} = 750\ \mathrm{MW}$
      vii. $a_0 = 0$
      viii. $a_1 = 0$
      ix. $H = 3.0$ seconds per machine
      x. $f_0 = 50\ \mathrm{Hz}$

   b. *Response is provided*
      i. $\Sigma P_{\mathrm{GEN(NR)}} = 53 \times 250 = 13{,}250\ \mathrm{MW}$
      ii. $\Sigma P_{\mathrm{GEN}n\mathrm{(R)}} = 30 \times 250 = 7500\ \mathrm{MW}$
      iii. $\Sigma P_{\mathrm{LOAD}n} = 20{,}000\ \mathrm{MW}$
      iv. $k_1 = 20{,}000 \times 0.02 + \left[\frac{100}{(4 \times 50)}\right] \times 7500 = 4150\ \mathrm{MW/Hz}$
      v. $f_n = 50\ \mathrm{Hz}$
      vi. $P_{\mathrm{TRIP}} = 750\ \mathrm{MW}$
      vii. $a_0 = 0$

viii. $a_1 = 0$
ix. $H = 3.0$ seconds (per machine)
x. $f_0 = 50$ Hz

4. *The full data in "per unit"*

The full data in "per unit" assuming an *MVA base* of 20,000/0.8 = 25,000 MVA and a *frequency base* of 50 Hz is

a. *No response is provided*
   i. $\Sigma P_{GEN(NR)} = 20,000\ \text{MW}/25,000\ \text{MVA} = 0.8$ pu
   ii. $\Sigma P_{GENn(R)} = 0$ pu
   iii. $\Sigma P_{LOADn} = 20,000\ \text{MW}/25,000\ \text{MVA} = 0.8$ pu
   iv. $k_1 = 400 \times 50/25,000 = 0.8$ pu
   v. $f_n = 50\ \text{Hz}/50\ \text{Hz} = 1.0$ pu
   vi. $P_{TRIP} = 750\ \text{MW}/25,000\ \text{MVA} = 0.03$ pu
   vii. $a_0 = 0$
   viii. $a_1 = 0$
   ix. $H = 3.0$ seconds
   x. $f_0 = 50\ \text{Hz}/50\ \text{Hz} = 1.0$ pu

b. *Response is provided*
   i. $\Sigma P_{GEN(NR)} = 13,250\ \text{MW}/25,000\ \text{MVA} = 0.53$ pu
   ii. $\Sigma P_{GENn(R)} = 7500\ \text{MW}/25,000\ \text{MVA} = 0.30$ pu
   iii. $\Sigma P_{LOADn} = 20,000\ \text{MW}/25,000\ \text{MVA} = 0.8$ pu
   iv. $k_1 = 4150 \times 50/25,000 = 8.3$ pu
   v. $f_n = 50\ \text{Hz}/50\ \text{Hz} = 1$ pu
   vi. $P_{TRIP} = 750\ \text{MW}/25,000\ \text{MVA} = 0.03$ pu
   vii. $a_0 = 0$
   viii. $a_1 = 0$
   ix. $H = 3.0\ \text{seconds} \times (13,250 + 7500)/20,000 = 3.1125$ seconds (system average)
   x. $f_0 = 50\ \text{Hz}/50\ \text{Hz} = 1.0$ pu

5. *The system event*

At time $t = 0$, 3 × 250 MW sets trip from the system simultaneously. Using the "Milne–Hamming" method, what is the subsequent evolution of the frequency with time if: (1) no response is provided and (2) response is provided by the deloaded sets?

7.11.3 The general solution to the problem

1. *Calculating the first four points*

To follow Eq. (7.42), we first must evaluate f_0, f_1, f_2, and f_3 by a method other than the "Milne–Hamming" method. We shall use the fourth-order "Runge–Kutta" method to do this because the "Runge–Kutta" method is of a comparable order to the "Milne–Hamming" method and because it is straightforward to use.

2. *Calculating the fifth and all subsequent points*
   To calculate the fifth and all subsequent points, we shall use Eq. (7.42) for the "predictor" step and Eq. (7.43) for the "corrector" step.
3. *The system equations*
   The system equations in both cases will be those given in Eqs. (7.41), (7.41a), and (7.41b).

7.11.4 The input data to the *MATLAB* code

1. *Generation loss with no response inactive*
   In this case the input data to the *MATLAB* code will be consist of:
   a. The total system generation running
   b. The total system demand (load)
   c. The amount of generation tripped at $t = 0$
   d. The system inertia constant
2. *Generation loss with response activated*
   a. Total output of nonresponsive generators
   b. Total output of responsive generators at nominal frequency
   c. Total system demand (load)
   d. The amount of generation tripped at $t = 0$
   e. The system inertia constant H
   f. The generator governor "droop" D
   g. The proportion of deloading p of the responsive generators

7.11.5 The output data from the *MATLAB* code

In each case the output will be the "system frequency" f for values of time t from 0 to 10 seconds. A graph of this output can be plotted if required.

7.11.6 The *MATLAB* codes

The following *MATLAB* codes are suggested:

1. *"No response is provided" case*
   In this case, we use the data listed in Section 7.11.2(4)(a).

```
% The Milne-Hamming Method: No response is provided.
%
% Type in the input data in "per unit" format:
%
  PNR = input('Please enter the non-responsive generation in per unit')
  PR = input('Please enter the responsive generation in per unit')
  PL = input('Please enter the system load in per unit')
  K1 = input('Please enter the value of the constant k1 in per unit')
  FN = input('Please enter the nominal frequency in per unit')
  PTRIP = input('Please enter the generation tripped in per unit')
  A_0 = input('Please enter the value of the constant a0 in per unit')
```

```
    A_1 = input('Please enter the value of the constant a1in per unit')
    H = input('Please enter the value of the H-constant in seconds')
    F0 = input('Please enter the value of the initial frequency in per unit')
%
% Calculate the power balance at nominal frequency as a parameter:
%
    PBAL = PNR+ PR − PL − PTRIP
%
% Start off the routine using the "Classic Fourth-Order Runge-Kutta"
% algorithm:
% Calculate the parameters for the "initial condition":
% Input the size of the time step for the "Classic Fourth-Order Runge Kutta"
  method:
%
    TSTEP = input('Please enter the value of the time step h in seconds')
%
% Calculate the value of the parameter k1 in "per unit":
%
    K_1 = 0.02 * 50 * PL
%
% Calculate the parameter k2 in "per unit":
%
          K_2 = PBAL+ K_1 + A_0
% Calculate the parameters for the "initial condition" of the "Classic
  Fourth-Order Runge-Kutta" algorithm:
          K1 = TSTEP*(−K_1*F0+ K_2)/(2*H*F0)
          K2 = TSTEP*(−K_1*(F0+ 0.5*K1)+ K_2)/(2*H*F0)
          K3 = TSTEP*(−K_1*(F0+ 0.5*K2)+ K_2)/(2*H*F0)
          K4 = TSTEP*(−K_1*(F0+ 0.5*K3)+ K_2)/(2*H*F0)
% Calculate the frequency F(1) for the first step:
          F(1) = F0+ (K1+ 2*K2+ 2*K3+ K4)/6
% Calculate the frequency for three time-steps:
      for k = 1,3
          % Calculate the coefficients for this time step:
          K1 = TSTEP*(−K_1*F(k)+ K_2)/(2*H*F0)
          K2 = TSTEP*(−K_1*(F(k)+ 0.5*K1)+ K_2)/(2*H*F0)
          K3 = TSTEP*(−K_1*(F(k)+ 0.5*K2)+ K_2)/(2*H*F0)
          K4 = TSTEP*(−K_1*(F(k)+ 0.5*K3)+ K_2)/(2*H*F0)
          % Calculate the frequency at the next time-step:
          F(k+1) = F(k)+ (K1+ 2*K2+ 2*K3+ K4)/6
      end
% The values f0, f1, f2 and f3 of the frequency required for starting the
% "Milne-Hamming" algorithm are now all known.
%
% Programming the "Milne-Hamming" algorithm:
%
      for k = 3, 199
          DFDT(k) = (−K_1*F(k)+ K_2)/(2*H*F0)
```

```
        DFDT(k-1) = (-K_1*F(k-1) + K_2)/(2*H*F0)
        DFDT(k-2) = (-K_1*F(k-2) + K_2)/(2*H*F0)
        PF(k+1) = F(k-3) + (4*TSTEP/3)*(2*DFDT(k) - DFDT(k-1)
                         + 2*DFDT(k-2))
        PDFDT(k+1) = (-K_1*PF(K+1) + K_2)/(2*H*F0)
        F(k+1) = (1/8)*(9*F(k) - F(k-2))
                         + (3*TSTEP/8)*(PDFDT(k+1) + 2*DFDT(k) - DFDT(k-1))
    end
% The results for F can now be plotted.
```

2. *The "Response is provided by downloading some generating machines" case*

In this case, we use the data listed in Section 7.11.2(4)(b).

```
% Milne-Hamming Method: Response is provided by down-loading sets.
%
% Type in the input data in "per unit" format:
%
  PNR = input('Please enter the total rated non-responsive generation in per
        unit')
  PR = input('Please enter the total rated responsive generation in per
       unit')
  P = input('Please enter the de-loading factor p for the responsive machines
      in per unit')
  PL = input('Please enter the total system load in per unit')
  FN = input('Please enter the nominal frequency in per unit')
  PTRIP = input('Please enter the generation tripped in per unit')
  A_0 = input('Please enter the value of the constant a0 in per unit')
  A_1 = input('Please enter the value of the constant a1 in per unit')
  H = input('Please enter the value of the H-constant in seconds')
  F0 = input('Please enter the value of the initial frequency in per unit')
  DPC = input('Please enter the value of the droop in per cent')
%
% Calculate the power balance at nominal frequency as a parameter:
%
  PBAL = PNR + P*PR - PL - PTRIP
%
% Calculate the value of the parameter k1 in "per unit":
%
          K_1 = 0.02*50*PL + (100/DPC)*PR
%
% Calculate the parameter k2 in "per unit":
%
          K_2 = PBAL + K_1 + A_0
% Calculate the parameters for the "initial condition" of the "Classic
  Fourth-
% Order Runge-Kutta" algorithm:
    K1 = TSTEP*(-K_1*F0 + K_2)/(2*H*F0)
    K2 = TSTEP*(-K_1*(F0 + 0.5*K1) + K_2)/(2*H*F0)
```

```
        K3 = TSTEP*(-K_1*(F0+ 0.5*K2) + K_2)/(2*H*F0)
        K4 = TSTEP*(-K_1*(F0+ 0.5*K3) + K_2)/(2*H*F0)
%  Calculate the frequency F(1) for the first step:
        F(1) = F0+ (K1+ 2*K2+ 2*K3+ K4)/6
%  Calculate the frequency for three time-steps:
        for k = 1,3
            %  Calculate the coefficients for this time step:
            K1 = TSTEP*(-K_1*F(k)+ K_2)/(2*H*F0)
            K2 = TSTEP*(-K_1*(F(k)+ 0.5*K1)+ K_2)/(2*H*F0)
            K3 = TSTEP*(-K_1*(F(k)+ 0.5*K2)+ K_2)/(2*H*F0)
            K4 = TSTEP*(-K_1*(F(k)+ 0.5*K3)+ K_2)/(2*H*F0)
            %  Calculate the frequency at the next time-step:
            F(k+1)  =  F(k)+ (K1+ 2*K2+ 2*K3+ K4)/6
        end
%  The values f0, f1, f2 and f3 of the frequency required for starting the
%  "Milne-Hamming" algorithm are now all known.
%
%  Programming the "Milne-Hamming" algorithm:
%
        for k = 3, 199
            DFDT(k) = (-K_1*F(k)+ K_2)/(2*H*F0)
            DFDT(k-1) = (-K_1*F(k-1)+ K_2)/(2*H*F0)
            DFDT(k-2) = (-K_1*F(k-2)+ K_2)/(2*H*F0)
            PF(k+1) = F(k-3)+ (4*TSTEP/3)*(2*DFDT(k) - DFDT(k-1)
                          + 2*DFDT(k-2))
            PDFDT(k+1) = (-K_1*PF(K+1)+ K_2)/(2*H*F0)
            F(k+1) = (1/8)*(9*F(k) - F(k-2))
                          + (3*TSTEP/8)*(PDFDT(k+1)+ 2*DFDT(k) - DFDT(k-1))
        end
%  The results for frequency F can now be plotted against time t.
```

7.11.7 Instructions for the user

In each case, enter the input data in per unit (pu) when prompted and process the output data in such a way that a graph of "system frequency" against time may be plotted out.

7.11.8 In the next section

In the next section, we look at the "Adams–Bashforth–Moulton" method.

7.12 The "Adams–Bashforth–Moulton" multistep predictor–corrector method

7.12.1 A brief background to the "Adams–Bashforth–Moulton" method

1. *Composition*

   The "Adams–Bashforth–Moulton" method is a "multistep predictor–corrector" method which is made up from an "Adams–Bashforth" formula for the "predictor" and an "Adams–Moulton" formula for the "corrector" (see Ref. [1], p. 181, for more details on this topic).
2. *Possible combinations*

   There are *seven each* of the "Adams–Bashforth" and "Adams–Moulton" formulas, but for solving the "swing equation," we shall be concentrating on the *fourth-order* formulas that have a *four-step* "Adams–Bashforth" *predictor* stage in combination with a *three-step* "Adams–Moulton" *corrector* stage.

7.12.2 The definition of the *fourth-order* "Adams–Bashforth–Moulton" method

The *fourth-order* "Adams–Bashforth–Moulton" method is defined as follows:

1. *The "predictor" for the "Adams–Bashforth–Moulton" fourth-order method*

   The "predictor" for the "Adams–Bashforth–Moulton" *fourth-order* method is defined as

$$py_{n+1} = y_n + \left(\frac{h}{24}\right) \cdot \left[55 \cdot y'_n - 59 \cdot y'_{n-1} + 37 \cdot y'_{n-2} - 9 \cdot y'_{n-3}\right] \tag{7.44}$$

   where p denotes, as usual, a "predicted" value.

   See Ref. [1], p. 182, for further details.
2. *The "corrector" for the "Adams–Bashforth–Moulton" fourth-order method*

   The "corrector" for the "Adams–Bashforth–Moulton" fourth-order method is defined as

$$y_{n+1} = y_n + \left(\frac{h}{24}\right) \cdot \left[9 \cdot py'_{n+1} + 19 \cdot y'_n - 5 \cdot y'_{n-1} + y'_{n-2}\right) \tag{7.45}$$

   where again p denotes a "predicted" value.

   Again, see Ref. [1], p. 182, for further details.

7.12.3 The "truncation errors" associated with the fourth-order "Adams–Bashforth–Moulton" method

1. *The "truncation error" of the "predictor" for the "Adams–Bashforth–Moulton" fourth-order method*

   The "truncation error" associated with the "predictor" of the *fourth-order* "Adams–Bashforth–Moulton" method is given by

$$T(x,h) = \left(\frac{251 \cdot h^5}{720}\right) \cdot y^{[5]}(\zeta) \tag{7.46}$$

   where the superscript [5] indicates the *fifth derivative* and $(x_{n-3} < \zeta < x_{n+1})$.
2. *The "truncation error" of the "corrector" for the "Adams–Bashforth–Moulton" fourth-order method*

   The "truncation error" associated with the "corrector" of the *fourth-order* "Adams–Bashforth–Moulton" method is given by

$$T(x,h) = -\left(\frac{19 \cdot h^5}{720}\right) \cdot y^{[5]}(\zeta) \tag{7.47}$$

   where the superscript [5] indicates the *fifth derivative* and $(x_{n-2} < \zeta < x_{n+1})$.

   See Ref. [1], p. 182, for further details on this topic.

7.12.4 The "stability" of the "Adams–Bashforth–Moulton" method

1. *Introduction*

   The "Adams–Bashforth–Moulton" method is one of the more stable methods (see Ref. [1], Fig. 4.19 on p. 187, for the "region of stability" in the complex $h\lambda$ plane for the *fourth-order* "Adams–Bashforth" plus "Adams–Moulton" combination).
2. *Comparative "stability" of the "Adams–Bashforth–Moulton" method*

   The "region of stability" of the "Adams–Bashforth–Moulton" method is about *one-third* the size of that of the *fourth-order* "Runge–Kutta" method (see Ref. [1], p. 120) and about *twice* that of the "Milne–Hamming" method (see Ref. [1], p. 179).
3. *The trade-off between "accuracy" and "stability"*

   Please note that increasing the "accuracy" of the "Adams–Bashforth–Moulton" method is of course achievable by reducing the "step size" h, but at the expense of decreasing its "region of stability" (see Ref. [1], Fig. 4.21 on p. 188). Therefore to achieve a balance between "accuracy" and "stability," it is wise to select an "order" somewhere in the middle of the range, rather than go too high or too low.

Hence, a *fourth-order* method appears to be a good compromise that achieves reasonable accuracy and stability.

7.12.5 Formulating the "Adams–Bashforth–Moulton" method for solving the "swing equation"

1. *Applying the "Adams–Bashforth–Moulton" method to the "swing equation"*

   In the following section, we shall be programming the "Adams–Bashforth–Moulton" method for solving the "swing equation." To do this, we need to go back to a version of the "swing equation" that is expressed in a suitable format.
2. *Reviewing the "standard version" of the "swing equation"*

   Let us first review the "standard version" of the "swing equation" defined by Eq. (7.1) that was given at the beginning of this chapter. For convenience, we repeat this equation below:

$$\Sigma P_{\text{GEN (NR)}} + \Sigma P_{\text{GEN } n \text{ (R)}} - \Sigma P_{\text{LOAD } n} - k_1 \cdot (f - f_n) - P_{\text{TRIP}} \cdot u(t - t_{\text{TRIP}}) + a_0 + a_1 t = 2 \cdot H \cdot f_0 \cdot \frac{\mathrm{d}f}{\mathrm{d}t} \tag{7.48}$$

   where

$$k_1 = \Sigma P_{\text{LOAD}n} \cdot \alpha + \Sigma P_{\text{GEN}n(\text{R})} \cdot k \tag{7.48a}$$

3. *Choosing a "particular form" of Eq. (7.40) to analyze with the "Adams–Bashforth–Moulton" method*

   In keeping with our choice for some of the previous numerical solution methods that we have considered in this book, we shall simplify Eq. (7.40) and Eq. (7.40a) by assuming (without loss of generality) that $t_{\text{TRIP}} = 0$ and by setting $k = 0$. The first assumption models the case when the generation/load loss occurs at time $t = 0$ seconds, while the second assumption assumes that we shall be modeling any frequency response by using the "ramp" combination $a_0 + a_1 \times t$ rather than modeling the generator speed governor explicitly.
4. *Putting Eq. (7.36) into "standard mathematical form"*

   To further maintain consistency, we shall adopt Eqs. (7.15), (7.15a), and (7.15b) as previously, which we repeat here for convenience:

$$\frac{\mathrm{d}f}{\mathrm{d}t} = \frac{1}{2 \cdot H \cdot f_0}[-k_1 \cdot f + a_1 \cdot t + k_2] \tag{7.49}$$

   where

$$k_1 = \Sigma\ P_{\text{LOAD}n} \cdot \alpha \quad \text{only} \tag{7.49a}$$

and

$$k_2 = \Sigma P_{\text{GEN(NR)}} + \Sigma P_{\text{GEN}n\text{(R)}} - \Sigma P_{\text{LOAD}n} - P_{\text{TRIP}} \cdot u(t) + k_1 \cdot f_n + a_0 \tag{7.49b}$$

5. *Using the "Adams–Bashforth–Moulton" algorithm to solve the "swing equation"*
   a. *The "Adams–Bashforth–Moulton" method for our case*

      Translating the "predictor" and "corrector" equations (Eqs. 7.32 and 7.33) to our case by setting $y = f$ and $x = t$, we obtain

$$pf_{n+1} = f_n + \left(\frac{h}{24}\right) \cdot \left[55 \cdot f'_n - 59 \cdot f'_{n-1} + 37 \cdot f'_{n-2} - 9 \cdot f'_{n-3}\right] \tag{7.50}$$

      for the "predictor" stage and

$$f_{n+1} = f_n + \left(\frac{h}{24}\right) \cdot \left(9 \cdot pf'_{n+1} + 19 \cdot f'_n - 5 \cdot f'_{n-1} + f'_{n-2}\right) \tag{7.51}$$

      for the "corrector" stage.

      We note that Eq. (7.50) spans four iterations (i.e., equivalently, four points in time) and Eq. (7.51) three iterations. The first four frequency values, as was the case for the "Milne–Hamming" method, need to be provided by an independent method before the main algorithm can be used. We note, however, that as with that method, the first point $f = f_0$ is known already since it is the initial frequency.
   b. *The "initial conditions"*

      As explained in Section 7.10.1(2) and in the same way as for the "Milne–Hamming" method, we need more than just a value for the initial frequency f_0 to be able to start off the "Adams–Bashforth–Moulton" method. If we look at Eq. (7.30), we see that we need to have values for f_0, f_1, f_2, and f_3 to be able to evaluate the right-hand side of the equation, that is, we need to have values for the first four points in the sequence.
   c. *How to find values for the first four points of the "Adams–Bashforth–Moulton" method*

      We note first that the first value is the "initial frequency" f_0, and this is a known quantity since it is specified as part of the input data. To find values for the three remaining points, f_1, f_2, and f_3, of the first four points in the sequence in the "Adams–Bashforth–Moulton" method, we may use a "single-step method" three times over as for the "Milne–Hamming" method.

      Suitable "single-step methods" include the "modified Euler's" method and the "Runge–Kutta" method. Although the "standard fourth-order Runge–Kutta" method as formulated in this book is more accurate than the (only) *second-order* "modified Euler's" method, if the time step h is chosen small enough which one is used should not matter too much. On the other hand the "modified Euler's" method is an "implicit" method as well as being a "predictor–corrector" method, so is more cumbersome to use than the "Runge–Kutta" method. We shall therefore use the "Runge–Kutta" method to provide the first four values needed to be able to start the "Adams–Bashforth–Moulton" method.

d. *Further iterations*

To calculate f_4 and the estimated frequency at higher order time steps the formulas (7.50) and (7.51) may now be used since the values of the frequency f at the required previous time steps are now known quantities.

Eq. (7.49) evaluated at the appropriate time step should be used to evaluate the derivative terms on the right-hand sides of Eqs. (7.50) and (7.51).

This completes the description of the "Adams–Bashforth–Moulton" method.

e. *The next section*

In the next section, we shall go through an example calculation in *MATLAB* using the "Adams–Bashforth–Moulton" four-step, fourth-order predictor–corrector method.

7.13 MATLAB example simulation 7: using the "Adams–Bashforth–Moulton" method to solve the "swing equation"

7.13.1 Introduction

We shall work through an example that we can program in *MATLAB*. This example will use the "Adams–Bashforth–Moulton" method to solve numerically the version of the "swing equation" given in Eq. (7.49). We shall use the fourth-order "Runge–Kutta" algorithm also given earlier to initialize the method.

7.13.2 A statement of the problem

1. *The overall objective*

Our overall aim is to solve the reduced form of the "swing equation" given by Eq. (7.49) with the parameter k_1 given by Eq. (7.49a) and the parameter k_2 given by Eq. (7.49b).

2. *The data*

   We shall use the same data for the "Adams–Bashforth–Moulton" method as we did for the "Milne–Hamming" method so that we shall be able to compare the results of the two directly. This data, we recall, contains both "responsive" and "nonresponsive" generation.

   We shall consider two cases: (1) The "responsive" generation is not activated for providing response and (2) the "responsive" generation is activated for providing response.

7.13.3 The general solution of the problem

1. *The form of the "swing equation"*

   Our reference point is the "swing equation" as represented by Eq. (7.49). We note that the parameter k_1 given by Eq. (7.49a) models the "load damping" effect via the parameter α but that the effect of the generator governor is not modeled in this equation. We are going to model the responsive output in part (2) using instead the parameters a_0 and a_1 to define a straight-line response profile.
2. *The starting algorithm: the "Runge–Kutta" method*

   To start the solution off, we shall use the "Runge–Kutta" algorithm given in Eqs. (7.34a)–(7.34e) and (7.35a)–(7.35e) to provide the first *four* points of Eq. (7.50) and the first *three* points of Eq. (7.51) that define the "predictor" and "corrector" stages of the "Adams–Bashforth–Moulton" method.
3. *Note*

   Whenever there is a term with a derivative with respect to t in Eqs. (7.50) and (7.51), such as, for example f_{n-1}', this indicates that Eq. (7.49) is to be used, with the appropriate subscript, to evaluate it.

7.13.4 The input data

The network data are the same as those for *MATLAB example simulation 6*. Again, we shall be considering two cases: (1) No response is provided; (2) but this time response is to be provided in the form of a "ramped" output modeled by the combination of terms "$a_0 + a_1 \times t$." The explicit governor model is not used this time.

7.13.5 The output data

As in all previous cases, the output data will be predicted values for the system frequency f at values of time t ranging from 0 to 10 seconds.

7.13.6 The MATLAB codes

Suggested MATLAB codes for the two cases are as follows:

```
% Adams-Bashforth-Moulton method Case (i): No response is provided
%
% In this case the parameters a0 and a1 are both zero.
% The data for this case are listed in Section 7.11.2 (d) (i) above.
%
% Enter the data in "per unit" format:
%
  PNR = input('Please enter the non-responsive generation in per unit')
  PR = input('Please enter the responsive generation in per unit')
  PL = input('Please enter the system load in per unit')
  K1 = input('Please enter the value of the constant k1in per unit')
  FN = input('Please enter the nominal frequency in per unit')
  PTRIP = input('Please enter the generation tripped in per unit')
  A_0 = input('Please enter the value of the constant a0in per unit')
  A_1 = input('Please enter the value of the constant a1in per unit')
  H = input('Please enter the value of the H-constant in seconds')
  F0 = input('Please enter the value of the initial frequency in per unit')
%
% Calculate the power balance at nominal frequency as a parameter:
%
          PBAL = PNR + PR - PL - PTRIP
%
% Start off the routine using the "Classic Fourth-Order Runge-Kutta"
% algorithm:
%
% Calculate the parameters for the "initial condition":
% Input the size of the time step for the "Classic Fourth-Order Runge
  Kutta" method:
%
          TSTEP = input('Please enter the value of the time step h in seconds')
%
% Calculate the value of the parameter k1 in "per unit":
%
          K_1 = 0.02*50*PL
%
% Calculate the parameter k2 in "per unit":
%
          K_2 = PBAL + K_1*FN + A_0
%
% Calculate the parameters for the "initial condition" of the "Classic
  Fourth-
%   Order Runge-Kutta" algorithm:
          K1 = TSTEP*(-K_1*F0 + K_2)/(2*H*F0)
          K2 = TSTEP*(-K_1*(F0 + 0.5*K1) + K_2)/(2*H*F0)
```

```
          K3 = TSTEP*(-K_1*(F0+ 0.5*K2) + K_2)/(2*H*F0)
          K4 = TSTEP*(-K_1*(F0+ 0.5*K3) + K_2)/(2*H*F0)
% Calculate the frequency F(1) for the first step:
          F(1) = F0+ (K1+ 2*K2+ 2*K3+ K4)/6
% Calculate the frequency for three time-steps:
          for k = 1,3
% Calculate the coefficients for this time step:
          K1 = TSTEP*(-K_1*F(k) + K_2)/(2*H*F0)
          K2 = TSTEP*(-K_1*(F(k)+ 0.5*K1) + K_2)/(2*H*F0)
          K3 = TSTEP*(-K_1*(F(k)+ 0.5*K2) + K_2)/(2*H*F0)
          K4 = TSTEP*(-K_1*(F(k)+ 0.5*K3) + K_2)/(2*H*F0)
% Calculate the frequency at the next time-step:
                    F(k + 1) = F(k) + (K1+ 2*K2+ 2*K3+ K4)/6
          end
% The values f0, f1, f2 and f3of the frequency required for starting the
% "Adams-Bashforth-Moulton" algorithm are now all known.
%
% Setting up the time vector:
%
  for k = 1,200
      T(k) = TSTEP*k
  end
%
% Programming the "Adams-Bashforth-Moulton" algorithm
%
  for k = 3,199
      DFDT(k) = (-K_1*PF(k) + A1*T(k) + K_2)/(2*H*F0)
      DFDT(k-1) = (-K_1*PF(k-1) + A1*T(k-1) + K_2)/(2*H*F0)
      DFDT(k-2) = (-K_1*PF(k-2) + A1*T(k-2) + K_2)/(2*H*F0)
      DFDT(k-3) = (-K_1*PF(k-3) + A1*T(k-3) + K_2)/(2*H*F0)
%
%     The "Predictor":
%
      PF(k + 1) = F(k) + (TSTEP/24)*(55*DFDT(k) -59*DFDT(k-1)
                        + 37*DFDT(k-2) - 9*DFDT(k-3))
%
      PDFDT(k + 1) = (-K_1*PF(k + 1) + A1*T(k + 1) + K_2)/(2*H*F0)
%
%     The "Corrector":
%
      F(k + 1) = F(k) + (TSTEP/24)*(9*PDFDT(k + 1) + 19*DFDT(k)
                        -5*DFDT(k-1) + DFDT(k-2))
  end
%
% The results for frequency F can now be plotted against time t.
%
% Adams-Bashforth Moulton method Case (ii): Ramped response provided
%
```

```
% In this case the parameters a0and a1are NOT both zero.
% The data for this case are listed in Section 7.11.2 (d) (ii) above.
%
% Enter the data in "per unit" format:
%
  PNR = input('Please enter the non-responsive generation in per unit')
  PR = input('Please enter the responsive generation in per unit')
  PL = input('Please enter the system load in per unit')
  K1 = input('Please enter the value of the constant k1 in per unit')
  FN = input('Please enter the nominal frequency in per unit')
  PTRIP = input('Please enter the generation tripped in per unit')
  A_0 = input('Please enter the value of the constant a0 in per unit')
  A_1 = input('Please enter the value of the constant a1 in per unit')
  H = input('Please enter the value of the H-constant in seconds')
  F0 = input('Please enter the value of the initial frequency in per unit')
%
% Calculate the power balance at nominal frequency as a parameter:
%
  PBAL = PNR + PR - PL - PTRIP
%
% Start off the routine using the "Classic Fourth-Order Runge-Kutta"
% algorithm:
%
% Calculate the parameters for the "initial condition":
% Input the size of the time step for the "Classic Fourth-Order Runge Kutta"
% method:
%
  TSTEP = input('Please enter the value of the time step h in seconds')
%
% Calculate the value of the parameter k1 in "per unit":
%
        K_1 = 1.0*PL
%
% Calculate the parameter k2 in "per unit":
%
        K_2 = PBAL + K_1*FN + A_0
%
% Calculate the parameters for the "initial condition" of the "Classic
  Fourth-Order
% Runge-Kutta algorithm:
        K1 = TSTEP*(-K_1*F0 + K_2)/(2*H*F0)
        K2 = TSTEP*(-K_1*(F0 + 0.5*K1) + K_2)/(2*H*F0)
        K3 = TSTEP*(-K_1*(F0 + 0.5*K2) + K_2)/(2*H*F0)
        K4 = TSTEP*(-K_1*(F0 + 0.5*K3) + K_2)/(2*H*F0)
% Calculate the frequency F(1) for the first step:
        F(1) = F0 + (K1 + 2*K2 + 2*K3 + K4)/6
% Calculate the frequency for three time-steps:
        for k = 1,3
```

```
% Calculate the coefficients for this time step:
                    K1 = TSTEP*(-K_1*F(k) + K_2)/(2*H*F0)
                    K2 = TSTEP*(-K_1*(F(k) + 0.5*K1) + K_2)/(2*H*F0)
                    K3 = TSTEP*(-K_1*(F(k) + 0.5*K2) + K_2)/(2*H*F0)
                    K4 = TSTEP*(-K_1*(F(k) + 0.5*K3) + K_2)/(2*H*F0)
% Calculate the frequency at the next time-step:
                    F(k+1) = F(k) + (K1 + 2*K2 + 2*K3 + K4)/6
          end
% The values f0, f1, f2 and f3 of the frequency required for starting the
% "Adams-Bashforth-Moulton" algorithm are now all known.
%
% Setting up the time vector:
%
  for k = 1, 200
     T(k) = TSTEP*k
  end
%
% Programming the "Adams-Bashforth-Moulton" algorithm
%
  for k = 3, 199
     DFDT(k) = (-K_1*PF(k) + A1*T(k) + K_2)/(2*H*F0)
     DFDT(k-1) = (-K_1*PF(k-1) + A1*T(k-1) + K_2)/(2*H*F0)
     DFDT(k-2) = (-K_1*PF(k-2) + A1*T(k-2) + K_2)/(2*H*F0)
     DFDT(k-3) = (-K_1*PF(k-3) + A1*T(k-3) + K_2)/(2*H*F0)
%
%    The "Predictor":
%
     PF(k+1) = F(k) + (TSTEP/24)*(55*DFDT(k) - 59*DFDT(k-1)
                    + 37*DFDT(k-2) - 9*DFDT(k-3))
%
     PDFDT(k+1) = (-K_1*PF(k+1) + A1*T(k+1) + K_2)/(2*H*F0)
%
%    The "Corrector":
%
     F(k+1) = F(k) + (TSTEP/24)*(9*PDFDT(k+1) + 19*DFDT(k)
                   -5*DFDT(k-1) + DFDT(k-2))
  end
%
% The results for frequency F can now be plotted out against time T.
%
```

7.13.7 Instructions for the user

Run the *MATLAB* programs for this example. In each case the input data as specified above are to be inputted when prompted. A graph of frequency against time can be plotted out at the end.

7.14 Advantages and disadvantages of the numerical methods

7.14.1 Advantages

Compared with the analytical methods, the numerical methods cope more easily with step changes in the power balance on the system being modeled. The new power balance can be simply introduced into the equation at the next time step, without having to be concerned about having to satisfy boundary conditions.

7.14.2 Disadvantages

The disadvantages of the numerical methods are (1) that the solutions generated are not exact and their accuracy depends upon the step size chosen; (2) there is no guarantee of numerical stability, although the risk of this can be minimized by choosing a method with a sufficiently large region of stability.

7.15 The next chapter: the "control diagram" method

In the next chapter, we investigate a method that can be used to solve the "swing equation" based on a "control diagram" approach.

References

Books

[1] L. Lapidus, J.H. Seinfeld, Numerical Solution of Ordinary Differential Equations, Academic Press, 1971 (Library of Congress Catalog Card Number: 73-127689).

[2] J.D. Lambert, Numerical Methods for Ordinary Differential Systems: The Initial Value Problem, Wiley, 1991. ISBN: 0471929905.

[3] R. Bronson, (Schaum's Outline Series) Theory and Problems of Differential Equations, second ed., McGraw-Hill Inc., New York, St. Louis, San Francisco, Auckland, Bogotá, Caracas, Lisbon, London, Madrid, Mexico City, Milan, Montreal, New Delhi, San Juan, Singapore, Sydney, Tokyo & Toronto, 1994. ISBN: 0-07-008019-4.

[4] K.K. Mishra, A Handbook on Numerical Technique Lab: MATLAB Based Experiments, second ed., I. K. International Publishing House, 2007. ISBN: 978-8189-866433. 2015.

Paper

[5] R.M. Warten, Automatic step-size control for Runge-Kutta Integration, IBM J. Res. Develop. 2 (1963) 340.

Chapter 8

The "control diagram" method

8.1 A short introduction to the "control diagram" method

8.1.1 A third approach to solving the "Swing Equation"

In addition to the "exact solution of the differential equation" and "Laplace transform" methods covered in Chapter 6, Analytical solutions, and the "numerical solution" methods of solving the "swing equation" covered in Chapter 7, Numerical methods for solving the "swing equation," of this book, there is a third possible approach available, which may be termed the "control diagram" method.

This approach couches the problem of the solution of the "swing equation" in the language of "control engineering" by using, instead of algebraic and differential equations, "block diagrams" or "control diagrams." For examples of this approach, please see Refs. [3] and [4] listed at the end of this chapter.

The problem is the same as for the other methods, namely solving Eq. (7.1), but the way the solution is approached is slightly different from that of the other methods.

8.1.2 The use of the "control diagram" to represent a differential equation

The idea of using a "control diagram" to represent a differential equation such as the "swing equation" is, of course, by no means new (please see, for example, Ref. [1], Figures 11.4 and 11.5 on p. 585 for a basic illustration of the "control diagram" representation of the elementary "swing equation" with the addition of a single feedback loop to model the effect of "load damping," that is, the variation of load with frequency).

However, it can be a useful and visually appealing means of better understanding the relationships between the various physical and logical elements in an algebraic-differential equation such as the "swing equation."

For example, it groups the sum of all the sources of generation minus the demand into one element: the input, and the change in frequency as the output. In this way it separates out "cause" and "effect" by representing the system frequency equations as "input" and "output," and it represents the damping-type terms all together as one feedback loop.

Modern Aspects of Power System Frequency Stability and Control.
DOI: https://doi.org/10.1016/B978-0-12-816139-5.00008-4

8.1.3 Development of the "control diagram" representation

The basic representation of the control diagram given in Ref. [1], Figures 11.4 and 11.5 on p. 585 can be extended to include the extra features contained in our "standard version" of the "swing equation" given in Eqs. (8.1a)–(8.1d). Thus in addition to the features contained already in Figures 11.4 and 11.5 on p. 585 of Ref. [1], namely the system power balance including the variation of load with frequency (load damping) and generator "governor response," we can also include further elements, such as the "general responsive profile", represented in Eqs. (8.1a)–(8.1d) by $a_0 + a_1\,t$. However, representing the unit step-change term $P_{\mathrm{TRIP}} \cdot u(t - t_{\mathrm{TRIP}})$ of Eqs. (8.1a)–(8.1d) may prove more challenging using the "control system" method.

8.2 Recalling the "swing equation"

8.2.1 The "standard form" of the "swing equation"

Let us quote, for convenience, and for further use in the present chapter, the "standard form" of the "swing equation" that was last given in the previous chapter as [Eq. (7.1)]:

$$\begin{aligned}\Sigma P_{\mathrm{GEN(NR)}} + \Sigma P_{\mathrm{GEN}\ n\ \mathrm{(R)}} - \Sigma P_{\mathrm{LOAD}\ n} - k_1 \cdot (f - f_n) - P_{\mathrm{TRIP}} \cdot u(t - t_{\mathrm{TRIP}})\\ + a_0 + a_1 t = 2 \cdot H \cdot f_0 \cdot \frac{\mathrm{d}f}{\mathrm{d}t}\end{aligned} \tag{8.1a}$$

8.2.2 A simplifying substitution

We can simplify [Eq. (8.1a)] further for our purposes in this chapter by making the substitution:

$$\Delta f \equiv f - f_n \tag{8.1b}$$

so that Eq. (8.1a) may be rewritten in a more convenient form as

$$\begin{aligned}2 \cdot H \cdot f_0 \cdot \frac{\mathrm{d}(\Delta f)}{\mathrm{d}t} + k_1 \cdot (\Delta f) = \Sigma P_{\mathrm{GEN\ (NR)}} + \Sigma P_{\mathrm{GEN}\ n\ \mathrm{(R)}} - \Sigma P_{\mathrm{LOAD}\ n}\\ - P_{\mathrm{TRIP}} \cdot u(t - t_{\mathrm{TRIP}}) + a_0 + a_1 t\end{aligned} \tag{8.1c}$$

where in addition we have taken advantage of the fact that

$$\frac{\mathrm{d}(\Delta f)}{\mathrm{d}t} = \frac{\mathrm{d}f}{\mathrm{d}t} \tag{8.1d}$$

which is true since we have $\mathrm{d}f_n/\mathrm{d}t = 0$.

All that we have done here is to change the dependent variable from f to Δf because this fits in better with the "block diagram" representation we are going to use.

8.2.3 In the next subsection

In the next subsection we are going to start to look at how to formulate the "control system" method, beginning with Eq. (8.1c). This involves first representing the "swing equation" in "state-space format." This is an intermediate step between the algebraic-differential representation we are starting with and the "control diagram" or "block diagram" representation we are going to finish up with.

8.3 The representation of the "swing equation" in "state-space" format

8.3.1 An introduction to "state-space" format

First, the equation [Eqs. (8.1a)–(8.1d)] can be expressed, in a completely equivalent way, in the following well-known "state-space" format, which may be defined as follows:

$$\dot{x} = A \cdot x + B \cdot u \tag{8.2a}$$

and

$$y = C \cdot x + D \cdot u \tag{8.2b}$$

(see, for example, Ref. [1], p. 704 for a more detailed description of this representation).

In Eqs. (8.2a) and (8.2b) the variable u is the "input variable," the variable y is the "output variable," and the variable x is the "system variable" or "state variable." The parameters A, B, C, and D are constants.

The "state-space" format has the in-built flexibility for choosing the variables and constants to suit a desired application—a flexibility that we shall be making use of in the next section.

8.3.2 Application of the "state-space" format to the "swing equation"

To represent the "swing equation," in the form represented by Eq. (8.1c), in the "state-space" format of Eqs. (8.2a) and (8.2b), we may select

$$x = y = \Delta f = f - f_n \tag{8.2c}$$

as the output variable. This represents the amount that the "system frequency" differs from the "nominal frequency."

Following Ref. [2], p. 753 and using their notation, our version of the "swing equation" given in Eq. (8.1c) may be written in the form:

$$a_0 \cdot \dot{y} + a_1 \cdot y = b_1 \cdot u \tag{8.3a}$$

so that, in their notation, we have

$$a_0 = 2 \cdot H \cdot f_0 \tag{8.3b}$$

$$a_1 = k_1 \tag{8.3c}$$

and we may take

$$b_1 = 1 \tag{8.3d}$$

For the *input variable u* we therefore have

$$u = \Sigma P_{\text{GEN (NR)}} + \Sigma P_{\text{GEN } n \text{ (R)}} - \Sigma P_{\text{LOAD } n} - P_{\text{TRIP}} \cdot u(t - t_{\text{TRIP}}) + a_0 + a_1 t \tag{8.3e}$$

Thus we have provided values for the three parameters A, B, and u of Eq. (8.2a) in addition to taking $C = 1$ and $D = 0$ in Eq. (8.2b) to complete the conversion of the "swing equation" into "state-space" format.

In the next section, we shall provide an interpretation of some of the terms in the above set of equations.

8.3.3 An interpretation of some of the terms in Eqs. (8.3a)–(8.3e) inclusive

1. *A comment on the information distribution of Eqs. (8.3a)–(8.3e)*

   We can see from the list of equations earlier that most of the information for the "state-space" format of the "swing equation" is contained in Eqs. (8.3c) and (8.3e). We now therefore proceed to look more closely at these two equations, taking them in turn.

2. *The information contained in Eq. (8.3c)*

   In Eq. (8.3c), there is information contained about the dependency of the "system frequency" f on itself since it multiplies y in Eq. (8.3a). It is therefore a kind of "feedback" term. We shall see that this assertion will be confirmed graphically later in our analysis.

   We recall from previous chapters, for example, in Eq. (7.1a), that the parameter k_1 is given by

   $$k_1 = \Sigma P_{\text{LOAD } n} \cdot \alpha + \Sigma P_{\text{GEN } n \text{ (R)}} \cdot k \tag{8.3f}$$

   where $\Sigma P_{\text{LOAD } n}$ represents the total MW load at nominal frequency, α is the "load-damping coefficient," $\Sigma P_{\text{GEN } n \text{ (R)}}$ represents the total MW generation of responsive synchronous machines at nominal frequency, and k is a parameter essentially equal to the reciprocal of the slope D of the synchronous machine governor characteristic [see, for example, Eq. (A.9) of Appendix A for the full definition of this parameter].

3. *The information contained in Eq. (8.3e)*

In Eq. (8.3e), we have all the information relating to the net *input* to the system: generation minus demand (or load: demand and load are the same here because we are ignoring I^2R losses).

We can divide Eq. (8.3e) into the following groups of terms:

a. *The basic "power balance" term at "nominal frequency"*

The term "$\Sigma P_{GEN\ (NR)} + \Sigma P_{GEN\ n\ (R)} - \Sigma P_{LOAD\ n}$" in Eq. (8.3e) represents the basic power balance at the "nominal frequency" of the system in the absence of any "trips" and before any "responsive output" has appeared on the system.

b. *In the absence of any responsive output only*

If "$\Sigma P_{GEN\ (NR)} + \Sigma P_{GEN\ n\ (R)}$" is *less than* "$\Sigma P_{LOAD\ n}$" then the frequency will settle down to a value *lower than* the "nominal frequency."

If "$\Sigma P_{GEN\ (NR)} + \Sigma P_{GEN\ n\ (R)}$" *equals* "$\Sigma P_{LOAD\ n}$" then the frequency will settle down *to the* "nominal frequency."

If "$\Sigma P_{GEN\ (NR)} + \Sigma P_{GEN\ n\ (R)}$" is *greater than* "$\Sigma P_{LOAD\ n}$" then the frequency will settle down to a value *higher than* the "nominal frequency."

c. *The "reference frequency" for both the "dependency of the load on frequency" ("load damping") term and the "governor response" term*

The term "$k_1 \cdot f_n$" in Eq. (8.1a) contains the "reference frequency" for both the "dependency of the load on frequency" ("load damping") term and the "governor response" term. In this instance the nominal frequency f_n is being applied to both. This does not *necessarily* have to be so since the synchronous generator governor response could have a reference or "target" frequency different from the nominal frequency. It is simpler to do it like this, though, for the purposes of analysis. However, the "load damping" term is invariably based on the nominal frequency f_n.

d. *The "instantaneous trip" term*

The term "$-P_{TRIP} \times u(t - t_{TRIP})$" in Eq. (8.3e) represents an "instantaneous trip" of *generation*. It can be made easily to represent an "instantaneous trip" of *load* by simply changing the sign from negative to positive. This term works as follows: when $t \leq t_{TRIP}$ the term = 0; when $t > t_{TRIP}$ the term = $-P_{TRIP}$.

e. *The "power ramp" term*

The term "$a_0 + a_1\ t$" in Eq. (8.3e) represents a kind of general "power ramp." It can be used to model an *increase* or *decrease* in either, or both, of *generation* or *demand* by the appropriate choice of values for the parameters "a_0" and "a_1." It is a very general representation of any response profile that is "piecewise linear." It can represent the responsive output of a synchronous generator during the "primary response" phase or perhaps the initial output of a general responsive device (e.g., a battery).

8.3.4 A note on the application of the "state-space" formulation to general "stability and control" problems

The earlier "state-space" representation, being equivalent in effect to an algebraic-differential representation, can be used for modeling and finding solutions to other frequency stability and control problems (please see, for example, Ref. [1], Chapter 12 for an application to small-signal stability analysis). We recall that it is being employed here in this book as a kind of stepping-stone to the "transfer function" format version we wish to set up for our version of the "swing equation."

8.4 Conversion from the "state-space" format to the "transfer function" format

8.4.1 An introduction to the "transfer function" format

There is a standard procedure by which an equation in the "state-space" format can be converted in a straightforward, although, in some cases, perhaps arduous, manner to an equivalent representation in the "transfer function" format (for example, please see Ref. [2]. p. 761 for more details). We now proceed to accomplish this conversion for our case of the "swing equation."

8.4.2 Converting the "state-space" format of the "swing equation" given in Eq. (8.3) to "transfer function" format

Now that we have obtained the "swing equation" in the "state-space" format in the form of Eqs. (8.3a)–(8.3f) inclusive, we are now ready to convert this representation to the "transfer function" format.

To accomplish this task, we are going to apply the formula in equation (11-2) on p. 753 of Ref. [2] to Eq. (8.3a).

Having done this, the resulting expression for the output $Y(s)$ divided by the input $U(s)$, using the same terminology as in Eq. (8.3a), is

$$\frac{Y(s)}{U(s)} = \frac{b_1}{a_0 s + a_1} \tag{8.4}$$

or

$$\frac{Y(s)}{U(s)} = \frac{1}{(2Hf_0 \cdot s + k_1)} \tag{8.5}$$

In calculating which we have also used the results from Eqs. (8.3b) to (8.3d).

8.4.3 The "gain" and "time constant" associated with the system described by Eq. (8.3)

Two further important pieces of information contained within Eq. (8.3) are the "gain K" and "time constant T." These can be deduced in a straightforward manner from Eq. (8.4).

If we introduce the "gain K" and the "time constant T" by adopting the standard formula:

$$\frac{Y(s)}{U(s)} = \frac{K}{1 + s \cdot T} \tag{8.6}$$

then by a straight comparison of Eq. (8.6) with Eq. (8.5) we therefore find that

$$K = \frac{1}{k_1} \tag{8.7a}$$

and

$$T = 2 \cdot H \cdot \frac{f_0}{k_1} \tag{8.7b}$$

or, equivalently,

$$K = \frac{1}{a_1} \tag{8.8a}$$

and

$$T = 2 \cdot H \cdot \frac{f_0}{a_1} \tag{8.8b}$$

by using the result from Eq. (8.3c).

Please see Ref. [1], p. 586 for further details on this topic.

8.5 Conversion from the "transfer function" format to the "control diagram" format

8.5.1 An introduction to the "control diagram" format

The "control diagram" format is simply the familiar "block diagram" that we can now construct easily, since we now know the details of the "transfer function" as given by Eq. (8.6).

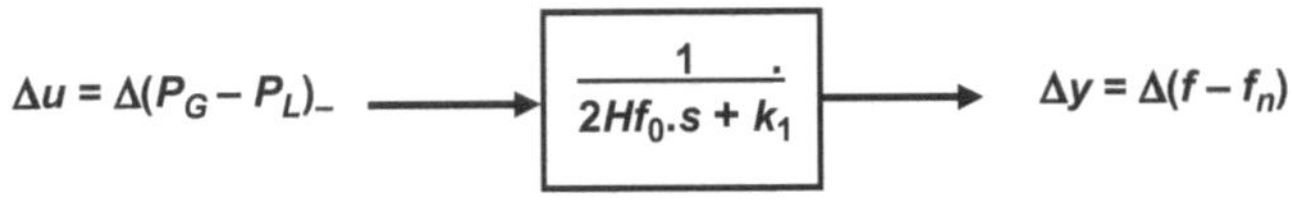

DIAGRAM 8.1 A simple control (block) diagram representation of the "standard swing equation" [Eqs. (8.1a)–(8.1d)].

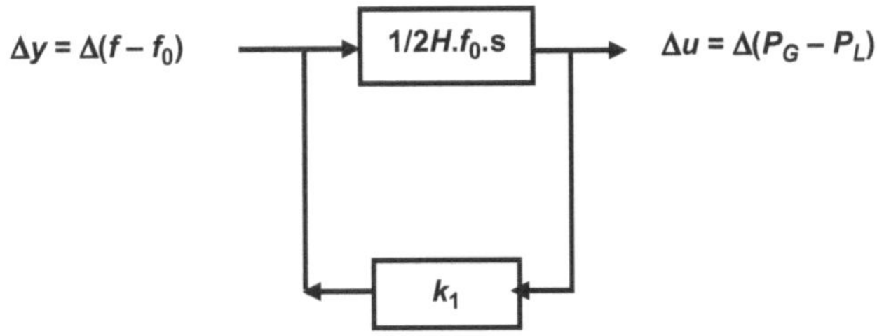

DIAGRAM 8.2 An alternative control diagram representation of the "standard swing equation" [Eqs. (8.1a)–(8.1d)].

8.5.2 Converting the "transfer function" representation of the "standard swing equation" of Eq. (8.5) to "control diagram" format

From Eq. (8.5), we may now draw the corresponding "control (block) diagram" as shown in Diagram 8.1.

We see graphically from Diagram 8.1 how a mismatch between *system generation* and *system demand (load)* translates by means of a "transfer function" expressed in Laplace transform format into the deviation of the power frequency from the nominal power frequency.

Note that in Diagram 8.1 the input variable $u =$ "$P_G - P_L$" is given by Eq. (8.3e), and the output variable $y = f - f_n = \Delta f$ is given by Eq. (8.2c).

8.5.3 An alternative "control diagram" format for the "standard swing equation" [Eqs. (8.1a)–(8.1d)]

A second way to represent the "swing equation" in "control diagram" format is to separate away the "k_1" term in Diagram 8.1 into a "feedback" loop, as in Diagram 8.2.

where again $u =$ "$P_G - P_L$" is given by Eq. (8.3e), and $y = f - f_n = \Delta f$ is given by Eq. (8.2c).

8.5.4 A comparison of Diagram 8.2 with Diagram 8.1

1. *The equivalence of*

   Diagrams 8.1 and 8.2 are equivalent to each other.

It is left as an exercise to the interested reader to prove that Diagrams 8.1 and 8.2 are logically equivalent to one another. [Hint: Let the "feed-forward branch" be called G and let the "feedback branch" be called H.]

2. *The advantage of Diagram 8.2 over Diagram 8.1*

   The advantage of the representation shown in Diagram 8.2 over the representation shown in Diagram 8.1 is that Diagram 8.2 clearly shows that the term "k_1" is a simple "feedback" term, whereas this fact is not as clear from looking at Diagram 8.1.

8.6 Iterating the values in the "control diagram" representation of the "swing equation" to find a solution for the "system frequency" as a function of time

8.6.1 The overall objective of the process

The overall objective of this process is to find out, by calculation, how much generation needs to be held in reserve (the "response requirement") to ensure that for a given loss of generation the ensuing path of the frequency does not transgress the statutory lower limit for the system frequency during the "primary response" period. This period extends from the initial loss about 10 seconds following the loss, although this timescale will vary from system to system.

8.6.2 The "initial condition" in the "control diagram" method

1. *An introduction to the "initial condition" of the "control diagram" method*

The "initial condition" of the "control diagram" method may clearly be either a "steady state" or "not a steady state." We shall consider these two possibilities in turn:

a. *If the "initial condition" of the "control diagram" method is a "steady state"*

   If the "initial condition" of the "control diagram" is a "steady state," then, by definition, at $t = 0$, *at whatever frequency the system is running at*, the system *generation* and *demand* are *exactly in balance at that frequency*, and the frequency therefore is momentarily unchanging. Further, if nothing on the system changes to disturb this equilibrium, *the system frequency will stay constant at its current value, whatever that happens to be.*

b. *If the "initial condition" of the "control diagram" method is "not a steady state"*

   If the "initial condition" of the "control diagram" is "not a steady state," then, by definition, at $t = 0$, *at whatever frequency, the system generation and demand are not in balance*. Therefore *the system frequency will change to some other value.*

8.6.3 How the "control diagram" reacts to the status of the "input"

1. *If the "input u" is in a "steady state"*

If the "input *u*" is in a "steady state" with $\Delta u = 0$ (that is, the system "generation" and "demand" are "in balance" with no change), then the "output *y*" will *not* change, that is, $\Delta y = 0$.

2. *If the "input u" is not in a "steady state"*

If the "input *u*" is *not* in a "steady state" with $\Delta u \neq 0$ (that is, the system *generation* and *demand* are *not* "in balance"), then the "output *y*" will change, that is, $\Delta y \neq 0$ as well.

8.6.4 How to perform a "single iteration" using the "control diagram" method

1. *The meaning of an iteration in the "control diagram" method.*

What we mean by an "iteration" in the "control diagram" method is as follows:

a. We start with the initial values for generation and demand.
b. We remove an amount of generation or demand to represent a "trip."
c. We increment or decrement as appropriate the response by an amount "Δu" according to an estimated value of the amount of "frequency response" that can be made available in one step-length Δt.
d. We apply the iteration procedure as described in the next paragraph to find the change in the frequency "Δf" corresponding to our applied change to the input "Δu."

2. *The iteration procedure*

To perform a "single iteration," or "step," using the "control diagram" representation of the "swing equation," as given in Diagram 8.1, we must go through the following process (please see Ref. [1], pp. 585–587 for an example of how to follow this process):

a. *Calculate "Δu" in "Laplace transform space," that is, find "$\Delta u(s)$"*

To calculate "$\Delta u(s)$" it is necessary to take the input power balance (which may or may not be zero) and divide it by s (Ref.[1], p. 586).

b. *Calculate "Δy" in "Laplace transform space," that is, find "$\Delta y(s)$"*

To calculate "$\Delta y(s)$" we take the input power balance "Δu" divided by s and then multiply it by the "transfer function" $K/(1 + sT)$ as defined earlier.

c. *Calculate the "inverse Laplace transform" of "$\Delta y(s)$" to find "$\Delta y(t)$"*

The final step of this single-step process is to calculate the "inverse Laplace transform" of "$\Delta y(s)$" obtained as described earlier to find "$\Delta y(t)$."

d. *Choose the time step*

Choose the time step Δt, for example, 0.05 seconds.

e. *Plotting the result for the single iteration*

Finally, we plot "$\Delta y(t)$" against "t" for one time step Δt.

8.6.5 How to perform "multiple iterations" of the "control diagram" method

1. *Repetition of the "single iteration"*

   To achieve "multiple iterations" of the "control diagram" method, we repeat the process described in Section 8.6.4 with updated values for the input "Δu." The updated value is the value from the previous iteration plus an increment calculated according to an assumed "responsive generator ramp rate R." This is an assumed amount in *MW/s* calculated according to the ramp rate for a single responsive machine of a selected size multiplied by the number of responsive machines that we estimate to be available.
2. *Adjusting the amount of response available*

   Once a full plot has been drawn up to a selected time (say 10 seconds), the plot is examined to determine whether the frequency is within the required bounds for all values of time within the plotting range.

   If the frequency is within the required bounds for all values of time within the plotting range, the process is terminated.

   If the frequency is not within the required bounds for all values of time within the plotting range, then the amount of frequency response available is updated, and the whole process repeated until the frequency is within the required bounds.
3. *The "convergence" of the calculation process*

   The process described in Section 8.6.3 is repeated until "convergence" is attained by some definition given in the preceding paragraph.
4. *Acceptable frequency behavior*

   To find the "frequency response requirements" for a system the "response provision" that can be included with the "input Δu" is varied until the resulting frequency behavior $f(t)$ becomes acceptable, that is, the "system frequency" does not rise too high or fall too low in a certain timescale that is defined by the relevant "operational standards" and within preselected tolerances.
5. *Choosing "convergence tolerances"*

   In our example a "convergence tolerance" for Δu or $\Delta f(t)$ (or both) can be defined such that when the change resulting from an iteration is smaller than one or both tolerances, the iterations are terminated. We can then say that we have "the solution" for the "response provision" required.

8.7 MATLAB example simulation 8: an example of the "control diagram" method: modeling a generator "ramped response"

8.7.1 Introduction

In this exercise, we shall be following the prescription laid out in Section 8.6.4 of this chapter to solve an example problem in which an initial power balance between generation and demand on a system is disturbed by a loss of generation from the system caused by an unexpected "trip."

We shall assume that there is a source of ramped response available which attempts to restrain the ensuing fall in frequency and to return the "system frequency" as far as possible to normal operating levels. This ramped response is represented by the two terms $a_0 + a_1 t$ using our standard notation.

At the end of our calculations, we shall create a plot of the "system frequency" f versus time t to show the time-evolution of the "system frequency."

8.7.2 A statement of the problem

Let us adopt the system of MATLAB example simulation 1 which is a small 50 Hz system with a demand/load of 3000 MW. Initially 10 identical generators, each with an "inertia constant" $H = 5.0$ seconds are running at their rated output of 300 MW each. At $t = 0$ second one of the generators trips off the system. We need to calculate the time-evolution of the "system frequency" up to a time $t = 10$ seconds.

8.7.3 The general solution of the problem

The general solution of the problem is as described in Sections 8.6.4 and 8.6.5. Section 8.6.4 explains a single iteration of the process, while Section 8.6.5 explains multiple iterations.

8.7.4 The input data

1. *General data*

We need to feed data to represent Eqs. (8.1a) and (8.3f) with $k = 0$ into the program. We list these two following equations again here for ease of reference:

$$\Sigma P_{\text{GEN (NR)}} + \Sigma P_{\text{GEN }n(R)} - \Sigma P_{\text{LOAD }n} - k_1 \cdot (f - f_n) - P_{\text{TRIP}} \cdot u(t - t_{\text{TRIP}}) + a_0 + a_1 t = 2 \cdot H \cdot f_0 \cdot \frac{\mathrm{d}f}{\mathrm{d}t} \tag{8.9}$$

$$k_1 = \Sigma P_{\text{LOAD }n} \cdot \alpha + \Sigma P_{\text{GEN }n\text{ (R)}} \cdot k \tag{8.10}$$

where as noted earlier, we shall take $k = 0$ because we are not modeling a generator governor response, but ramped response from some other (electronic) source using the combination of parameters $a_0 + a_1\ t$ to describe the ramp.

With reference to Eqs. (8.9) and (8.10) the data we shall need is as follows:

$\Sigma P_{\text{GEN (NR)}} = 3000\,\text{MW}$; $\Sigma P_{\text{GEN}\,n\,\text{(R)}} = 0\,\text{MW}$; $f_n = 50\,\text{Hz}$; $P_{\text{TRIP}} = 300\,\text{MW}$; $t_{\text{TRIP}} = 0$ s; $H = 5.0$ s; $\Sigma P_{\text{LOAD}\,n} = 3000$ MW; $\alpha = 0.02$ pu/Hz.

2. *Data for the responsive ramp*

   We need to choose some values for the "ramp parameters" a_0 and a_1 to model some provision of response following the generation loss. Let us assume for this example that the response is instantaneous, that is, there is no time delay between the generator trip and the start of the responsive ramp. In this limiting case we have

$$a_0 = 0 \tag{8.11}$$

   Let us also assume that we have 150 MW of a responsive source available which can change its output from 0 to 150 MW in 5 seconds. In this case we have

$$a_1 = 30\ \text{MW}/s \tag{8.12}$$

   for our second "ramp parameter."

3. *Calculation of the input parameter u*

   Since we are framing the problem in terms of the response of an output y to an input u, we need to begin by finding values for the input variable u. This variable is given by Eq. (8.3e). We repeat this following equation here for ease of reference:

$$u = \Sigma P_{\text{GEN (NR)}} + \Sigma P_{\text{GEN}\,n\,\text{(R)}} - \Sigma P_{\text{LOAD}\,n} - P_{\text{TRIP}} \cdot u(t - t_{\text{TRIP}}) + a_0 + a_1 t \tag{8.13}$$

   This expression may now be evaluated by substituting the data from Section 8.7.4 (1) into Eq. (8.13).

   Doing this, we get

$$u = 3000\ \text{MW} + 0\ \text{MW} - 3000\ \text{MW} - 300 \cdot u(t) + 0 + \left(30\ \text{MW}/s\right) \cdot t$$

   or

$$u = -\,300 \cdot u(t) + \left(30\ \text{MW}/s\right) \cdot t \tag{8.14}$$

4. *Choosing the "time-step" Δt*

   Let us divide the ramping-up time of 5 seconds by 100 to give a time step of

$$\Delta t = 0.05\ \text{s} \tag{8.15}$$

   We shall use this time-step to increment the input u and hence the output y.

8.7.5 The output data

The output data will consist of values of $\Delta y = \Delta(f - f_n)$ that correspond to increments of the input $\Delta u = \Delta(P_G - P_L)$ as shown in Diagrams 8.1 and 8.2. The output values for each step are obtained by following the procedure described in Sections 8.6.4 and 8.6.5. Values of $f - f_n$ can then be plotted against time t. Alternatively, values of f may be plotted against t.

8.7.6 The suggested *MATLAB* code

The suggested *MATLAB* code to solve the problem is as follows:

```
%       Solving the "Swing Equation" by the "control diagram" method.
%
%       Enter the input data in physical units:
%
        PNR_P = input('Please enter the total output of the non-
        responsive generation in MW')
        PR_P = input('Please enter the total rated output of the
        responsive generation In MW')
        FN_P = input('Please enter the system nominal frequency in
        Hz')
        PTRIP_P = input('Please enter the amount of generation
        tripped in MW')
        TTRIP = input('Please enter the time of the generation trip
        in seconds')
        H = input('Please enter the machine inertia constant in
        seconds')
        PLOAD_P = input('Please enter the total system load in MW')
        ALPHA_P = input('Please enter the "load damping" constant α
        in per cent per Hz')
%
%       Now enter the "responsive ramp" data:
%
        A0_P = input('Please enter the value of the "ramp" constant
        a0 in MW')
        A1_P = input('Please enter the value of the "ramp" constant
        a1 in MW per second)
%
%       Convert the data as necessary to "per unit" by defining the
        MVA and f bases:
%
        MVA_BASE = 3000/0.8
        F_BASE = FN_P
%
%       Now convert all quantities to "per unit" as necessary:
```

```
%
         PNR = PNR_P/MVA_BASE
         PR = PR_P/MVA_BASE
         FN = FN_P/F_BASE
         PTRIP = PTRIP_P/MVA_BASE
         PLOAD = PLOAD_P/MVA_BASE
         ALPHA = ALPHA_P*F_BASE/100
         A0 = A0_P/MVA_BASE
         A1 = A1_P/MVA_BASE
%
%       Now enter the size of the time step in seconds:
%
         TSTEP = input('Please enter the value of the time-step in
         seconds')
%
%       Calculate the values of the parameters K and T of the transfer
        function using
%       (Eq 8.8a) and (Eq 8.8b):
%
         K = 1/A1
         T = 2*H*F0/A1
%
%       In Laplace space Δu(s) = Δu/s. We need to multiply this by
        K/(1 + s.T)
%       to get Δy(s), of which we can then take the inverse transform
        to get Δy(t).
%
%       For the first time increment only:
%
         DELTA _OMEGA(1) = PTRIP * K * (EXP(-TSTEP/T) – 1)
         DELTA_F(1) = 2 * PI * DELTA_OMEGA(1)
%
%       Now perform iterations to generate the successive speed
        changes
%       up to a time limit of 5 seconds:
%
         NITER = 5/TSTEP
%
         for k = 2, NITER
                T = k * TSTEP
                DELTA_OMEGA(k) = A1 * T * K * (EXP(-TSTEP/T) – 1)
                DELTA_F(k) = 2 * PI * DELTA_OMEGA(k)
         end
%
%       A graph of the results may now be printed out.
```

8.7.7 Instructions for the user

Enter the required input data when prompted by the MATLAB code. The output will be a graph of "system frequency" against time for time $t = 0-10$ seconds.

8.8 Comparison of the "control diagram" approach with some of the other methods available

8.8.1 Introduction

The "control diagram" approach is a sort of hybrid between the "analytical methods" and the "numerical methods": it contains elements of both. We describe these similarities in turn next.

8.8.2 Similarity of the "control diagram" approach to the "analytical methods"

The "control diagram" approach has similarities with the "analytical methods." There are elements of the "Laplace transform" method because Laplace transforms are needed as part of the "control diagram" approach as well to calculate the change in frequency output as a response to a change in the power balance input.

8.8.3 Similarity of the "control diagram" approach to the "numerical methods"

There is also an element of the "numerical methods" in the "control diagram" approach because iterations are required in both methods to find the solution.

8.8.4 Uniqueness of the "control diagram" approach

The "control diagram" approach is unlike the other two groups of methods in its use of the "block diagram" and in the "state-space" approach.

8.9 The next chapter: some important practical applications

8.9.1 Introduction

In the next chapter, we look at some important practical applications. They are as follows:

- Calculation of the "rate of change of frequency" (RoCoF) following a loss of *generation* or *demand*
- Calculation of the "system response requirements" following a loss of *generation* (the "low frequency excursion")

- Calculation of the "system response requirements" following a loss of *demand* (the "high frequency excursion")
- Calculation of the "system response requirements" during "normal operation" (relatively small variations in the "system frequency")

These calculations are all important tasks performed regularly every day by control engineers at national control centers around the world. Because of their importance, we give short descriptions of each in preparation for the next chapter as follows:

8.9.2 *[I]* The calculation of the "rate of change of frequency" following a loss of *generation* or *demand*

1. *RoCoF relays*
   a. *The equipping of generators with "RoCoF relays"*
      Some generators are equipped with "RoCoF relays".
   b. *Why are "RoCoF" relays installed on generators?*
      "RoCoF" relays may be installed on generators for several reasons:
   i. *Loss of grid/loss of mains/islanding protection*
      One potential use of having "RoCoF" relays on generators is to detect a "loss of grid" condition (also known as "loss of mains" or "islanding"). The purpose of the "RoCoF" relays in this situation is to protect the generator. When the "RoCoF" relay detects a "loss of grid" condition, the generator is "tripped," that is, disconnected from the main system.
   ii. *The detection of "under-frequency" and "over-frequency" conditions*
      "RoCoF relays" may also be installed on generators to detect both "under-frequency" and "over-frequency" conditions. The purpose of the "RoCoF relay" here is to protect the generator by tripping it before the frequency becomes too low or too high, thus avoiding potentially costly damage to the generator.
2. *The control of the RoCoF*
   The "RoCoF" on the power system can be controlled by the scheduling of generation to *ensure that there is enough inertia on the system to limit the RoCoF to below generator tripping threshold levels.*
3. *Calculating the "RoCoF"*
   Calculating the "RoCoF" is achieved straightforwardly by evaluating the right-hand side of the "swing equation," since "RoCoF" is simply "d*f*/d*t*."

8.9.3 *[II]* The calculation of the "system response requirements" following a loss of *generation* (a "low frequency excursion")

1. *What is a "low frequency excursion"?*

By "low frequency excursion" we mean that the "system frequency" has deviated, perhaps significantly, from the "normal operating range" and is *lower* than where it started.

2. *Calculating the "system response requirements" following a loss of generation (a "low frequency excursion")*

   Calculating the "system response requirements" following a *loss of generation* (a "low frequency excursion") may be achieved by using the "swing equation" to model sources of *generation* and *demand* including "frequency responsive plant" and iterating one of the solution methods in Chapters 6–8 following a *loss of generation* to find the amount of "frequency response" needed to maintain the "system frequency" to *above* the *low voltage tolerance* required by the "statutory requirements" for that size of *generation loss.*

8.9.4 *[III]* The calculation of the "system response requirements" following a *loss of demand* (a "high frequency excursion")

1. *What is a "high frequency excursion"?*

   By "high frequency excursion" we mean that the "system frequency" has deviated, perhaps significantly, from the "normal operating range" and is *higher* than where it started.

2. *Calculating the "system response requirements" following a loss of demand (a "high frequency excursion")*

   Calculating the "system response requirements" following a *loss of demand* (a "high frequency excursion") may be achieved by using the "swing equation" to model sources of *generation* and *demand* including "frequency responsive plant" and iterating one of the solution methods in Chapters 6–8 following a *loss of demand* to find the amount of "frequency response" needed to maintain the "system frequency" to *below* the *high voltage* tolerance required by the "statutory requirements" for that size of *demand loss.*

8.9.5 *[IV]* The calculation of the "system response requirements" during "normal operation" ("relatively small variations" in the "system frequency")

1. *What is "normal operation"?*

   By "normal operation" we mean that the "system frequency" is contained within a narrow band around the "nominal frequency." There are no "low frequency excursions" or "high frequency excursions."

2. *Calculating the "system response requirements" during "normal operation" (relatively small variations in the "system frequency")*

   Calculating the "system response requirements" during "normal operation" (relatively small variations in the "system frequency") may be achieved by using the "swing equation" to model sources of *generation* and *demand* including "frequency responsive plant" and iterating one of the solution methods in Chapters 6–8 following a *selected small size of loss of generation or demand* to find the amount of "frequency response" needed to maintain the "system frequency" *both above* the *low voltage* tolerance *and below* the *high voltage* tolerance defined by the "normal band of operation" around the "nominal frequency."

References

Books

[1] P. Kundur, 22nd reprint Power System Stability and Control. Power System Engineering Series, Electric Power Research Institute, McGraw-Hill Education, 2017.

[2] K. Ogata, Modern Control Engineering, fourth ed., Prentice Hall/Pearson Education International, 2002.

[3] D.R. Towill, Transfer Function Techniques for Control Engineers, Iliffe Books Ltd, 1970.

[4] J.G. Truxal, Control Engineers' Handbook, McGraw-Hill Book Company, Inc., 1958.

Chapter 9

Some important practical applications

9.1 An introduction to some important practical applications

9.1.1 Different uses of the "swing equation"

In this chapter, we look at some important practical applications of the "swing equation" and different solutions of it. These applications are important because they are (or should be) daily calculations performed at power system control centers by control engineers to protect the integrity of the whole power system. Without these calculations, it would not be possible to ensure, for all times of the daily load cycle, that enough sources of frequency response were available on the system to compensate for the "reasonably expected" (also known as "credible") unplanned losses of generation and/or demand.

9.1.2 The rate of change of frequency calculation

The first calculation we shall look at is that of the "rate of change of frequency" (RoCoF). It is naturally the first calculation that we should look at because following the loss of generation and/or demand from the system it is the first problem chronologically that we must contend with. Further, if we do not contend satisfactorily with it, we may be faced with significant and increasing imbalances on the system, possibly leading to the loss of part of the system or, in extreme circumstances, the whole system.

9.2 The system "rate of change of frequency" assessment

9.2.1 *[I]* The calculation of the "rate of change of frequency" following a loss of *generation or demand*

1. "RoCoF relays"
   a. *The equipping of generators with "RoCoF relays"*
      Some generators are equipped with "RoCoF relays." They detect and measure the RoCoF following a loss of generation or demand from the system.

Modern Aspects of Power System Frequency Stability and Control.
DOI: https://doi.org/10.1016/B978-0-12-816139-5.00009-6

b. *Why are "RoCoF" relays installed on generators?*
"RoCoF" relays may be installed on generators for several reasons:

i. *Loss of grid/loss of mains/islanding protection*
One potential use of having "RoCoF" relays on generators is to detect a "loss of grid" condition (also known as "loss of mains" or "islanding"). The purpose of the "RoCoF" relays in this situation is to protect the generator. When the "RoCoF" relay detects a "loss of grid" condition, the generator is deliberately "tripped," that is, it is deliberately disconnected from the main system, that is, from the grid and/or supergrid.

ii. *The detection of "under-frequency" and "over-frequency" conditions*
"RoCoF relays" may also be installed on generators to detect both "under-frequency" and "over-frequency" conditions. The purpose of the "RoCoF relay" here is to protect the generator by tripping it before the frequency becomes too low or too high, thus avoiding potentially costly damage to the generator caused by these conditions.

2. *The control of the "RoCoF"*
The "RoCoF" on the power system can be controlled by the scheduling of enough generation in reserve to *ensure that there is enough power and inertia on the system to limit the RoCoF to below generator tripping threshold levels.*

3. *Calculating the "RoCoF"*
Calculating the "RoCoF" is achieved in a straightforward way by evaluating the right-hand side of the "swing equation," since "RoCoF" is the same as "d*f*/d*t*."

9.2.2 A procedure for calculating the "rate of change of frequency"

1. *Introduction*
Using a suitable form of the "swing equation," such as Eq. (8.1a), we can say that

$$\frac{\mathrm{d}f}{\mathrm{d}t} = \frac{\Sigma P_{\mathrm{GEN(NR)}} + \Sigma P_{\mathrm{GEN}\ n\mathrm{(R)}} - \Sigma P_{\mathrm{LOAD}\ n} - k_1\cdot(f - f_n) - P_{\mathrm{TRIP}}\cdot u(t - t_{\mathrm{TRIP}}) + a_0 + a_1 t}{2\cdot H\cdot f_0} \tag{9.1}$$

$$k_1 = k\cdot\Sigma P_{\mathrm{GEN}\ n\mathrm{(R)}} + \alpha\cdot\Sigma P_{\mathrm{LOAD}\ n} \tag{9.2}$$

where k and α are *positive* constants of proportionality, the subscript "GEN (R)" denotes "responsive generation," and the subscript "GEN (NR)" denotes "nonresponsive generation." These terms respectively correspond to those generators that are in active response mode and those that are not.

2. *A note on the correction term to the load for the dependency of load on frequency ("load damping").*

   The following correction term may be *added to* the *total load measured at* "nominal frequency":

$$\Sigma P_{\mathrm{LOAD}\ n} \cdot \alpha \cdot (f - f_n) \tag{9.3}$$

   We note from the form of Eq. (9.3) that we are treating *all* the load as being frequency responsive with an average rate α and consequently that all the load must undergo a correction in line with the prevailing system frequency. There is no correction brought about by this term if the frequency is the nominal frequency.

3. *A note on the correction term to the generation for the effect of "governor action":*

   The following correction described by Eq. (9.4) is *subtracted from* the amount of "responsive generation" *only measured at* "nominal frequency":

$$\Sigma P_{\mathrm{GEN}\ n(\mathrm{R})} \cdot k(f - f_n) \tag{9.4}$$

   We note that the "responsive generation" *only* gets corrected by this process. The "unresponsive generation" does not get corrected.

4. *The sense of the dependencies of the load and the generation on frequency*

   We note that from Eq. (9.3) *the load increases with increasing frequency*, but from Eq. (9.4) that *the generation decreases with increasing frequency*. In fact, both these terms are stabilizing for the system: the first because increasing the load with frequency reduces the deficit between generation and demand, thus tending to reduce the frequency excursion, and decreasing the generation with increasing frequency has the same effect. It should be emphasized that both formulas are not contrivances but faithfully model actual observed phenomena, the first due to loads and the second due to response-activated generation.

5. *The full expressions for the "system load P_L" and the "system generation P_G"*

   From the earlier reasoning, we may define the following expressions for the "total system load $P_{\mathrm{L(TOT)}}$" and the "total system generation $P_{\mathrm{G(TOT)}}$" as functions of the prevailing system frequency f:

$$P_{\mathrm{L(TOT)}}(f) = \Sigma P_{\mathrm{LOAD}\ n}[1 + \alpha \cdot (f - f_n)] \tag{9.5}$$

   and

$$P_{\mathrm{G(TOT)}}(f) = \Sigma P_{\mathrm{GEN(NR)}} + \Sigma P_{\mathrm{GEN}\ n(\mathrm{R})} \cdot [1 - k(f - f_n)] \tag{9.6}$$

   We note that these are variables dependent on frequency, because of the "dependency of load on frequency" effect and the effect of the "governor characteristic" in the case of those generators that have been selected to be in "frequency-responsive mode."

6. *Rewriting Eq. (9.1) in a more compact form*

Using the definitions given in Eqs. (9.5) and (9.6), we may now rewrite Eq. (9.1) in the following more compact form:

$$\left(\frac{\mathrm{d}f}{\mathrm{d}t}\right)(f,t) = \frac{P_{\mathrm{G(TOT)}}(f) - P_{\mathrm{L(TOT)}}(f) - P_{\mathrm{TRIP}}\cdot u(t - t_{\mathrm{TRIP}}) + a_0 + a_1 t}{2\cdot H\cdot f_0} \tag{9.7}$$

We see from the formulation in Eq. (9.7) that "d*f*/d*t*" is a function of both the prevailing power frequency "*f*" and the time "*t*."

7. *The terms in Eq. (9.7) relevant for a "RoCoF" calculation*

We may ask the question, which terms in Eq. (9.7) are relevant for a "RoCoF" calculation?

Since a "RoCoF" calculation applies to the first second or so following a system trip, the said trip being either of generation or of demand, we are then only interested in those terms of Eq. (9.7) that are relevant during this time period.

If we assume that the trip occurs at $t=0$, an appropriate use of Eq. (9.7) for calculating the "RoCoF" is to take $t_{\mathrm{TRIP}}=0$ and $a_0=a_1=0$ leaving therefore only

$$\left(\frac{\mathrm{d}f}{\mathrm{d}t}\right)(f,t) = \frac{P_{\mathrm{G(TOT)}}(f) - P_{\mathrm{L(TOT)}}(f) - P_{\mathrm{TRIP}}\cdot u(t)}{2\cdot H\cdot f_0} \tag{9.8}$$

(We have assumed that no "general ramped response" $a_0 + a_1 t$ occurs during the chosen time of calculation.)

8. *An assumption about the "generator response characteristic" during the "RoCoF" period*

Since the period over which "RoCoF" is measured (about 1 second, say) is shorter than the average initial delay time (the "dead band") before responsive generators begin to react (about 2 seconds following the initial incident), in Eq. (9.6) we may safely ignore the variation of power output with frequency during the "RoCoF period."

Therefore instead of the variation of output with frequency for the power generation during the "RoCoF period" we have simply

$$P_{\mathrm{G\ TOT}}(f_0) = \Sigma P_{\mathrm{GEN(NR)}} + \Sigma P_{\mathrm{GEN}\ n\mathrm{(R)}} \tag{9.9}$$

which is independent of frequency during the "RoCoF" period.

9. *A comment about the "variation of the load with frequency" during the "RoCoF" period*

The load varies with frequency in a way that depends on the time constants for devices, such as induction motors, that form part of the load.

On the assumption that the time constants for induction machines are generally much shorter than the corresponding time constants for HV-connected synchronous machines, because they are much smaller in size, d*f*/d*t* as given by Eq. (9.8) *changes during the* "RoCoF" *period because it cannot be assumed justifiably that the load does not change with frequency during this period.*

We would ideally like to account for the modification in the value of the load during the "RoCoF" period of approximately ($0<t<1$ second) according to Eq. (9.5), which we repeat below for convenience:

$$P_{L(TOT)}(f) = \Sigma P_{LOAD\ n}[1 + \alpha \cdot (f - f_n)] \tag{9.10}$$

With $P_{L\ (TOT)}\ (f)$ as given by Eq. (9.10), Eq. (9.8) becomes

$$\left(\frac{df}{dt}\right)(f,t) = \frac{P_{G(TOT)}(f) - \Sigma P_{LOAD\ n}[1 + \alpha \cdot (f - f_n)] - P_{TRIP} \cdot u(t)}{2 \cdot H \cdot f_0} \tag{9.11}$$

We can see that the right-hand side of Eq. (9.11), except for the trip at $t=0^+$, is a function of *f*. Therefore the "correct" approach to calculating the "RoCoF" for the whole of the first second following the trip is to solve Eq. (9.11) as a differential equation in *f* for ($0^+ < t < 1$ second).

Alternatively, an approximate value, correct only at $t=0^+$, may be obtained by ignoring the variation of the load with frequency effect in the first second.

10. *A note on the "initial frequency f_0"*

We state here, for the record, that the simulations we shall do, which start at time $t = 0$ do *not* assume that the "initial frequency" is necessarily equal to the nominal frequency f_n but is, in fact, *any* general initial frequency f_0.

11. *A note on finding the "generation output at nominal frequency" from the "initial generation output"*

To find the nominal generation output from the initial generation output we can use Eq. (9.6) by putting $f=f_0$ (the initial frequency) and rearranging to get

$$P_{G(TOT)}(f_n) = P_{G(TOT)}(f_0) + P_{G(R)}(f_0) \cdot \frac{k \cdot (f_0 - f_n)}{[1 - k \cdot (f_0 - f_n)]} \tag{9.12}$$

in which

$$P_{G(TOT)}(f_n) \equiv \Sigma P_{GEN(NR)} + \Sigma P_{GEN\ n(R)} \tag{9.13}$$

and

$$P_{G(R)}(f_0) \equiv \Sigma P_{GEN(R)}(f_0) \tag{9.14}$$

As a commonsense check on the reasonableness of Eq. (9.12), we see that putting $f_0 = f_n$ on the right-hand side then $P_{G\ (TOT)}\ (f_n) = P_{G\ (TOT)}\ (f_0)$ as expected.

12. *A note on finding the nominal load from the initial load*

To find the load at nominal frequency from the initial load we can use Eq. (9.5) in a similar way by setting $f = f_0$ (the initial frequency) and rearranging it to get

$$P_{L(TOT)}(f_n) = \frac{P_{L(TOT)}(f_0)}{[1 + \alpha \cdot (f_0 - f_n)]} \tag{9.15}$$

in which $P_{L\ (TOT)}\ (f_n) \equiv \Sigma P_{LOAD\ n}$.

As a commonsense check on the reasonableness of Eq. (9.14), we see that putting $f_0 = f_n$ on the right-hand side then $P_{L\ (TOT)}\ (f_n) = P_{L\ (TOT)}\ (f_0)$ as expected.

a. We now go on to describe the proposed "RoCoF procedure" itself.

13. *A summary of the proposed "RoCoF procedure"*

We note here the steps that we propose should be followed:

Step 1: Make a note of/calculate the system inertia constant "H."

Step 2: Make a note of the "initial frequency f_0."

Step 3: Make a note of the "total initial generation $P_{G\ (TOT)}\ (f_0)$"

Step 4: Make a note of the "total initial load $P_{L\ (TOT)}\ (f_0)$"

Step 5: Make a note of the "generation to be tripped (alternatively, the "load to be tripped" etc.) P_{TRIP}" at $t = 0^+$.

Step 6: Calculate the "RoCoF" using Eq. (9.8) evaluated at $f = f_0$ and $t = 0^+$, that is,

$$\left(\frac{df}{dt}\right)(f_0, 0^+) = \frac{P_{G(TOT)}(f_0) - P_{L(TOT)}(f_0) - P_{TRIP}}{2 \cdot H \cdot f_0} \tag{9.16}$$

Note: We recall that we are assuming that Eq. (9.16) is approximately valid throughout the "RoCoF" period of ($0 < t < 1$ second). Please see paragraphs (8) and (9) for a summary of the rationale behind this assumption.

14. *In the next section*:

In the next section, we move on to work through an example "RoCoF" calculation that can be run in *MATLAB*.

9.3 MATLAB example simulation #9: the "rate of change of frequency" calculation

9.3.1 Introduction

In this section, we shall be using Eq. (9.16) to give an example "RoCoF" calculation. To assist with completing the task we shall be providing some appropriate code that can be run in *MATLAB*. This code is merely a suggestion and may, of course, be changed by the user to suit his or her own needs.

9.3.2 The statement of the problem

A 30 GW power system operating at 50 Hz nominal frequency comprises 30 identical generators, each having a rated power output of 1000 MW, an *H*-constant of 2.5 seconds, and a rated power factor of 0.8 lagging.

1. If the system load is30,000 MW, calculate the initial "RoCoF" in Hz/s following the trip of one of the generators if the system frequency just before the trip occurred was (1) 50.0, (2) 50.5, (3) 49.5 Hz. (Ans: −0.267, −0.264, and −0.269 Hz/s).
2. Repeat the earlier calculation for the same three initial frequencies, in turn, if the system load is instead only 15,000 MW when (1) all machines used are *fully loaded* (power output = 100%), and (2) all machines used are *half loaded* (power output = 50%). (Ans: (1) −0.533, −0.528, and −0.539 Hz/s; (2) −0.267, −0.264, and −0.269 Hz/s).
3. What can we learn from this exercise?

9.3.3 The general solution to the problem

We use Eq. (9.16)

$$\left(\frac{\mathrm{d}f}{\mathrm{d}t}\right)(f_0, 0^+) = \frac{P_{\mathrm{G(TOT)}}(f_0) - P_{\mathrm{L(TOT)}}(f_0) - P_{\mathrm{TRIP}}}{2 \cdot H_{\mathrm{sys}} \cdot f_0} \tag{9.16}$$

which we note is in "per unit" form.

For the system in the question, we can select a *system MVA base = System demand (MW)/0.8 (MVA)* and a frequency base of 50 Hz.

Assuming an initially balanced system, Eq. (9.16) becomes

$$\begin{aligned}\frac{\mathrm{d}f}{\mathrm{d}t}(\mathrm{pu}) &= \frac{-P_{\mathrm{TRIP}}(\mathrm{MW})/\left[\mathrm{Sys.\ Dem.(MW)}/0.8\right]}{2 \cdot H_{\mathrm{sys}} \cdot \left[f_0(\mathrm{Hz})/50\right]} \\ &= \frac{-20 \cdot P_{\mathrm{TRIP}}(\mathrm{MW})}{\mathrm{Sys.\ Dem.(MW)} \cdot H_{\mathrm{sys}} \cdot f_0(\mathrm{Hz})}.\end{aligned} \tag{9.17}$$

To convert d*f*/d*t* (pu) into units of Hz/s

$$\begin{aligned}\frac{\mathrm{d}f}{\mathrm{d}t}(\mathrm{Hz}/s) &= 50\ \mathrm{Hz} \times \frac{\mathrm{d}f}{\mathrm{d}t}(\mathrm{pu}) \\ &= \frac{-1000 \cdot P_{\mathrm{TRIP}}(\mathrm{MW})}{\mathrm{Sys.\ Dem.(MW)} \cdot H_{\mathrm{sys}} \cdot f_0(\mathrm{Hz})}\mathrm{Hz}/s\end{aligned}$$

that is,

$$\frac{df}{dt}(\mathrm{Hz}/s) = \frac{-P_{\mathrm{TRIP}}(\mathrm{MW})}{\mathrm{Sys.Dem.(GW)} \cdot H_{\mathrm{sys}}(\mathrm{s}) \cdot f_0(\mathrm{Hz})} \mathrm{Hz/s} \tag{9.18}$$

This equation will be programed into MATLAB for use in this exercise.

Note 1: In part (a) all 30 machines out of 30 are used, so $H_{\mathrm{ave}=}(30/30) \times$ 2.5 seconds = 2.5 seconds. In part (b) only 15 machines out of 30 are used but $H_{\mathrm{ave}=}(15/15) \times 2.5$ seconds = 2.5 seconds, the same number.

Note 2: The "suspiciously neat-looking" formula in Eq. (9.17) is neat only by chance because $f_{n=}50$ Hz and the machine rated $pf = 0.8$ making $pf \times \{f_n(\mathrm{Hz})\}^2 = 2000$. The formula is the same whenever $pf = 2000/\{f_n(\mathrm{Hz})\}^2$. For example, if instead $f_n(\mathrm{Hz}) = 60$ Hz, with a pf of $2000/(60)^{2=}0.5556$ would generate the same formula, although this latter pf is not a realistic number.

9.3.4 The input data to the MATLAB code

The input data are P_{TRIP} (MW), Sys. Size (GW), H_{sys} (second), and f_0 (Hz).

9.3.5 The output datum from the MATLAB code

The output datum is the *initial RoCoF in Hz/s.*

9.3.6 The suggested *MATLAB* code

A suggested way to program the earlier example in MATLAB is as follows:

Suggested MATLAB code #9: Rate of Change of Frequency (RoCoF) Calculation

```
    HSYS = input('Please enter a value for the inertia constant H for
            the system in seconds')
% The system inertia constant H in seconds for the system has now been
assigned a
% value
    F0 = input('Please enter a value for the initial frequency in Hz')
% The initial frequency in Hz has now been assigned a value.
    GEN0 = input('Please enter a value for the total initial
            generation in GW')
% The total initial generation in GW has now been assigned a value,
    PL0 = input('Please enter a value for the initial load/demand in
          GW')
% The total initial load in GW has now been assigned a value.
    PTRIP = input('Please enter a value for the amount of generation
            or load tripped in MW')
% The total amount of generation tripped has now been assigned a value.
```

```
            RoCoF = - PTRIP/(PLO*HSYS*FO);
% The corresponding value of RoCoF has been calculated but not
printed out.
            RoCoF
% The corresponding value of RoCoF has been printed out.
```

9.3.7 Notes

This is a particularly straightforward example.

9.3.8 General instructions for the user

Assign values to P_{TRIP} *in MW, Sys. Dem. in GW, average* H_{sys} *in second,* and f_0 *in Hz* and run the program to generate the *initial "RoCoF" in Hz/s.*

9.4 The "system response requirements" for a "low frequency excursion"

9.4.1 *[II]* The calculation of the "system response requirements" following a loss of generation (a "low frequency excursion")

1. *What is a "low frequency excursion"?*

   By "low frequency excursion" we mean that the "system frequency" has deviated, perhaps significantly, from the "normal operating range" and is *lower* than its starting point or initial value.
2. *Calculating the "system response requirements" following a loss of generation (a "low frequency excursion")*

   Calculating the "system response requirements" following a *loss of generation* (a "low frequency excursion") may be achieved by using the "swing equation" to model sources of *generation* and *demand* including "frequency-responsive plant" and iterating one of the solution methods in Chapter 6, Chapter 7 or Chapter 8 following a *loss of generation* to find the amount of "frequency response" needed to maintain the "system frequency" to *above* the *low voltage tolerance* required by the "statutory requirements" pertaining to that particular size of *generation loss.*

9.4.2 A procedure for calculating the "system response requirements" following a *loss of generation* (a "low frequency excursion")

1. *The need to provide (schedule) partly loaded generating sets*

   To be able to provide frequency response for a traditional synchronous generator-fed system, it is necessary to include some generators that are "part-loaded" to provide some headroom if the system frequency were to fall, in addition to including some generators that are "fully loaded."

2. *The problem to be solved: finding the "response requirements"*

   For a given network, we need to find out how much plant needs to be "part-loaded" to ensure that the system frequency stays within statutory limits following a "credible loss" of generation from the system.

3. *The inertia of an individual "part-loaded" generator*

   Let us remind ourselves at this point of the definition of the "H-constant" for a single generating set:

$$H \equiv \frac{(1/2)J \cdot \omega_n^2}{S_n} \tag{9.19}$$

Where ω_n denotes the rated speed of the machine, and S_n denotes the "rated MVA" of the machine.

Since, by definition, we have

$$S_n = \frac{P_n}{pf_n} \tag{9.20}$$

where P_n denotes the "rated power output," and "pf_n" denotes the "rated power factor" of the machine, we can rewrite Eq. (9.19) in the equivalent alternative form:

$$H = \frac{pf_n \cdot (1/2)J \cdot \omega_n^2}{P_n} \tag{9.21}$$

Since all quantities in Eq. (9.21) are "nominal," this expression holds true for a single synchronous machine, irrespective of its loading.

In other words a "part-loaded" machine has the same inertia as a "fully loaded" machine. The material difference is in its different "power output," which, however, has no effect on its "inertia."

4. *The "inertia" of a system containing "part-loaded" machines*

   In *MATLAB example simulation #9*, we assumed, for the sake of simplicity, that all machines used were "fully loaded." However, this is never the case in practice because some machines are always run "part-loaded" to provide frequency response.

5. *How generators are "loaded up" for running in a "real" power system*

   a. *The "merit order"?*

      In a real power system, "loading up" of generators, that is, the selection of the power output for each generator, is usually done on the following basis: for a given system demand, the main principle followed is to select the least expensive generator first, that is, to follow a "merit order." In the past in the United Kingdom, this was literally done in this way when all generation was under central control. Now there is a pseudo-market set up whereby generators compete to run by submitting prices, with "bids" and "offers" to have their outputs increased or decreased.

b. *Two "merit orders"*

However, since we also need to provide response, in practice there are two "merit orders" in operation at the same time: one for "fully loaded" generators and the other for "partly loaded" generators.

c. *The degree of part-loading*

In theory a generator may be downloaded to its minimum stability limit, for example, 30% of full output, the precise figure depending on the make and type of machine. In practice, most generators used for response provision will be downloaded to some much higher loading, typically 90% of its rated output.

6. *A simple "generator scheduling" procedure for estimating frequency response requirements*

In practice the selection of generation to meet demand is a kind of multiple optimization process involving many parameters, but to demonstrate the main principles of "inertia" and "downloading for response" and how they affect the subsequent behavior of system frequency, in the next exercise we shall be making an initial estimated mix of "fully loaded" and "partly loaded" machines to meet the demand and estimated response requirements and then refining this estimate to find the best ratio that just satisfies the statutory frequency requirements for the size of the loss involved.

7. *An assumption about the nature of generator response*

In the next exercise, we shall be assuming that all the generators available for responsive duty are traditional synchronous machines possessing a "ramped response profile."

8. *Using the "exact solution" of the "swing equation"*

To help us complete the next exercise (*MATLAB example simulation #10*: calculating the "response requirements" for a "low frequency excursion"), we shall be using the "exact solution" of the "swing equation" as given at the beginning of this chapter:

$$\frac{\mathrm{d}f}{\mathrm{d}t} = \frac{\Sigma P_{\mathrm{GEN(NR)}} + \Sigma P_{\mathrm{GEN}\ n(\mathrm{R})} - \Sigma P_{\mathrm{LOAD}\ n} - k_1 \cdot (f - f_n) - P_{\mathrm{TRIP}} \cdot \mathrm{u}(t - t_{\mathrm{TRIP}}) + a_0 + a_1 t}{2 \cdot H \cdot f_0} \tag{9.22}$$

$$k_1 = k \cdot \Sigma P_{\mathrm{GEN}\ n(R)} + \alpha \cdot \Sigma P_{\mathrm{LOAD}\ n} \tag{9.23}$$

in the form given in Eq. (6.27) of Chapter 6, Analytical methods for solving the "swing equation":

$$f(t) = A_1 + B_1 \cdot t + C_1 \cdot \exp(D_1 \cdot t) \tag{9.24}$$

with parameters A_1, B_1, C_1, and D_1 given by Eqs. (6.35), (6.34), (6.36), and (6.37), respectively, which we reproduce here for convenience.

$$A_1 = f_n + \frac{\left(a_0 + \Sigma P_{\text{GEN(NR)}} + \Sigma P_{\text{GEN } n\text{(R)}} - \Sigma P_{\text{LOAD } n}\right)}{k_1} - 2 \cdot H \cdot f_0 \cdot \frac{a_1}{k_1^2} \tag{9.25}$$

$$B_1 = \frac{a_1}{k_1} \tag{9.26}$$

$$C_1 = f_0 - f_n - \frac{\left(a_0 + \Sigma P_{\text{GEN(NR)}} + \Sigma P_{\text{GEN } n\text{(R)}} - \Sigma P_{\text{LOAD } n}\right)}{k_1} + 2 \cdot H \cdot f_0 \cdot \frac{a_1}{k_1^2} \tag{9.27}$$

$$D_1 = \frac{-k_1}{2 \cdot H \cdot f_0} \tag{9.28}$$

where

$$k_1 = \Sigma P_{\text{LOAD } n} \cdot \alpha \cdot + \Sigma P_{\text{GEN } n\text{(R)}} \cdot k \tag{9.29}$$

in which the constant "α" is defined implicitly by the equation

$$\Sigma P_{\text{LOAD}} = \Sigma P_{\text{LOAD } n}[1 + \alpha \cdot (f - f_n)] \tag{9.30}$$

and the constant "k" is defined by the equation

$$\Delta P_{\text{RESP}} = -P_{\text{RESP}} \cdot k \Delta f \tag{9.31}$$

where

$$D(\%) = \frac{100}{k(\text{pu})} \tag{9.32}$$

is the "generator governor droop" in "%" and $\Delta f \equiv f - f_n$. Some typical values are $\alpha = 2\%$ per Hz and $D = 4\%$.

9.5 MATLAB example simulation #10: calculating the "response requirements" for a "low frequency excursion"

9.5.1 Introduction

The objective of this exercise is to calculate how much response must be held in total on system generators to cover for the loss of one of the generators. This involves finding out how much responsive generation needs to be held by de-loading some machines to less than their rated MW outputs.

9.5.2 The statement of the problem

As a practical example, let a power system operating at 50 Hz nominal frequency has a total of 50 identical generators available to run, each having a rated power output of 1000 MW and a rated power factor of 0.8 lagging.

The problem is to find the system response requirements for the loss of one of the generators when the system loading is 30 GW and the H (inertia) constant of each generator is (1) 2.5, (2) 5.0, and (3) 1.25 seconds if the system frequency is not permitted to fall below (i) 49.2 and (ii) 49.5 Hz.

Assume that all generators that are selected for response are run at *90%* output preincident and that the "load-damping coefficient" $\alpha = 0.02$ pu/Hz.

9.5.3 The general solution of the "low frequency excursion" problem

1. *The general method*

   In general terms the problem is solved by assuming an initial proportion of responsive generation to be de-loaded and iterating until the amount required to allow the frequency to fall just above the permitted lower level is obtained.

2. *The details*

   Starting from Eqs. (9.1) and (9.2) which we restate here, we have

$$\frac{df}{dt} = \frac{\Sigma P_{\mathrm{GEN(NR)}} - \Sigma P_{\mathrm{LOAD}\ n} - k_1 \cdot (f - f_n) - P_{\mathrm{TRIP}} \cdot u(t - t_{\mathrm{TRIP}}) + a_0 + a_1 t}{2 \cdot H \cdot f_0} \tag{9.33}$$

$$k_1 = k \cdot \Sigma P_{\mathrm{GEN}\ n(R)} + \alpha \cdot \Sigma P_{\mathrm{LOAD}\ n} \tag{9.34}$$

   which has the general solution

$$f(t) = A_1 + B_1 \cdot t + C_1 \cdot \exp(D_1 \cdot t) \tag{9.35}$$

   in which the parameters A_1, B_1, C_1, and D_1 are given by Eq. (6.35) were stated earlier in this chapter in Eq. (9.25) and the equations immediately following it.

3. *A convenient assumption*

   Since the generator governor operates with a time delay of a few seconds, to avoid adding more complexity we can simulate an ideal "ramped response" using the "$a_0 + a_1 t$" term in Eq. (9.1) while at the same time setting, $k = 0$ in Eq. (9.2).

   Using this approach, Eqs. (9.1) and (9.2) become

$$\frac{df}{dt} = \frac{\Sigma P_{\mathrm{GEN(NR)}} - \Sigma P_{\mathrm{LOAD}\ n} - k_1 \cdot (f - f_n) - P_{\mathrm{TRIP}} u(t) + a_0 + a_1 t}{2 \cdot H \cdot f_0} \tag{9.36}$$

   and

$$k_1 = \alpha \cdot \Sigma P_{\mathrm{LOAD}\ n} \quad \text{only} \tag{9.37}$$

Note that we have also simplified the formulation further by choosing $t_{TRIP=}0$, that is, we have assumed that the generation trip occurs at time $t = 0$.

4. *The value of the "load-damping" coefficient* α

   We have chosen for our present purposes the "load-damping" coefficient α to have the value0.02 pu/Hz. This means in practical terms that for a 50 Hz system a 0.5 Hz drop in frequency is commensurate with a 1% drop in demand.

5. *Initially balanced system*

   We shall assume, for the purposes of this exercise, that initially the system generation equals the system demand (or very nearly so) so that the starting frequency is one per unit or very close to it.

6. *The number of sets required to be in "responsive mode"*

   The point of this exercise is to find out how many generators are needed to be in "responsive mode," that is, whose output initially needs to be reduced by 10% and are available to "ramp up" if any generation were to be lost unexpectedly from the system.

7. *How the mathematics works out*

   Since the initial system is to be balanced (or nearly so), we need to find the closest balance possible when the number of responsive (10% de-loaded) machines is 0, 1, 2, 3, 4, ... and so on.

   In our example, all the 50 machines supplied are of 1000 MW≡1 GW size. The system demand is 30 GW or in other words 30 times the size of an individual machine.

   Let the number of de-loaded machines be denoted by "n," a nonnegative integer.

   Each de-loaded machine has a rated output of 900 MW. Hence, if there are "n" de-loaded machines being run at rated output, their total output is $900 \times n$ MW.

   Since the system demand is 30,000 MW, the total output of unresponsive machines would therefore be

$$30,000 - 900\, n \text{ MW} \tag{9.38}$$

In tabular form, for different values of "n," we have therefore for our example system.

We can see from Table 9.1 that in practice it is rarely possible to provide exactly the right combination of generators to meet the system demand exactly with whole generating sets, that is, generating sets operating at their rated outputs (unresponsive machines) or de-loaded by 10% (responsive machines). In fact, of those cases listed in Table 9.1, only when $n = 0$ or $n = 10$ is this condition true if we are to use "whole sets" throughout to represent the unresponsive generation. In all other cases, there is a mismatch between generation and demand, which

TABLE 9.1 Theoretically possible loadings of generators.

n	P (resp, MW)	P (unresp, MW)	Actual UR (MW)	$P_{gen}-P_{dem}$ (MW)
0	0	30,000	30,000	0
1	900	29,100	29,000	−100
2	1800	28,200	28,000	−200
3	2700	27,300	27,000	−300
4	3600	26,400	26,000	−400
5	4500	25,500	26,000	+500
6	5400	24,600	25,000	+400
7	6300	23,700	24,000	+300
8	7200	22,800	23,000	+200
9	8100	21,900	22,000	+100
10	9000	21,000	21,000	0
11	9900	20,100	20,000	−100
12	10,800	19,200	19,000	−200

means consequently that the initial frequency will necessarily be different from 1 pu.

There is, however, a way round this "initial balance" problem available to us. If we schedule enough "unresponsive" generating sets, we can download some or all of them sufficiently to enable the initial frequency to be equal to, or very nearly equal to, the nominal frequency. We shall cover an example of this balancing process a little later.

8. *The values of the ramp parameters a_0 and a_1*

In our example the values we select for the "ramp parameters" a_0 and a_1 depend on the value of n chosen. To take an example, if $n = 1$ and we want to ramp up linearly the responsive machine's remaining 10% output from 0% at $t = 2$ seconds to 10% at $t = 10$ seconds, then from Eq. (9.36) for $t \geq 2$ seconds we will need to solve simultaneously the equations

$$a_0 + a_1 \cdot 2 = 0 \tag{9.39}$$

and

$$a_0 + a_1 \cdot 10 = \frac{0.1 \times 1000\,\text{MW}}{S_{sys}} \tag{9.40}$$

where S_{sys} is the *MVA rating* that has been chosen for the system.

(We continue to bear in mind that all our equations are expressed in pu unless otherwise stated.)

We note that for a general integral value of "n" that Eq. (9.39) still stands but that Eq. (9.40) should be replaced by

$$a_0 + a_1 \cdot 10 = \frac{0.1 \times n \times 1000\ \text{MW}}{S_{\text{sys}}} \tag{9.41}$$

where we are assuming that all the "n" responsive machines "perform in unison," that is, they "ramp up" together in an identical fashion.

Solving Eq. (9.39) with Eq. (9.41) we obtain

$$a_0 = -50 \cdot n \cdot H \cdot \frac{f_0}{S_{\text{sys}}} \tag{9.42}$$

and

$$a_1 = 25 \cdot n \cdot H \cdot \frac{f_0}{S_{\text{sys}}} \tag{9.43}$$

as the values of a_0 and a_1 as functions of the number of responsive sets n.

9. *The system MVA rating S_{sys}*

We can derive a value for the system MVA rating S_{sys} knowing the assumed *system size* in MW and by using a nominal generating machine-rated power factor of 0.8.

Taking the "system MW rating" to be the same thing as the system size of 50,000 MW in our example, we therefore obtain

$$S_{\text{sys}} = 50,000/0.8 = 62,500\ \text{MVA} \tag{9.44}$$

as the MVA base for our calculations.

10. *Putting it all together*

We now have enough information to find numerical values for the parameters A_1, A_2, A_3, and A_4 as defined by Eqs. (9.25)–(9.28) inclusive.

11. *The solution*

We have now reached the solution for $f(t)$ which, repeating Eq. (9.35), is

$$f(t) = A_1 + B_1 \cdot t + C_1 \cdot \exp(D_1 \cdot t) \tag{9.45}$$

where the parameters A_1, B_1, C_1, and D_1 are given by Eqs. (9.25)–(9.28), respectively.

12. *An Example with $n = 1$*

Let us look in detail at the case when $n = 1$, that is, when only one generating set is put into "responsive mode" by the action of reducing its prefault output by 10% from 1000 to 900 MW.

13. *Unbalanced initial condition*

Looking at Table 9.1, we see that for $n = 1$ the initial condition is not balanced because we have reduced the output of one of the generating sets by 100 MW and cannot replace it exactly with any of the machines at our disposal, all of which are rated at 1000 MW.

14. *The closest match between generation and demand*

The closest match between generation and demand that we can achieve with "whole sets" is obtained by selecting 29 × 1000 MW sets to be in unresponsive mode, that is, 29 generating sets to run at 1000 MW in addition to the responsive set of 900 MW.

15. *The initial condition is unbalanced at the nominal frequency*

In practice the system is usually not exactly balanced when the initial frequency f_0 is the nominal frequency f_n (e.g., 50 Hz). Since in our example here with $n = 1$ the generation is slightly less than the demand, the system would be expected to be balanced at an initial frequency of slightly less than the nominal frequency.

16. *Balancing the frequency of a system unbalanced at the nominal frequency*

Following our simple example, let $n = 1$ so that the one generator selected to be in responsive mode is to be run at an output of 900 MW. Since from Table 9.1 we require 29,100 MW of nonresponsive generation to balance the system for a load of 30,000 MW, we can do this by selecting 30 *nonresponsive generators*, each running at an output of

$$\frac{29,100}{30,000} \times 1000\ \text{MW} = 970\ \text{MW} \tag{9.46}$$

This is a simple way of ensuring an initial power balance and therefore a starting frequency of 50 Hz. However, if we solve the problem like this, we must remember to correct the system inertia H by a factor of 31/30 since we now have 31 machines instead of 30.

17. *Introducing the trip of a generator from the system*

Typically, one generating set will trip from the system in a single event, although this is not always the case in practice, where sometimes more than one generator can trip almost simultaneously.

In our example, if we assume that the responsive set, that is, the one running at 900 MW output, does not itself trip, this means that in this event 970 MW will be lost in a single incident.

18. *Modeling the trip of a generator from the system*

To model the trip of a generator from the system, we can subtract its output from the initial power balance given in Eq. (9.36).

19. *What to do next*

The next thing to do is to calculate the values of all the parameters used in Eqs. (9.36) and (9.37). These parameter values are then inserted into Eqs. (9.25)–(9.28) inclusive to determine the corresponding values of the parameters A_1, B_1, C_1, and D_1. Knowing these four parameters, the solution $f(t)$ for a selected value of t is then given by using Eq. (9.45).

20. *Drawing graphs*

By repeating this process for different values of n and for a suitable range of values for the time t, graphs of f against t may be plotted.

21. *Choosing the "optimal" value for "n"*

The "optimal" value for n is the smallest value of n that achieves the objective of the system frequency f not falling below the stipulated statutory value (e.g., 49.2 Hz for a 50 Hz system) during an allotted period, for example, the first 10 seconds, which can be regarded as the de facto "primary response" period.

9.5.4 The input data

The input data for the problem are as stated in Section 9.5.2. We have made the reasonable assumption that those generating sets that are de-loaded for response are all de-loaded to 90% of their full output.

9.5.5 The output data

The output data is the amount of de-loaded plant necessary to achieve compliance with the minimum permitted frequency level for the system.

9.5.6 The MATLAB code

We shall use the "exact solution" of the "swing equation" method to help us to calculate for each iteration of n the lowest system frequency following the loss of one of the 1000 MW as detailed in Section 9.5.2.

A suggested example *MATLAB* code to solve our problem might be structured something like the following

MATLAB example simulation #10

Calculating the minimum response requirements for the loss of a generating set.

Calculating the parameters for the problem.

First, we calculate the case where no generating sets are de-loaded, that is the amount of response provided is zero.

Case 1: No response is provided (no responsive generating sets are provided)

In this case, since we are providing no reserve, 30 × 1 GW unresponsive sets are selected to run and there are no responsive sets scheduled (in other words, $n = 0$). One of the sets is the "trip set."

So, we have the following system of equations that must be evaluated

$$f(t) = A_1 + B_1 \cdot t + C_1 \cdot \exp(D_1 \cdot t) \tag{9.47}$$

where

$$A_1 = f_n + \frac{\left[\Sigma P_{\mathrm{GEN(NR)}} + \Sigma P_{\mathrm{GEN}\ n(\mathrm{R})} - \Sigma P_{\mathrm{LOAD}\ n} + a_0\right]}{k_1} - 2 \cdot H \cdot \frac{f_0 a_1}{k_1^2} \tag{9.48}$$

and

$$B_1 = \frac{a_1}{k_1} \tag{9.49}$$

and

$$C_1 = f_0 - f_n - \frac{\left[\Sigma P_{\mathrm{GEN(NR)}} + \Sigma P_{\mathrm{GEN}\ n(\mathrm{R})} - \Sigma P_{\mathrm{LOAD}\ n} + a_0\right]}{k_1} + 2 \cdot H \cdot \frac{f_0 a_1}{k_1^2} \tag{9.50}$$

and

$$D_1 = \frac{-k_1}{2} \cdot H \cdot f_0 \tag{9.51}$$

according to the standard analytical solution given in Chapter 6, Analytical methods for solving the "swing equation" by *Eqs. (6.45)–(6.48)* and reproduced in Eqs.(9.24)–(9.27) provided earlier in this chapter.

Case 2: Response is provided in the form of "n" de-loaded generating sets

In this case we are providing "*n*" generating sets that are de-loaded to 90% of their rated output where $n = 1, 2, 3, 4$, and so on. In all these cases we use Eqs. (9.42) and (9.43), respectively, for the parameters a_0 and a_1. The parameters a_0 and a_1 describe the linear output profile of the responsive machines. As mentioned earlier, we want to find the optimum (best) value for *n* for the given system parameters and loss size. This is done by a process of trial and error until the smallest value of *n* that ensures that the postfault frequency stays above the statutory minimum during the *primary response* period, which we are taking here to be the first 10 seconds postfault.

```
%   SUGGESTED MATLAB® CODE FOR THE LOW FREQUENCY EXCURSION PROBLEM
%
%   We suggest the following MATLAB® code for solving the problem of
    finding the
%   correct amount of responsive generation to cater for the loss of
    a single
%   generator from the system in the example system described above.
```

```
%
%   Read in and calculate the input data
%
%   The input data is with reference to the main equations (Eq 9.36)
    and (Eq 9.37)
%   above. First read in the data in "physical units" and then
    convert them to "per
%   unit" quantities for the purposes of our subsequent calculations.
%
    N = input('Please enter the number of de-loaded machines to be
    considered')
    P = input('Please enter the loading factor for de-loaded
    machines')
%
%   Note: A typical loading factor to apply to the de-loaded
    machines is p = 0.9
%
    PLOAD_P = input('Please enter the total system load in MW')
%
%   Note: In our example the total system load is 30,000 MW.
%
    UNIT_SIZE_P = input('Please enter the rated generator size in MW')
%
%   Note: in our example UNIT_SIZE_P = 1000 MW.
%
    PR_P = UNIT_SIZE_P * N
    PNR_P = PLOAD_P – P * N * UNIT_SIZE_P
%
    PTRIP_P = input('Please enter the amount of generation to be
    tripped in MW')
    F0_P = input('Please enter the initial frequency in Hz')
%
    A0_P = input('Please enter the value of the constant a0 in MW')
    A1_P = input('Please enter the value of the constant a1 in MW/Hz')
%
    H_P = input('Please enter the value of the H-constant per
    machine in seconds')
    ALPHA_P = input('Please enter the value of the "load damping"
    constant α in
              per cent per Hz')
%
    K1_P = ALPHA_P * PLOAD_P
%
%   Now convert the input data from "physical" to "per unit" values
    MVA_BASE = PLOAD_P/0.8
    FN_P = input('Please enter the nominal frequency in Hz')
    F_BASE = FN_P
```

```
%
       PLOAD = PLOAD_P/MVA_BASE
       UNIT_SIZE =UNIT_SIZE_P/MVA_BASE
       PR = PR_P/MVA_BASE
       PNR = PNR_P/MVA_BASE
       PTRIP = PTRIP_P/MVA_BASE
       F0 = F0_P/F_BASE
       A0 = A0_P/MVA_BASE
       A1 = A1_P/MVA_BASE
%
%      Now calculate the average system H-constant:
%
       NUM_R_SETS = PR/UNIT_SIZE
       NUM_NR_SETS = input('How many non-responsive sets are there?')
       FACTOR = (NUM_R_SETS+ NUM_NR_SETS)/(PLOAD/UNIT_SIZE)
       H = H_P * FACTOR
%
%      Now calculate the constant K1 in "per unit":
%
       K1 = K1_P * F_BASE/(100*MVA_BASE)
%
%      Now calculate the values of the constants A1, B1, C1and D1:
%
%      Calculate the parameter PBAL:
%
       PBAL = A0 + PNR + PR - PLOAD - PTRIP
%
%      Now calculate the four constants:
%
       A_1 = FN + PBAL/K1 - 2*H*F0*A1/K1^2
       B_1 = A1/K1
       C_1 = F0 - A_1
       D_1 = - K1/(2*H*F0)
%
```

9.5.7 Instructions for the user

Run the MATLAB code earlier to generate a graph of frequency against time for successive choices of the amount of response held on the system (values of *N*). Terminate the iterations when the minimum frequency asked for has been achieved. Note the total amount of de-loaded MW required to achieve this. This is then the response requirement for the given system and the given loss of generation.

9.6 *[III]* The "system response requirements" for a "high frequency excursion"

9.6.1 Calculation of the "system response requirements" following a *loss of demand* (a "high frequency excursion")

1. *What is a "high frequency excursion"?*

   By a "high frequency excursion" we mean that the "system frequency" has deviated, perhaps significantly, from the "normal operating range" and is higher than where it started.

2. *Calculating the "system response requirements" following a loss of demand (a "high frequency excursion")*

   Calculating the "system response requirements" following a loss of demand (a "high frequency excursion") may be achieved by using the "swing equation" to model sources of generation and demand including "frequency-responsive plant" and iterating one of the solution methods in Chapter 6, Chapter 7 or Chapter 8 following a *loss of demand* to find the amount of "frequency response" needed to maintain the "system frequency" to *below* the *high frequency* tolerance required by the "statutory requirements" for that size of *demand loss*.

9.6.2 A procedure for calculating the "system response requirements" following a *loss of demand* (a "high frequency excursion")

1. *The need to provide (schedule) generating sets for high-frequency events*

   By contrast to low-frequency events, high-frequency events cause the power output of frequency-responsive sets to decrease rather than increase. Hence, it is not necessary to de-load generating sets prefault to allow for high-frequency incidents. However, there do need to be enough responsive machines scheduled on the system to allow for both low- and high-frequency events.

2. *The problem to be solved: finding the "response requirements"*

   In practice a certain magnitude of load loss for any given system needs to be catered for. This will generally be indicated by the largest block of demand that is likely to be at risk of disconnection. This information will generally be known in advance by the *system operator* and so can be planned for.

3. *Inertia considerations in high-frequency event scenarios*

   As for low-frequency events, the system inertia, sometimes spoken of in terms of the sum of the "MVA × *H*" values of all the generators in service on the system at a given time, is an important factor in the postevent performance of the system frequency. Whether or not the system frequency can be maintained within the statutory frequency limits is determined largely by the system inertia irrespective of whether generation or load has been lost. If there is insufficient inertia on the system then this may not be possible.

4. *Generator loading considerations in high-frequency event scenarios*

   As mentioned earlier, it is not particularly critical what the loading of a responsive generator selected for high-frequency response is provided the minimum stable generation limit of any generator is not transgressed postfault.

5. *Assumptions about the nature of generator response during high frequency*

   For simplicity, it may be assumed that the trajectory of the power outputs of responsive sets with time following the loss of a block of demand from the system is linear.

6. *Using the "exact solution" of the "swing equation"*

   We can again use the "exact solution" of the "swing equation" to analyze the performance of the system frequency following the loss of a demand block.

9.7 *MATLAB example 11*: calculating the "response requirements" for a "high frequency excursion"

9.7.1 Introduction

By contrast with the preceding analysis that we did for a low frequency excursion event, the objective of this exercise is to calculate how much response must be held on system generators to cover for the loss of a block of load (demand). In general terms the size of a typical block of load is smaller in magnitude than the size of a typical supergrid connected generating set. The consequence of this is that typically when a block of load is lost from the system, the frequency deviates less from the starting frequency than when a major generating set is lost, albeit the change of frequency is in the opposite direction.

9.7.2 A statement of the high-frequency problem

Let us take the same basic scenario that we studied in the low-frequency event section: there are 50 identical generators available to run, each having a rated power output of 1000 MW and a rated power factor of 0.8 lagging. Assume in addition that the largest block of demand that can be lost from the system in a single event is 500 MW.

As before, let the system loading be 30 GW and the H (inertia) constant of each generator be in turn (1) 2.5 seconds, (2) 5.0 seconds, and (3) 1.25 seconds.

This time take two different high-frequency limits of (1) 50.5 and (2) 50.25 Hz, each of which is to be studied with the three different inertias given in the preceding section.

Finally, assume that all generators run at 100% of rated output prefault and that the "load-damping" coefficient α is again 0.02 pu/Hz (2%/Hz).

9.7.3 The general solution of the "high frequency excursion" problem

1. *The general method*

   In general terms the problem is solved by assuming a starting proportion of responsive generation (this time not de-loaded) and iterating until the amount required to allow the frequency to rise to just below the permitted level is obtained.

2. *The details*

   Again, starting from Eqs. (9.1) and (9.2) which we restate here, we have

$$\frac{df}{dt} = \frac{\Sigma P_{\mathrm{GEN(NR)}} + \Sigma P_{\mathrm{GEN}\ n(R)} - \Sigma P_{\mathrm{LOAD}\ n} - k_1 \cdot (f - f_n) - P_{\mathrm{TRIP}} \cdot u(t - t_{\mathrm{TRIP}}) + a_0 + a_1 t}{2 \cdot H \cdot f_0} \tag{9.52}$$

$$k_1 = k \cdot \Sigma P_{\mathrm{GEN}\ n(R)} + \alpha \cdot \Sigma P_{\mathrm{LOAD}\ n} \tag{9.53}$$

   which has the general solution

$$f(t) = A_1 + B_1 \cdot t + C_1 \cdot \exp(D_1 \cdot t) \tag{9.54}$$

   in which the parameters A_1, B_1, C_1, and D_1 are given by Eq. (6.35) and the equations immediately following it that were stated earlier in this chapter in Eq. (9.25) and the equations following it.

3. *A convenient assumption*

   Again, since the generator governor operates with a time delay of a few seconds, to avoid adding more complexity we can simulate an ideal "ramped response" using the "$a_0 + a_1\ t$" term in Eq. (9.1) while at the same time setting $k = 0$ in Eq. (9.2). To do this we need to introduce a *time delay of 2 seconds*.

   Using this approach, Eqs. (9.1) and (9.2) become

$$\frac{df}{dt} = \frac{\Sigma P_{\mathrm{GEN(NR)}} + \Sigma P_{\mathrm{GEN}\ n(R)} - \Sigma P_{\mathrm{LOAD}\ n} - k_1 \cdot (f - f_n) - P_{\mathrm{TRIP}} u(t) + a_0 + a_1 t}{2 \cdot H \cdot f_0} \tag{9.55}$$

   and

$$k_1 = \alpha \cdot \Sigma P_{\mathrm{LOAD}\ n} \tag{9.56}$$

Note that we have also simplified the formulation further by choosing $t_{TRIP}=0$, that is, we have assumed that the generation trip occurs at time $t=0$.

4. *The value of the "load-damping" coefficient α and its effect on rising frequency*

   We have again chosen the value $\alpha = 2\%/Hz$ for the *load-damping coefficient*. The effect of this parameter for rising frequencies is again stabilizing, as it was for falling frequencies. This can be seen in the form of the mathematical expression $\alpha \times (f - f_n)$ that models this effect. When $f < f_n$ (falling frequency), the load is reduced, which tends to increase the system frequency. On the other hand, when $f > f_n$ (rising frequency), the load is increased, which has the effect of reducing the system frequency. Therefore the action of α is to stabilize the system frequency in both cases.

5. *Initially balanced system*

   We shall again for the purposes of our exercise assume that the frequency when a block of load is lost from the system is close to nominal. This is a good starting point for our analysis.

6. *The number of sets required to be in "responsive mode"*

   We are aiming to find out how many generating sets are required to be in "responsive mode" to ensure that following a block of demand from the system that the system frequency is prevented from going too high. We remind the reader that when an amount of demand is lost from the system that initially the power system frequency will rise and therefore needs to be brought back down again. This process is achieved through the action of automatic speed governors on those generating units that are selected to be in "responsive mode." When the system frequency rises, the automatic speed governors act to pull it back down again after an initial time delay of about 2 seconds. As indicated earlier, we shall assume that this governing action operates on a straight-line profile in time.

7. *How the mathematics works out*

   In the case of high-frequency studies, we do not have the problem of de-loading that we must consider in low-frequency studies. In the illustrative examples for high frequency, we can justifiably start out with all our generating sets fully loaded to meet the system demand of 30 GW.

   The question to be answered remains, how many generating sets need to be placed in "responsive mode" to meet the stipulated frequency performance following the demand loss specified?

8. *The values of the ramp parameters a_0 and a_1*

   A clue to the answer to our question lies in the values that we assign to the parameters a_0 and a_1. The values of these parameters determine the "slope" and "set point" of the responsive profile, in other words how much responsive power is generated for a given change in the system frequency and at what frequency the responsive output is zero.

Since the responsive generators we are using are also to be used for low-frequency response when de-loaded, we can assume the same slope but a set point of 100% of rated output, that is, at rated output the responsive element is zero, but the slope is the same. We assume the same time delay of 2 seconds so that, as with the low-frequency analysis, we may take

$$a_0 = -50 \cdot n \cdot H \cdot \frac{f_0}{S_{sys}} \tag{9.57}$$

and

$$a_1 = 25 \cdot n \cdot H \cdot \frac{f_0}{S_{sys}} \tag{9.58}$$

as in Eqs. (9.42) and (9.43).

9. *The solution*

We use, as before, the solution for $f(t)$ given by Eq. (9.35)

$$f(t) = A_1 + B_1 \cdot t + C_1 \cdot \exp(D_1 \cdot t) \tag{9.59}$$

with once again the parameters A_1, B_1, C_1, and D_1 given by Eqs. (9.25)–(9.28) inclusive.

9.7.4 The input data

The input data for the problem are as stated in Section 9.7.2. As previously stated, for these high-frequency problems we are assuming that all generating sets are initially loaded to their rated output.

9.7.5 The output data

The output data for each of the subproblems are the number of generating sets that need to be put into "frequency-responsive" mode to ensure compliance with the frequency standards. Alternatively, the output data can be the total MW decrease on the responsive sets required to achieve this compliance objective.

9.7.6 The suggested MATLAB code

We shall use the "exact solution" of the "swing equation" method to help us to calculate the number of frequency-responsive sets required to meet the postfault frequency requirements.

It should be noted that the problem of finding the response requirements for a "high frequency excursion" is mathematically very similar to the problem of finding the response requirements for a "low frequency excursion." The essential differences are that "P_{TRIP}" is now negative rather than positive and the responsive "ramp" $a + a_1 \times t$ has now a *negative* rather than a *positive* slope. In addition the "responsive" generating sets are not in this case de-loaded prefault as they were in the "low frequency excursion" example.

A suggested example MATLAB code to solve the problem might be structured something like the following:

```
%    MATLAB® example simulation #11
%
%    Calculating the Minimum Response Requirements for the loss of a
     block of
%    load (demand)
%
%    Read in and calculate the input data
%
%    The input data is with reference to the main equations (Eq 9.36)
     and (Eq 9.37)
%    above. We again first read in the data in "physical units" and
     then convert
%    them to "per unit" quantities for the purposes of our subsequent
     calculations.
%
     N = input('Please enter the number of responsive machines to be
     considered')
%
     PLOAD_P = input('Please enter the total system load in MW')
%
%    Note: In our example the total system load is 30,000 MW.
%
     UNIT_SIZE_P = input('Please enter the rated generator size in MW')
%
%    Note: in our example UNIT_SIZE_P = 1000 MW.
%
     PR_P = UNIT_SIZE_P * N
     PNR_P = PLOAD_P - N * UNIT_SIZE_P
%
     PTRIP_P = input('Please enter the amount of generation to be
     tripped in MW')
     F0_P = input('Please enter the initial frequency in Hz')
%
     A0_P = input('Please enter the value of the constant a0 in MW')
     A1_P = input('Please enter the value of the constant a1 in MW/Hz')
%
     H_P = input('Please enter the value of the H-constant per
     machine in seconds')
```

```
ALPHA_P = input('Please enter the value of the "load damping"
constant α in
                per cent per Hz')
%
K1_P = ALPHA_P * PLOAD_P
%
%   Now convert the input data from "physical" to "per unit" values
MVA_BASE = PLOAD_P/0.8
FN_P = input('Please enter the nominal frequency in Hz')
F_BASE = FN_P
%
PLOAD = PLOAD_P/MVA_BASE
UNIT_SIZE = UNIT_SIZE_P/MVA_BASE
PR = PR_P/MVA_BASE
PNR = PNR_P/MVA_BASE
PTRIP = PTRIP_P/MVA_BASE
F0 = F0_P/F_BASE
A0 = A0_P/MVA_BASE
A1 = A1_P/MVA_BASE
%
%   Now calculate the constant K1 in "per unit":
%
K1 = K1_P * F_BASE/(100*MVA_BASE)
%
%   Now calculate the values of the constants A1, B1, C1and D1:
%
%   Calculate the parameter PBAL:
%
PBAL = A0 + PNR + PR - PLOAD - PTRIP
%
%   Now calculate the four constants:
%
A_1 = FN + PBAL/K1 - 2*H*F0*A1/K1^2
B_1 = A1/K1
C_1 = F0 - A_1
D_1 = - K1/(2*H*F0)
%
```

9.7.7 Instructions for the user

Run the MATLAB code earlier to generate a graph of frequency against time for successive choices of the amount of response held on the system. Terminate the iterations when the maximum frequency asked for has been achieved. Note the number of sets required to be in "frequency-responsive mode" to achieve this. This then constitutes the "high-frequency response requirement" for the given system parameters and the given loss of demand (load).

9.8 The "system response requirements" during "normal operation"

9.8.1 *[IV]* Calculation of the "system response requirements" during "normal operation" ("relatively small variations" in the "system frequency")

1. *What is "normal operation"?*

   By "normal operation" we mean that the "system frequency" is contained within a narrow band around the "nominal frequency." In other words, there are no "low frequency excursions" or "high frequency excursions."

   During "normal operation," relatively small changes to the system frequency occur due to load variation, minor switching events, and the like. There are no large-scale changes to the system.

2. *Calculating the "system response requirements" during "normal operation" (relatively small variations in the "system frequency")*

   Calculating the "system response requirements" during "normal operation" (relatively small variations in the "system frequency") may be achieved by using the "swing equation" to model sources of generation and demand including "frequency-responsive plant" and iterating one of the solution methods in Chapter 6, Chapter 7 or Chapter 8 following a *selected small size of loss of generation or demand* to find the amount of "frequency response" needed to maintain the "system frequency" both above the low voltage tolerance and below the *high voltage* tolerance defined by the "normal band of operation" around the "nominal frequency."

9.8.2 A procedure for calculating the "response requirements" during "normal operation"

There are several ways potentially of calculating the amount of response to be held for "normal operation," but the one to be described in this book will follow the same logic of the calculations of response for low and high frequency excursions described earlier.

We shall consider the bands of "normal operational frequency" that straddle the nominal frequency (see Diagram 5.1) and ask ourselves the question, "how much response needs to be held to contain the frequency to within this double band?."

We shall try to answer this question by again using the "exact solution" of the "swing equation" given by Eq. (9.47) and its associated equations and calculating the amount of response to be held on this basis.

9.9 *MATLAB example 12*: calculating the "response requirements" during "normal operation"

9.9.1 Introduction

1. *How large are "small" changes in the system power balance?*
   To be able to calculate the response requirements during "normal operation" by our standard method, we need first to know just what size the regular "small" changes in the system power balance are.
2. *The complexity of measurement*
   Direct measurement is always difficult from the "system operator's" perspective because load changes on the system are mixed up with continuous and automatic changes from generator governor systems and general noise from switching and continuous load variations.
3. *A possible solution*
   A potential way to measure actual changes to loads and power outputs on the system may be to perform a Fourier decomposition of the power balance profile. Then, knowing the components from the Fourier decomposition, various statistical techniques could be employed to produce such data as means and standard deviations. Then, for example, the mean change to the power balance could be used as a representative of a "typical change" during normal operation.
4. *An alternative approach*
   An alternative approach is simply to make an estimate of a typical power balance change during "normal operation" based on one's experience of one's own system. This method is more likely to be used in practice.

9.9.2 A statement of the problem

Supplied with a known or estimated "typical power change" during "normal operation" on our power system of interest, how much frequency response (if any) is needed to keep the system power frequency during "normal operation" between defined operational limits, either absolutely or for a prescribed minimum percentage of the time?

9.9.3 The general solution of the problem

A straightforward approach using the "exact solution" of the "swing equation" method is to take the representative typical operational power change; however, this has been calculated, to define the trip of an equivalent generator for the lower frequency limit and the negative of it for the trip of an equivalent block of demand for the upper frequency limit.

9.9.4 The input data

1. *Solution method*

   A system with a given size and inertia with given generating machine sizes is supplied, the parameters of which form elements of the data needed to define the input data to the "exact solution" of the "swing equation" method in the usual way.
2. *The "normal band of operation"*

   An additional, and vitally, part of the input data is the information concerning the assumed "normal band of operation" for the system frequency. This may be, for example, *±0.4%* of the power frequency (which happens to be the figure that has been used in Britain in the past).

   Let the nominal frequency of the system be *50 Hz*, so that with the foregoing definition, the band of acceptable frequency operationally will be *49.8–50.2 Hz* in the absence of any "faults" or "events" on the system.
3. *The assumed typical load (demand) variation*

   Assume for the purposes of this example that the typical range for the load (demand) variation is *±250 MW*.
4. *Other input data*

   For the rest of the input data in this example, we shall choose the network data that we used for *MATLAB example simulation #1* given in Chapter 6, Analytical methods for solving the "swing equation", together with some minor modifications, as follows:

   The input data in per unit:

   a. $\Sigma P_{GEN(NR)} = 3000\ MW/3750\ MVA = 0.8$ pu
   b. $\Sigma P_{GEN\ n(R)} = 0$ pu
   c. $\Sigma P_{LOAD\ n} = 3000\ MW/3750\ MVA = 0.8$ pu
   d. $k_1 = 60\ MW/Hz \times 50\ Hz/3750\ MVA = 0.8$ pu
   e. $f_n = 50/50\ Hz = 1.0$ pu
   f. $P_{TRIP} = \pm 250\ MW/3750\ MVA = \pm 0.06667$ pu
   g. $a_0 = 0$
   h. $a_1 = 0$
   i. $H = 5$ seconds(the average H for the system)
   j. $f_0 = 50/50\ Hz = 1.0$ pu

   Note 1: Response is modeled by choosing nonzero values for a_0 and a_1 as needed. A positive value of a_1 is required when P_{TRIP} is positive, and a negative value of a_1 is required when P_{TRIP} is negative. These choices correspond to positive and negative slopes for the responsive "ramp."

   Note 2: Different values of the inertia H may be chosen to see the effect on the resulting response requirements.

9.9.5 The main output data

The main output data will be (1) the amount of response required to cover a load (demand) increase of 250 MW and (2) the amount of response required to cover a load decrease of 250 MW, in both cases keeping the system frequency within the acceptable range defined earlier.

9.9.6 The suggested *MATLAB* code

We suggest the following *MATLAB* code for solving the problem of how much response, both positive and negative, is needed during the period of "normal operation" of the grid system.

```
% Inputting the data for the problem in "per unit" :
    PNR = input('Please enter the non-responsive generation in per unit')
    PR = input('Please enter the responsive generation in per unit')
    PL = input('Please enter the system load in per unit')
    FN = input('Please enter the nominal frequency in per unit')
    PTRIP = input('Please enter the generation tripped in per unit')
    A_0 = input('Please enter the value of the constant a0in per unit')
    A_1 = input('Please enter the value of the constant a1in per unit')
    H = input('Please enter the value of the H-constant in seconds')
    F0 = input('Please enter the value of the initial frequency in per
          unit')
% Calculate the power balance at nominal frequency as a parameter:
    PBAL = PNR + PR - PL - PTRIP
% Calculate the value of the parameter k1:
    K1 = 0.02*FN*PL
% Calculate the values of the constants A1, B1, C1and D1:
    A1 = FN + (A_0 + PBAL)/K1
    B1 = A_1/K1
    C1 = F0 - (A_0 + PBAL)/K1 + 2*H*F0*A_1/(K1^2)
    D1 = -K1/(2*H*F0)
% Calculate the solution for the frequency f at a sequence of values
  of t from 0 to 10
% seconds at intervals of 0.1 second:
    for k = 1,100
        t = k/10
        f(k) = A1 + B1*t + C1*EXP(D1*t)
    end
% Note that the entries of the vector f are in "per unit" and so need to
  be modified to
% be in Hz.
```

9.9.7 General instructions for the user

Run the given MATLAB code with the data suggested above the code listing given earlier. Run the program twice, once with P_{TRIP} positive and once with P_{TRIP} negative and plot the resulting frequency results against time to see if the frequency stays within the required band about the nominal frequency or not. Remember to choose appropriate values of a_0 and a_1 to provide appropriate response in each case. Find the best values to use to just satisfy the requirements.

9.10 The next chapter: the challenges of operating systems with high penetrations of renewables (low-inertia systems)

In the next chapter, we shall be looking at the issues that are encountered when there is not much inertia on the system. This can be caused, for example, by high penetrations of renewable energy sources that have little or no inertia of their own. We shall look at some possible options for addressing this increasingly common operational situation.

Chapter 10

Challenges of operating systems with high penetrations of renewables (low-inertia systems)

10.1 An introduction to low-inertia systems

10.1.1 The original concept of the alternating current transmission system

When large-scale "alternating current (AC)" transmission networks, such as that in Britain, were first designed and built, it was, perhaps with every justification at that time, not foreseen that we would eventually have the type of power systems that we have today, with their many and varied types of power source that were unknown on the system in the 1950s and 1960s when these large-scale "AC" transmission systems were first being developed in Britain.

Examples of some of the more important innovations to transmission systems today include numerous high-voltage direct current (HVDC) links, wind turbines, and solar photovoltaic (solar PV) cell technologies. We shall focus on these three major technologies in this chapter of the book.

10.1.2 The developing challenge of the system inertia

When AC transmission systems were first designed, it was taken for granted that the great majority, if not all, of the generators feeding the system would be "AC" machines with massive, rotating parts that provided, among other characteristics, adequate *inertia* to the system to assist in counteracting various system disturbances. One might say, with some justification, that "AC" generating machines and "AC" power systems went "hand in hand."

Modern Aspects of Power System Frequency Stability and Control.
DOI: https://doi.org/10.1016/B978-0-12-816139-5.00010-2

10.1.3 The frequency stability of "alternating current" machines in power systems

The frequency stability of "AC" machines under system disturbance conditions depends mainly on following two factors:

1. the ability of selected machines to replace power that has been lost from the system due to the tripping of large generators
   and
2. the total inertia of the rotating machines connected to the system.

The first of these characteristics can be provided by just about any sort of electrical generator, given suitable controls in place, and even by batteries, but the second characteristic can only be provided (in its true physical sense) by machines possessing massive rotating parts.

However, both these characteristics are needed to ensure the frequency stability of the system. Indeed, it has been shown in system studies that, for a given total system demand, there is a lower limit of "system inertia" below which a power system will become transiently unstable when the system is disturbed (see, for example, Ref. [1]).

10.1.4 The "erosion" of inertia from power systems

In the case of some countries (e.g., most of those in mainland Europe), a lack of system inertia is not felt because the interconnections with the power systems of surrounding systems are so strong that frequency control is not a major concern. These connections are strong because they are "AC." However, if these connections were "DC" they would not be as strong.

However, in more isolated geographical regions, such as Britain or Ireland, for example, these systems are less strong (=have a lower "fault level"), and frequency control is correspondingly more important. In these latter countries, therefore, the lowering or "erosion" of system inertia is more of a challenge.

10.1.5 The closure of "fossil-fuel" power stations

In some of the more electrically isolated countries the closure of "fossil-fuel" power stations that are subsequently replaced in the generation schedule by renewable sources that have no inertia in the traditional sense of the word has become more prevalent in recent years. Thus there has been a need to find ways to compensate for what could be termed "inertia erosion."

10.1.6 Synthetic inertia

One of the proposed approaches to "inertia erosion" has been the proposal of a concept known as "synthetic inertia." As the name suggests, this concept refers to the *simulation* of inertia by electronic means in the absence of actual real inertia (see, for example, Ref. [2]).

If this idea works in practice, it will have to be as good as inertia in the traditional meaning of the word, and thus, among other things, it must ensure the frequency stability of the more isolated power systems, such as those of Britain or Ireland, if used there.

10.1.7 Fast frequency response

Another recent concept is "fast frequency response (FFR)" (see, for example, Ref. [3]). It is not the same as "synthetic inertia" although the two are easily confused (for disambiguation see, for example, Ref. [4]).

The basic idea of "FFR" is to respond very quickly by electronic means in microsecond timescales to compensate in MW terms for the unplanned loss of generation from a power system. For these purposes a device, such as an "HVDC" link, may be used if supported by a suitable energy source.

A possible drawback associated with using FFR is that if the speed of response is not calculated appropriately for the location on the system then instabilities such as small-signal instability could be triggered.

10.1.8 The "virtual synchronous machine"

The "virtual synchronous machine" (VSM) is a concept related to the idea of "synthetic inertia" that was mentioned in Section 10.1.6. The basic idea here is that the overall properties of a "synchronous machine," such as power output, fault current, governor "droop," "ramp rate," and inertia, are simulated in such a way so that from the terminals of the "VSM" its behavior reproduces almost exactly that of a "synchronous machine." Research work is currently being undertaken in this area (see, for example, Ref. [5]).

10.1.9 Energy storage devices

"Energy storage devices," such as HV-connected batteries, are being increasingly used in grid systems. They can be used to provide frequency response by providing them with a suitable controller.

A quite recent research idea has been the coupling of a doubly fed induction generator (DFIG)-type wind turbine with lithium-ion supercapacitors to provide a means of control of "virtual inertia" (see Ref. [6] for more details on this).

10.1.10 In the next section

In the next section, we shall examine in more detail the case of the addition of multiple "HVDC" links to a power system and the consequences for system inertia and for the frequency stability of the system (if any).

10.2 Low-inertia systems: the modeling of high-voltage direct current links in system frequency studies

10.2.1 An introduction to "high-voltage direct current links"

1. *A definition of an "HVDC link"*

   An "HVDC link" may be defined as a means of connecting two "AC" systems by means of "DC" cables with two "AC" to "DC" converter stations, there being one at each end of the link.
2. *The economy of "HVDC links"*

   "HVDC links" are more economical than "AC" connections for connections over longer distances. Therefore they are invariably used for international links or "interconnectors" and in larger countries, such as the United States, are also used internally.
3. *External "HVDC links"*

   External "HVDC" links are those from one bulk system to another, for example, those from Britain to France or that from Britain to the Netherlands (consult Ref. [7] for more details on these links).
4. *Internal "HVDC links"*

   Internal "HVDC links" are those that connect locations, both of which are situated in the same bulk system. There are examples of these in the United States where distances between demand centers can be large compared with those in Western Europe.
5. *Monopolar "HVDC links"*

   A monopolar "HVDC link" connects the converters at each end of the link with one "DC" cable only. This type of link is not commonly used nowadays.
6. *Bipolar "HVDC links"*

   The majority of "HVDC links" are of the bipolar kind. In this kind of link, there are two "DC" cables that are arranged in parallel, each connecting the two converter stations.

10.2.2 The effect of "high-voltage direct current links" on the "system inertia"

1. *The status of an "HVDC link" in the "intact system"*

   In the "intact system," that is, before any fault outages have occurred, an "HVDC link" appears from one end either as a generation infeed or as a demand. In the case of "external" links a generation infeed in this respect is termed "importing" and conversely if the link appears as a demand it is termed "exporting."
2. *The effects of the tripping of an "importing" "HVDC link"*

   The effect of the loss of an "importing" "HVDC link" is numerically the same as losing the MW output of a conventional generator, but no inertia is lost since the "HVDC" link does not have any to lose.

3. *The effects of the tripping of an "exporting" "HVDC link"*

   The effect of the loss of an "exporting" "HVDC link" is numerically the same as losing a block of demand. Again, there is no change in the total inertia of the system since again none is involved.

10.2.3 The provision of "frequency response" by "high-voltage direct current links" themselves

The control systems of "HVDC links" can be programmed to act in a frequency responsive mode and indeed there are real examples of their being contracted to operate in just this way. This works in practice in the following way.

1. *The "importing" "HVDC link"*

   In the case of an "importing" "HVDC link," when some generation in the system is lost, accompanied by the usual fall in the system frequency, the link can be set to "overshoot" by a certain percentage of its rated import amount, typically by something like 5% or 10%. In this way it has provided "frequency response" in MW terms although the increase is very fast compared with the ramped response of synchronous generation. This process is of course equivalent in commercial terms by buying more MWs from the provider at the "exporting" end.
2. *The "exporting" "HVDC link"*

   In the case of an "exporting" "HVDC link," when a block of load is lost from the system instead, accompanied by a rise in the system frequency, the link can be set to "undershoot" by a similar amount, thus providing "frequency response" for a loss of load.

10.2.4 In the next section

In the next section, we shall go through a MATLAB simulation to demonstrate the effect of an "HVDC link" on the "system frequency" when the grid system is disturbed.

10.3 *MATLAB example simulation #13*: the effect of "high-voltage direct current links" on the system frequency

10.3.1 Introduction

10.3.1.1 Importing and exporting cases

In this example, we shall investigate both the effect of (1) losing an "importing link" and (2) losing an "exporting link."

10.3.1.2 The "importing link": a summary

In the case of a transmission system that is fed by an "importing link," power is being injected into the system from outside. This power does not add any inertia to the transmission system since any inertia associated with it resides in the external system. Hence, when the "importing link" is interrupted the net effect on the "home" transmission system is one of losing an amount of generated power but not inertia.

10.3.1.3 The "exporting link": a summary

In the case of a transmission system that is fed by an "exporting link," power is generated from within the transmission system but transferred to the external system. This power is provided by generators from within the transmission system, some of which possess inertia. When an "exporting link" is interrupted the net effect on the transmission system is one of losing an amount of load (demand) but not inertia.

10.3.2 A statement of the problem

Let us consider a system of 30,000 MW demand operating at 50 Hz which incorporates *two HVDC links: one importing of size* 1000 MW and *one exporting of size* 1000 MW. Let the average inertia constant of the transmission system in either case be 4.0 seconds.

What is the effect on the system frequency when if (1) the "importing link" alone is lost or (2) the "exporting link" alone is lost? Assume that the initial frequency is 50 Hz. Assume also that no response is provided and plot the time evolution of the frequency for the first 5 seconds.

10.3.3 The general solution of the problem

The general solution to the problem is to use any of the methods described earlier in the book to calculate the time evolution of the "system frequency" in each case. For clarity and convenience, we shall use the "direct solution of the differential equation" method described in the first part of Chapter 6, Analytical solutions, to solve the problem. For ease of reference we repeat the relevant equations here:

$$\begin{aligned} &\Sigma P_{\mathrm{GEN(NR)}} + \Sigma P_{\mathrm{GEN}\ n\mathrm{(R)}} - \Sigma P_{\mathrm{LOAD}\ n} - k_1 \cdot (f - f_n) \\ &- P_{\mathrm{TRIP}} \cdot u(t - t_{\mathrm{TRIP}}) + a_0 + a_1 t = 2 \cdot H \cdot f_0 \cdot \frac{\mathrm{d}f}{\mathrm{d}t} \end{aligned} \tag{10.1}$$

Eq. (10.1) is the basic form of the "swing equation" that we shall solve. The parameter k_1 is given by

$$k_1 = \Sigma P_{\text{LOAD } n} \cdot \alpha \cdot + \Sigma P_{\text{GEN } n(\text{R})} \cdot k \quad (10.2)$$

where α is the "load-damping" constant we shall assume to have the value 2% per Hz and k is the constant associated with the "droop" of the speed governor of responsive synchronous plant. Since we shall not be considering response in this example, the second term on the right-hand side of Eq. (10.2) we shall take to be zero.

From Chapter 6, Analytical solutions, the solution of Eq. (10.1) with Eq. (10.2) is given by

$$f(t) = A_1 + B_1 \cdot t + C_1 \cdot \exp(D_1 \cdot t) \quad (10.3)$$

where

$$A_1 = f_n + \frac{\Sigma P_{\text{GEN(NR)}} + \Sigma P_{\text{GEN } n(\text{R})} - \Sigma P_{\text{LOAD } n} + a_0}{k_1} - \frac{2 \cdot H \cdot f_0 a_1}{k_1^2} \quad (10.4)$$

$$B_1 = \frac{a_1}{k_1} \quad (10.5)$$

$$C_1 = f_0 - f_n - \frac{\Sigma P_{\text{GEN(NR)}} + \Sigma P_{\text{GEN } n(\text{R})} - \Sigma P_{\text{LOAD } n} + a_0}{k_1} + \frac{2 \cdot H \cdot f_0 \cdot a_1}{k_1^2} \quad (10.6)$$

and

$$D_1 = -\frac{k_1}{2 \cdot H \cdot f_0} \quad (10.7)$$

are the expressions for the four coefficients that appear in the solution.

10.3.4 The input data

The input data for the problem are as follows:

1. *Loss of the "importing interconnector"*
   a. *Input data in "physical units"*

      The input data for the "importing interconnector" case in physical units are

      $\Sigma P_{\text{GEN(NR)}} = 30,000$ MW; $\Sigma P_{\text{GEN } n(\text{R})} = 0$ MW; $\Sigma P_{\text{LOAD } n} = 30,000$ MW; $P_{\text{TRIP}} = 1000$ MW; $t_{\text{TRIP}} = 0$ second; $a_0 = 0$; $a_1 = 0$; $H = 4.0$ seconds; $f_0 = 50$ Hz

   b. *Input data in "per unit" quantities*

      We need to convert the foregoing data to "per unit," so that Eqs. (10.3)–(10.7) are correct. To accomplish this, we need to define some "base quantities." A suitable MVA base is $S_{\text{base}} = 30,000/0.8 = 37,500$ MVA, and a suitable frequency base is 50 Hz. Using these bases, we obtain the following "per unit" values for the input data for the "importing interconnector" loss case:

$\Sigma P_{GEN(NR)} = 0.8$ pu; $\Sigma P_{GEN\ n(R)} = 0$; $\Sigma P_{LOAD\ n} = 0.8$ pu; $P_{TRIP} = 0.0333$ pu; $t_{TRIP} = 0$ second; $a_0 = 0$; $a_1 = 0$; $H = 4.0$ seconds; $f_0 = 1.0$ pu

2. *Loss of the "exporting interconnector"*
   The input data for the "exporting interconnector" case are the same as the data for the "importing interconnector" case except for P_{TRIP}. For the "exporting interconnector" case we have $P_{TRIP} = -0.0333$ pu which is of the opposite sign since the loss of the "exporting interconnector" is equivalent to the loss of a load.

10.3.5 The output data

In both cases the output data will consist of the solution *f*(*t*) of the equation. We shall plot the results of "system frequency" *f* against time *t* for the first 5 seconds following the loss of the interconnector.

10.3.6 The suggested *MATLAB* code

We suggest the following *MATLAB* code to examine the above cases of (1) loss of an "importing interconnector" and (2) loss of an "exporting interconnector." Both cases may be analyzed using the straightforward coding method used in *MATLAB example simulation #1* with the parameters changed as necessary to suit our needs here.

```
%       Loss of (a) importing, and (b) exporting interconnectors
%
%       The input data in per unit:
                (i)     Σ P_GEN(NR) = 30,000 MW / 37500 MVA = 0.8 pu
                (ii)    Σ PGEN n (R) = 0 pu
                (iii)   Σ PLOAD n = 30,000 MW / 3750 MVA = 0.8 pu
                (iv)    fn = 50 Hz / 50 Hz = 1.0 pu
                (v)     PTRIP = 1000 MW / 37500 MVA = 0.02667 pu
                (vi)    a0 = 0
                (vii)   a1 = 0
                (viii)  H = 4.0 s (the average H for the system)
                (ix)    f0 = 50 Hz / 50 Hz = 1.0 pu
% Inputting the data for the problem in "per unit" :
  PNR = input('Please enter the non-responsive generation in per unit')
  PR = input('Please enter the responsive generation in per unit')
  PL = input('Please enter the system load in per unit')
  FN = input('Please enter the nominal frequency in per unit')
  PTRIP = input('Please enter the generation tripped in per unit')
  A_0 = input('Please enter the value of the constant a0 in per unit')
```

```
    A_1 = input('Please enter the value of the constant a1 in per unit')
    H = input('Please enter the value of the H-constant in seconds')
    F0=input('Please enter the value of the initial frequency in per unit')
% Calculate the power balance at nominal frequency as a parameter:
    PBAL = PNR + PR – PL – PTRIP
% Calculate the value of the parameter k1:
    K1 = 0.02*FN*PL
% Calculate the values of the constants A1, B1, C1 and D1:
    A1 = FN + (A_0 + PBAL)/K1
    B1 = A_1/K1
    C1 = F0 – (A_0 + PBAL)/K1 + 2*H*F0*A_1/(K1^2)
    D1 = -K1/(2*H*F0)
% Calculate the solution for the frequency f at a sequence of values
of t from 0 to 10
% seconds at intervals of 0.1 second:
    for k = 1,100
        t = k/10
        f(k) = A1 + B1*t + C1*EXP(D1*t)
    end
% Note that the entries of the vector f are in "per unit" and so need to
be modified to
% be in Hz if this is desired when plotting the output.
```

10.3.7 Instructions for the user

The user should input the relevant data in "per unit" for each case when prompted by the program. A graph of the output may be plotted. Note that in case (a) "importing interconnector" T_{TRIP} is positive and in case (b) "exporting interconnector" T_{TRIP} is negative.

10.4 Low-inertia systems: the modeling of "wind farms" in system frequency studies

10.4.1 An introduction to "wind farms"

1. *A definition of a "wind farm"*

   A "wind farm" is an arrangement in which multiple wind turbine generators (WTGs) are constructed close to one another and pool their MW outputs to a combined connection point onto the HV power grid (Diagram 10.1).
2. *Types of WTGs*

   WTGs come in many different shapes and sizes. Some are mounted horizontally, and some are mounted vertically. However, a useful classification of the main types for our purposes here relating to the study of system frequency is given in Ref. [8].

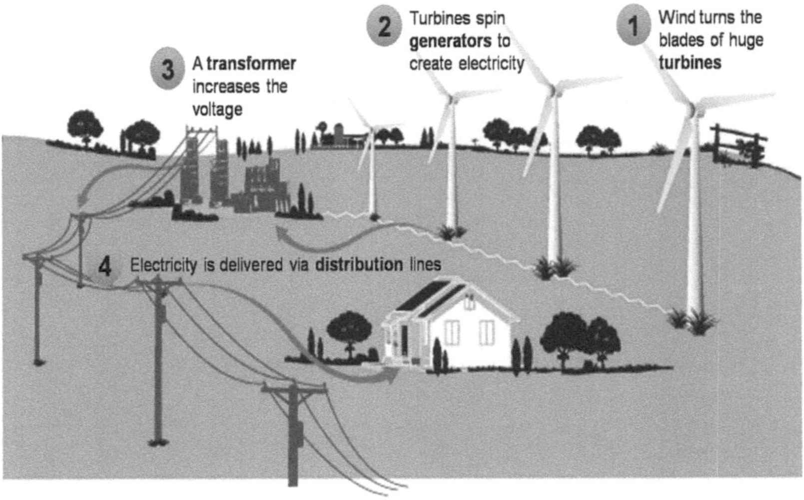

DIAGRAM 10.1 Schematic diagram of a wind farm. *elsalvadorla.org.*

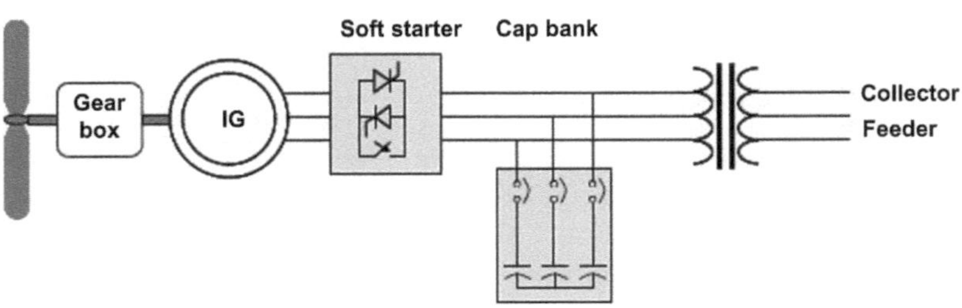

DIAGRAM 10.2 Schematic diagram of a Type 1 wind turbine generator. *IEEE PES Wind Plant Collector System Design Working Group.*

We can summarize the main types as follows:

a. *Limited variable speed (Type 1)*

This type of WTG is a relatively simple design in which the turbine blades are connected via a short shaft through a gearbox to a simple induction generator and then basically connected after that in a straightforward way to the grid system (Diagram 10.2).

The induction generator is an asynchronous rotating-shaft machine that generates electricity onto the grid whenever its rotor speed is greater than the electrical frequency of the system. A capacitor bank is included in this design to compensate for the reactive power (Mvar) absorption from the grid at all speeds of operation associated with this kind of machine.

b. *Limited variable speed (Type 2)*

This type of WTG has a design basically the same as that of Type 1 of the preceding section except a variable resistor is also included in the rotor circuit (Diagram 10.3).

The effect of the addition of the variable resistor is to modify the operating characteristics of the machine, so that the machine can deliver nearly constant power output even during gusty wind conditions and can also improve the machine's performance during disturbances on the grid side.

c. *Variable speed with partial power electronics conversion (Type 3)*

This type of WTG is known as the "DFIG." In this design the variable resistor of the Type 2 design of the previous section has been replaced by variable frequency AC excitation provided by a back-to-back power electronic converter connected from the rotor up to the grid (Diagram 10.4).

The converters provide much greater control of the output than with the Type 2 design. The "DFIG" can operate in two modes

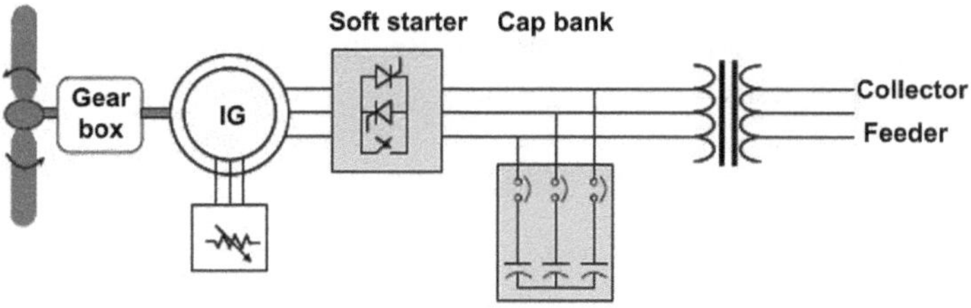

DIAGRAM 10.3 Schematic diagram of a Type 2 wind turbine generator. *IEEE PES Wind Plant Collector System Design Working Group.*

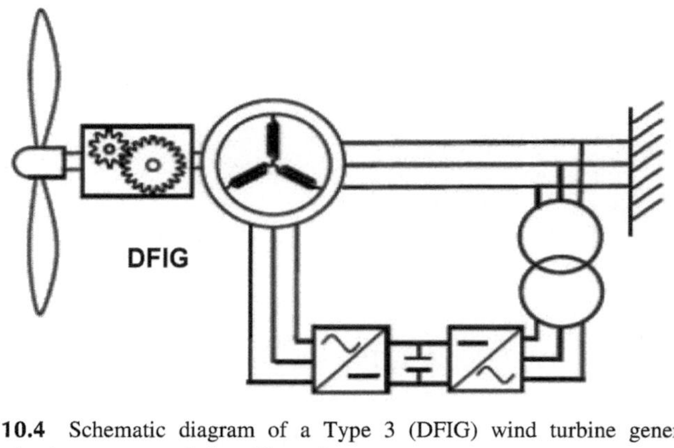

DIAGRAM 10.4 Schematic diagram of a Type 3 (DFIG) wind turbine generator. *DFIG*, Doubly fed induction generator. *kalyan07.wordpress.com.*

according to whether its rotor speed is greater than the grid frequency (generating mode) or less than the grid frequency (motoring mode).

d. *Variable speed with full power electronics conversion (Type 4)*

This type of WTG has great flexibility in both design and operation as the machine is connected entirely in series through back-to-back converters to the grid (Diagram 10.5).

The machine used can be either a synchronous machine like that seen in hydroelectric plants or an asynchronous "squirrel cage" induction generator. Also, the machine may be constructed with or without a gearbox.

e. *Speed/torque converter plus synchronous generator (Type 5)*

This type of WTG uses a variable speed drive train in series with a speed/torque converter to deliver a constant output shaft speed to a synchronous generator in series with the grid (Diagram 10.6).

The synchronous generator component can be designed to operate at any desired speed and voltage.

f. For the purposes of assessing frequency control, we are interested principally in whether the inertia of the rotating mass is isolated from the main transmission system by power electronic converters or not. In the next two sections, we therefore consider this aspect in relation to the five designs described earlier.

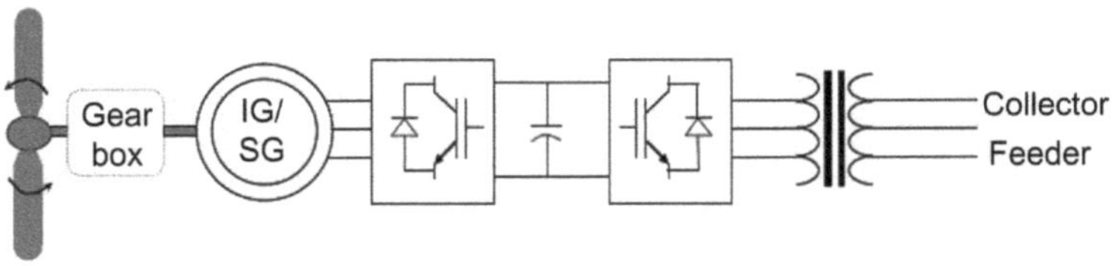

DIAGRAM 10.5 Schematic diagram of a Type 4 wind turbine generator. *IEEE PES Wind Plant Collector System Design Working Group.*

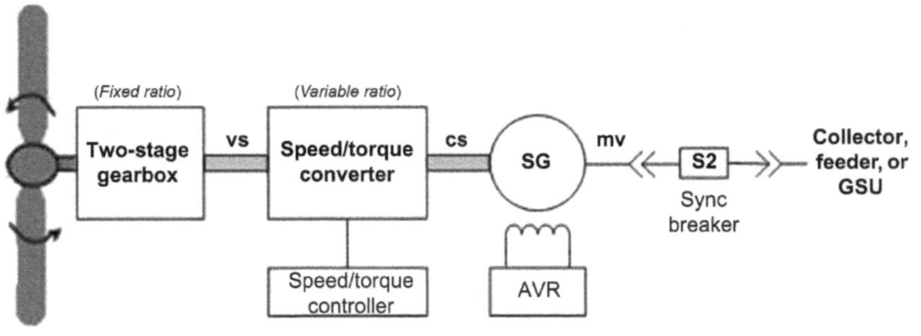

DIAGRAM 10.6 Schematic diagram of a Type 5 wind turbine generator. *cs*, Constant speed; *mv*, medium voltage; *vs*, variable speed. *IEEE PES Wind Plant Collector System Design Working Group.*

3. *"Nonconverter-based" WTGs*

This category includes the Type 1 and Type 2 WTGs described earlier. Therefore the inertia of the rotors of the machines of these two types is not "cutoff" from the grid and therefore *does* contribute to the total inertia of the whole system. This inertia helps in stabilizing the grid frequency during a disturbance or fault.

4. *"Converter-based" WTGs*

This group includes the remaining categories, *Type 3*, *Type 4*, and *Type 5*.

a. *Type 3—the "DFIG"*

Although *Type 3*, the "DFIG," is a hybrid system with feeds both directly from the system and through a converter, nevertheless the converter controls the frequency of the rotor of the induction machine. Within the design range of the converter, therefore, the inertia of the rotor of the machine is "decoupled from the grid."

b. *Type 4—the "full converter"*

Type 4—the "full converter" type—has the generator completely in series with the grid connection terminal, and because of this, within the useable range of the converters, there is no coupling of the inertia of the rotor with the grid. Therefore the inertia of this kind of machine is not useful for the control of frequency on the grid.

c. *Type 5—the "speed/torque converter type"*

In *Type 5* which has a speed/torque converter between the rotor/gearbox on one side and the synchronous generator through to the grid system on the other, there is a decoupling of the inertia of the blades and drive train from the grid side because the speed/torque converter maintains the shaft of the synchronous machine at constant speed (within the operating range of the converter). Therefore most of the inertia of this kind of "WTG" is decoupled from the grid and cannot help to stabilize the system in the event of a system disturbance or fault.

10.4.2 The effect of "wind farms" on the "system inertia"

1. *The status of "wind farms" in the intact system*

In the "steady state" the power output of a "wind farm" in general behaves like that of any other generator except it is more variable because of the ever-present changes in the speed of the wind. Since a wind turbine cannot function when the wind speed is either too low or too high, it is designed to "cut out" when either of these conditions is present.

2. *The effects of tripping a "nonconverter-based" "wind farm"*

The effects of tripping a "nonconverter-based" "wind farm" are that the MW output of the farm is lost, and in addition, its inertia becomes unavailable to the grid system. The loss of both these items is seen by the control center.

3. *The effects of tripping a "converter-based" "wind farm"*
   The effects of tripping a "converter-based" "wind farm" is that, since the machines' inertias were unavailable to the grid system in the first place, only a loss of MW output is seen by the control center.

10.4.3 The provision of "frequency response" by "wind farms" themselves

Wind farms composed of converter-based WTGs such as *Type 3*, the "DFIG," have the capacity to be programmed to provide frequency response (see Ref. [9] for more details on this). They can be made to respond in a similar way to synchronous machines in terms of the change in MW output when there is a sudden, unplanned fall or rise of the system frequency, but they have no inertia in the conventional sense of the word. Therefore local stabilization of the postfault frequency must necessarily be achieved in other ways.

10.4.4 In the next section

In the next section, we look at the effect on the "system frequency" by looking at a simple example of how the response of a power system changes under conditions of the increased penetration of "wind farms."

10.5 MATLAB example simulation #14: the effect of "wind farms" on the "system frequency"

10.5.1 Introduction

1. *Losing a "wind farm" from the system*
   From the viewpoint of the "grid," that is, the transmission system, a "wind farm" appears to be a large generator whose power output is constantly changing with time. If we imagine there to be a fault on the transformer that connects the wind farm to the "grid," the whole "wind farm" could be unexpectedly cutoff from the "grid." The effect on the "system frequency" would be the same in its effect as losing a large generator.
2. *The objective of the simulation*
   In this exercise, we shall simulate the loss of a wind farm that is composed of convertor-based turbines. Thus as discussed earlier, losing the wind farm entails the loss of generated MW but not the loss of any inertia due to the shielding effect of the back-to-back AC–DC converters.

10.5.2 A statement of the problem

1. *Details of the system*
   Let us consider a 50 Hz system of total demand (load) 30,000 MW with an average *H-constant* of 4.0 seconds as in *MATLAB example*

simulation #13. Take the parameters from the "importing interconnector" case of that simulation, except for the initial (starting) frequency and the size of the loss. This time we are going to have a starting frequency of 49.95 Hz rather than the 50.0 Hz of that example. Also, this time we are going to take the size of the loss to be 1500 MW, rather than 1000 MW. This, for example, might represent a wind farm composed of 50 turbines, each of 30 MW.

2. *Two cases*

We shall look at two cases:

a. No response is made available.

b. 1000 MW of response is made available such that the response *ramps up linearly from 2 to 1000 MW in 10 seconds.* Model this ramp-up using the parameters a_0 and a_1 (general responsive profile).

10.5.3 The general solution of the problem

1. *The solution method*

We could address the problem with any of the methods discussed in Chapter 6, 7, or 8, The "control diagram" method, but this time we shall use the "direct solution of the differential equation" method described in Chapter 7, Numerical methods for solving the "swing equation," as in our previous example.

2. *Case (b): the two periods*

The solution should be divided into two periods: (1) $0 \leq t < 2$ seconds and (2) $2 \leq t \leq 10$ seconds. This is to cover the two periods: (1) before the responsive ramp has started its output and (2) after the responsive ramp has started its output.

3. *Values of the parameters a_0 and a_1*

To run the program, we need to know values for the two ramp parameters a_0 and a_1. As discussed earlier, these two parameters are to describe a power ramp that has *zero* output from $t=0$ to 2 seconds and then has a linear increase from *zero* to 1000 MW in the remaining 8 seconds.

We shall calculate a_0 and a_1 in (a) physical units and (b) "per unit."

a. *a_0 and a_1 in "physical units"*

We shall model the period from 0 to 2 seconds with $a_0 = a_1 = 0$.

This ignores the so-called inertial response and results in the response requirement calculated being somewhat of an overestimate. However, this simplifies our calculation.

In the period from 2 to 10 seconds, we have the following formula:

$$P_{\text{resp}} = a_0 + a_1 t \tag{10.8}$$

Applying in turn the boundary conditions at $t = 2$ and 10 seconds, we obtain the simultaneous equations:

$$0 = a_0 + 2 \cdot a_1 \tag{10.9}$$

and

$$1000 = a_0 + 10 \cdot a_1 \tag{10.10}$$

Solving the above pair of equations gives

$$a_0 = -250 \text{ MW} \tag{10.11}$$

and

$$a_1 = 125 \text{ MW}/s \tag{10.12}$$

b. *a_0 and a_1 in "per unit"*

To convert a_0 and a_1 to "per unit" values, we divide each of them by the MVA base 30,000/0.8 = 37,500 MVA to get

$$a_0 = -\frac{250}{37,500} = -6.667 \times 10^{-3} \text{ pu} \tag{10.13}$$

$$a_1 = \frac{125}{37,500} \text{ s}^{-1} = 3.333 \times 10^{-3} \text{ pu}/s \tag{10.14}$$

10.5.4 The complete input data

The complete input data are given as follows:

1. *The input data in "physical units"*
   a. $0 \le t < 2$ seconds

   $\Sigma P_{\text{GEN(NR)}} = 30,000$ MW; $\Sigma P_{\text{GEN }n\text{(R)}} = 0$ MW; $\Sigma P_{\text{LOAD }n} = 30,000$ MW; $P_{\text{TRIP}} = 1000$ MW; $t_{\text{TRIP}} = 0$ second; $a_0 = 0$; $a_1 = 0$; $H = 4.0$ seconds; $f_0 = 50$ Hz

   b. $2 \le t \le 10$ seconds

   $\Sigma P_{\text{GEN(NR)}} = 30,000$ MW; $\Sigma P_{\text{GEN }n\text{(R)}} = 0$ MW; $\Sigma P_{\text{LOAD }n} = 30,000$ MW; $P_{\text{TRIP}} = 1000$ MW; $t_{\text{TRIP}} = 0$ second; $a_0 = -250$ MW; $a_1 = 125$ MW/s; $H = 4.0$ seconds; $f_0 = 50$ Hz

2. *The input data in "per unit"*
   a. $0 \le t < 2$ seconds

   $\Sigma P_{\text{GEN(NR)}} = 0.8$ pu; $\Sigma P_{\text{GEN }n\text{(R)}} = 0$; $\Sigma P_{\text{LOAD }n} = 0.8$ pu; $P_{\text{TRIP}} = 0.0333$ pu; $t_{\text{TRIP}} = 0$ second; $a_0 = 0$; $a_1 = 0$; $H = 4.0$ seconds; $f_0 = 1.0$ pu

b. $2 \le t \le 10$ seconds

$\Sigma P_{\text{GEN(NR)}} = 0.8$ pu; $\Sigma P_{\text{GEN } n\text{(R)}} = 0$; $\Sigma P_{\text{LOAD } n} = 0.8$ pu; $P_{\text{TRIP}} = 0.0333$ pu;
$t_{\text{TRIP}} = 0$ second; $a_0 = -6.667 \times 10^{-3}$ pu;
$a_1 = 3.333 \times 10^{-3}$ pu/s; $H = 4.0$ seconds;
$f_0 = 1.0$ pu

10.5.5 The output data

The output data from the simulation will consist of values of the "system frequency" for equally spaced time values for the first 10 seconds following the loss of generation. A graph may be plotted to show the time evolution of the "system frequency."

10.5.6 The suggested *MATLAB* code

We again suggest that the code for *MATLAB example simulation #1* that was given in Chapter 6, Analytical solutions, be used again with suitable changes to the parameters to meet our requirements in this example.

```
%    MATLAB® Example: the frequency following the loss of a Wind Farm
%
% Inputting the data for the problem in "per unit" :
    PNR = input('Please enter the non-responsive generation in per unit')
    PR = input('Please enter the responsive generation in per unit')
    PL = input('Please enter the system load in per unit')
    FN = input('Please enter the nominal frequency in per unit')
    PTRIP = input('Please enter the generation tripped in per unit')
    A_0 = input('Please enter the value of the constant a0 in per unit')
    A_1 = input('Please enter the value of the constant a1 in per unit')
    H = input('Please enter the value of the H-constant in seconds')
    F0 = input('Please enter the value of the initial frequency in per
unit')
% Calculate the power balance at nominal frequency as a parameter:
    PBAL = PNR + PR - PL - PTRIP
% Calculate the value of the parameter k1:
    K1 = 0.02*FN*PL
% Calculate the values of the constants A1, B1, C1 and D1:
    A1 = FN + (A_0 + PBAL)/K1
    B1 = A_1/K1
    C1 = F0 - (A_0 + PBAL)/K1 + 2*H*F0*A_1/(K1^2)
    D1 = -K1/(2*H*F0)
% Calculate the solution for the frequency f at a sequence of values
of t from 0 to 10
```

```
% seconds at intervals of 0.1 second:
    for k = 1,100
        t = k/10
        f(k) = A1 + B1*t + C1*EXP(D1*t)
    end
% Note that the entries of the vector f are in "per unit" and so need to
be modified to
% be in Hz.
```

10.5.7 Instructions for the user

Data in "per unit" should be entered when prompted to do so by the program. The graph of "system frequency" against time may be plotted either automatically by the program by adding suitable code or by hand in, for example, Excel.

Note that the solution is divided into two time periods; and therefore, that the code needs to be run in two stages with the two sets of data provided earlier.

10.6 Low-inertia systems: the modeling of "solar photovoltaic farms" in frequency studies

10.6.1 A definition of a "solar photovoltaic farm"

A "solar PV farm" may be defined as an extended array of solar PV panels (solar PV) that is connected locally into a distribution system, or directly via suitable transformation to a transmission system.

10.6.2 Secondary applications of "solar photovoltaic farms"

1. *The provision of "voltage regulation" at night*

   "Solar PV farms" may be set up to generate MWs during sunlight hours and Mvars at night to provide voltage support. See Ref. [10] for further details on this topic; this is a comparatively early paper.
2. *The combination with "storage"*

   "Solar PV farms" may also be set up in combination with electricity "storage" devices. See Ref. [11] for further details on this topic; this is a much more recent paper than Ref. [10].

   In this arrangement, instead of supplying electrical current instantaneously from the solar energy impinging on the PV panels, electricity may be stored by, for example, a battery or capacitor for later use. In this way power can be produced when it is most needed.

10.6.3 The status of a "solar photovoltaic farm" in the intact system

A "solar PV farm" in an intact system appears to the system control center as a generator that is either switched on or switched off. In this respect, it looks like other forms of generation feeding into the system. However, like HVDC links and most wind farms, it makes no contribution to the system inertia.

10.6.4 The effect of tripping a "solar photovoltaic farm"

A "solar PV farm" does not have any conventional inertia because it does not have any massive rotating parts. Therefore the effect of the tripping (loss) of a "solar PV farm" involves only a loss of MW output: its inertia remains zero.

10.6.5 The provision of "frequency response" by "solar photovoltaic farms" themselves

At the present time the concept of "solar PV farms" themselves providing "frequency response" is not a common notion, but Chaiyathan and Ngamroo in their 2017 IEEE Systems Journal paper describe a possible method of improving the transient stability of a power system using a "PID controller" incorporated within a "solar PV farm" (see Ref. [12] for more details on this idea).

10.6.6 In the next section

In the next section, we work through an illustrative example in MATLAB about the possible effects of "solar farms" on the "system frequency."

10.7 MATLAB example #15: the effect of "solar photovoltaic farms" on the "system frequency"

10.7.1 Introduction

Let us imagine, for the purposes of argument, that we have a "solar farm" that is generating electricity on a sunny afternoon in the summer. What kinds of issues does the control engineer have to think about? Will the total system demand be high or low? Will there be enough inertia available in the system to cope with possible unplanned losses of generation or load? What would happen if the whole of the output of one of the wind farms was to be lost? Would this loss be better or worse to deal with than the loss of a traditional synchronous machine from the system?

10.7.2 A statement of the problem

1. *General description*

   We shall consider a 50 Hz grid system with a demand of 30,000 MW, which, because half of this demand is being met by inertia-less generation, the average system inertia is only 2.5 seconds. The synchronous generators available are all of size 500 MW and have an inertia constant of 5.0 seconds.

2. *The event*

   Without any warning, two of the synchronous generators trip off the system in quick succession. In this exercise, we shall treat these trips as being simultaneous.

3. *Modification of the "system inertia" postevent*

   How does the average inertia of the system change postevent? What is its new average value postevent?

4. *The frequency response requirement*

   How much response is needed postevent if the "system frequency" is not permitted to fall below 49.0 Hz?

10.7.3 The general solution of the problem

1. *Opening comment*

   This is an example of "inertia erosion" from a system due to the presence of substantial amounts of inertia-less generation. Only one half of the system's generation possesses inertia, and the other half has none.

2. *The change in the average system inertia*

   a. *The loss of generators having inertia*

      Following the loss of the two 500 MW synchronous generators, the average system inertia decreases. For the postevent system, this means that its frequency will fall lower than it would otherwise have done if there had been no change in the system frequency.

   b. *The loss of generators without inertia*

      There would have been no change to the system inertia if, for example, a "solar PV" installation of equivalent power output had been lost. In this latter case the challenge would have been reduced to simply replacing the lost power generation. In this example, however, we have lost a substantial amount of inertia along with the power generation.

3. *The numerical value of the postevent system inertia*

   a. *Preevent*

      In preevent, we have 15,000 MW of generation with an average inertia of 5 seconds and 15,000 MW of generation with an average inertia of 0 second giving an overall system average inertia of 2.5 seconds. The system demand is 30,000 MW.

   b. *Postevent*

      In postevent, we have 14,000 MW of generation with an average inertia of 5 seconds and 15,000 MW of generation with an average inertia of 0 second. The system demand is still 30,000 MW. This means that the new average system inertia works out to be

$$\frac{(14,000 \times 5 + 15,000 \times 0)}{30,000} = 2.33 \text{ } seconds$$

This is a slight but noticeable reduction which nevertheless is a quite sizeable reduction when one considers how small the average inertia was prefault.

The reduction in percentage terms is:

$$\frac{(2.5 - 2.33) \times 100}{2.5} = 7\%.$$

This constitutes a sizeable proportion of the original value, and its effect is noticeable in the postevent frequency behavior of the system.

4. *Calculating the frequency response requirement*

The frequency response requirement is calculated by solving the "swing equation" with the help of one of the methods explained in Chapters 6–8. In this instance, we shall use the "direct solution of the differential equation" method.

5. *The downloading of generators for response*

In our calculations, it will be assumed that responsive sets are synchronous generators of size 500 MW each, and that each is downloaded by 10% of its nominal power output. Thus if a synchronous generator is selected for response purposes, it is set to run preevent at an output of 450 MW.

The above affects our calculations as follows:

a. *The number of machines running*

If one or more machines are to run de-loaded by 10%, then we shall require more generators to run than if all generators were running fully loaded.

Since we are assuming that 15,000 MW of the total system demand is being met by nonsynchronous sources, then the remaining 15,000 MW needs to be met, in general, with a combination of sets running at their full output of 500 MW, sets running for response at 450 MW plus in some cases a set, or sets being downloaded to ensure a total of exactly 15,000 MW of synchronous generation in every case.

b. *The effect of downloading on average system inertia*

A 500 MW set running at 500 MW and a 500 MW set running at 450 MW have the same inertia since inertia depends on the MVA rating of a machine, not on its MW output. Put another way, a system that includes de-loaded sets has more inertia than a system made up solely of sets running at full output. Therefore the more downloaded sets there are in our solution, the higher will be the average system inertia, and the more stable the system will be regardless of the ramped response being provided.

10.7.4 The input data

10.7.4.1 The selection of some sets to be downloaded for response

The full details of the input data depend on how many generating sets have been selected to be in responsive mode and therefore downloaded by 10%. Let this number be n_r.

10.7.4.2 Generating capacity left over after selecting the responsive sets

To deal with the usual eventuality that after selecting an integral number of sets to be downloaded for response, there will be MW capacity still to be made up to meet the system demand, this will be added to the amount of nonresponsive generation to achieve the initial balance.

10.7.4.3 The limit on the number of sets selectable for response

We are assuming that a fixed amount (15,000 MW) of inertia-less generation is always present in the generation mix. When no generating sets are de-loaded for response, there are 15,000/500 = 30 responsive sets operating at their full load of 500 MW each. Remember that two generating sets with inertia 5.0 seconds each have been tripped from the system. Therefore the maximum value of response available is $28 \times 50\ \text{MW} = 1400\ \text{MW}$ since between 0 and 28 responsive sets are available in this problem.

10.7.4.4 Detailed input data

1. *Physical units*

   Thus in physical units, we have

   a. $0 \leq t < 2$ seconds

   $$\Sigma P_{\text{GEN(NR)}} = (15{,}000 + 50 \cdot n_r{}^*)\ \text{MW}; \quad \Sigma P_{\text{GEN}\ n\text{(R)}} = 15{,}000\ \text{MW};$$
   $$\Sigma P_{\text{LOAD}\ n} = 30{,}000\ \text{MW}; \quad P_{\text{TRIP}} = 1000\ \text{MW}; \quad t_{\text{TRIP}} = 0\ \text{second};$$
   $$a_0 = 0; \quad a_1 = 0; \quad H = 2.5\ \text{seconds}; \quad f_0 = 50\ \text{Hz}$$

   Note that the amount $50 \times n_r$ MW of generation is assumed to have an inertia constant of 0 second, that is, it has no inertia.

   Note also that the *total nominal amount* of responsive generation remains fixed at *15,000 MW less 1000 MW tripped* even though its *total MW output* is less than this figure once any generating sets have been selected for de-loading.

   b. $2 \leq t \leq 10$ seconds

   $$\Sigma P_{\text{GEN(NR)}} = (15{,}000 + 50 \cdot n_r)\ \text{MW}; \quad \Sigma P_{\text{GEN}\ n\text{(R)}} = 15{,}000\ \text{MW};$$
   $$\Sigma P_{\text{LOAD}\ n} = 30{,}000\ \text{MW}; \quad P_{\text{TRIP}} = 1000\ \text{MW}; \quad t_{\text{TRIP}} = 0\ \text{second};$$
   $$a_0 = -250\ \text{MW}; \quad a_1 = 125\ \text{MW/s}; \quad H = 2.3333\ \text{seconds};$$
   $$f_0 = 50\ \text{Hz}$$

2. *Per unit*

The earlier input data in "per unit," assuming a *base MVA* of 30,000/ 0.8 = 37,500 MVA, are

a. $0 \le t < 2$ seconds

$$\Sigma P_{\text{GEN(NR)}} = \left(\frac{0.8 + 50 \cdot n_r}{37,500}\right) \text{pu}; \Sigma P_{\text{GEN } n\text{(R)}} = 0.4 \text{ pu}; \Sigma P_{\text{LOAD } n} = 0.8 \text{ pu};$$

$P_{\text{TRIP}} = 0.026667$ pu; $t_{\text{TRIP}} = 0$ second; $a_0 = 0; a_1 = 0; H = 2.5$ seconds; $f_0 = 1.0$ pu

b. $2 \le t \le 10$ seconds

$$\Sigma P_{\text{GEN(NR)}} = \left(\frac{0.8 + 50 \cdot n_r}{37,500}\right) \text{pu}; \Sigma P_{\text{GEN } n\text{(R)}} = 0.4 \text{ pu}; \Sigma P_{\text{LOAD } n} = 0.8 \text{ pu};$$

$P_{\text{TRIP}} = 0.02667$ pu; $t_{\text{TRIP}} = 0$ second; $a_0 = -6.6667 \times 10^{-3}$ pu;
$a_1 = 3.3333 \times 10^{-3}$ pu/s; $H = 2.3333$ seconds; $f_0 = 1.0$ pu

We now move on to define the output data for this example.

10.7.5 The output data

The output data consist of "system frequency" against *time* value pairs that can subsequently be plotted as a graph.

10.7.6 The suggested MATLAB code

We suggest that the code for *MATLAB example simulation #1* given in Chapter 6, Analytical solutions, is again used. We describe this next.

```
%            The effect of Solar PV Farms on the system frequency
%
% Inputting the data for the problem in "per unit" :
    PNR = input('Please enter the non-responsive generation in per unit')
    PR = input('Please enter the responsive generation in per unit')
    PL = input('Please enter the system load in per unit')
    FN = input('Please enter the nominal frequency in per unit')
    PTRIP = input('Please enter the generation tripped in per unit')
    A_0 = input('Please enter the value of the constant a0 in per unit')
    A_1 = input('Please enter the value of the constant a1 in per unit')
    H = input('Please enter the value of the H-constant in seconds')
    FO = input('Please enter the value of the initial frequency in per
unit')
% Calculate the power balance at nominal frequency as a parameter:
    PBAL = PNR + PR - PL - PTRIP
% Calculate the value of the parameter k1:
    K1 = 0.02*FN*PL
```

```
% Calculate the values of the constants A1, B1, C1and D1:
    A1 = FN + (A_0 + PBAL)/K1
    B1 = A_1/K1
    C1 = F0 – (A_0 + PBAL)/K1 + 2*H*F0*A_1/(K1^2)
    D1 = -K1/(2*H*F0)
% Calculate the solution for the frequency f at a sequence of values
of t from 0 to 10
%    seconds at intervals of 0.1 second:
    for k = 1,100
        t = k/10
        f(k) = A1 + B1*t + C1*EXP(D1*t)
    end
% Note that the entries of the vector f are in “per unit” and so need to
be modified to
% be in Hz.
```

10.7.7 Instructions for the user

The input data in “per unit” should be entered when prompted. The program will help us to plot a graph of the evolution of the frequency postfault for the case when no response is made available and the case when either the amount of response required to keep the postfault frequency above 49.0 Hz is provided or the maximum amount available is provided.

10.8 The next chapter: “smart grids”

The next chapter is devoted to a discussion of “smart grids” and their possible consequences for frequency stability on the grid system. We shall cover the following topics:

1. a brief introduction to “smart grids”;
2. the potential effect of “smart grids” on the system frequency; and
3. MATLAB example simulation #16: “smart grids” and the “system frequency.”

References

[1] M. Yu, A. Dyśko, A. Roscoe, C. Booth, R. Ierna, H. Urdal, J. Zhu, Effects of swing equation-based inertial response (SEBIR) control on penetration limits of non-synchronous generation in the GB power system, in: International Conference on Renewable Power Generation (RPG 2015), 2015.

[2] P. Fairley, Can synthetic inertia from wind power stabilize grids?, IEEE Spectrum (November 7, 2016).

[3] K. Dallmer-Zerbe, E. Spahic, G. Kuhn, R. Morgenstern, G. Beck, Fast frequency response in the UK grid—challenges and solution, in: 13th IET International Conference on AC and DC Power Transmission (AC DC 2017), 2017.

[4] R. Eriksson, N. Modig, K. Elkington, Synthetic inertia versus fast frequency response: a definition, IET Renewable Gen. 12 (5) (2018) 507–514.

[5] Q.-C. Zhong, Virtual synchronous machines: a unified interface for grid integration, IEEE Power Electron. Mag. 3 (4) (2016) 18–27.

[6] J. Zhu, J. Hu, W. Hung, C. Wang, X. Zhang, S. Bu, et al., Synthetic inertia control strategy for doubly-fed induction generator wind turbine generators using lithium-ion supercapacitors, IEEE Trans. Energy Convers. 33 (2) (2018) 773–783.

[7] National Grid UK, Electricity Ten Year Statement (ETYS), National Grid UK, 2017.

[8] IEEE Power & Energy Society, Wind power plant collector design, IEEE Power & Energy Society, August 2016.

[9] D. Ochoa, S. Martinez, Fast frequency response provided by DFIG wind turbines and its impact on the grid, IEEE Trans. Power Syst. 32 (5) (2017) 4002–4011.

[10] R.K. Varma, V. Khadkikar, R. Seethapathy, Night-time application of PV solar farm to regulate grid voltage, IEEE Trans. Energy Convers. 24 (4) (2009) 983–985.

[11] Y. Ghiassi-Farrokhfal, F. Kazhamiaka, C. Rosenberg, S. Keshav, Optimal design of solar PV farms with storage, IEEE Trans. Sustain. Energy 6 (4) (2015) 1586–1593.

[12] T. Chaiyathan, I. Ngamroo, Improvement of power system transient stability by PV farm with fuzzy gain scheduling of PID controller, IEEE Syst. J. 11 (3) (2017) 1684–1691.

Chapter 11

"Smart grids" and the "system frequency"

11.1 A brief introduction to "smart grids"

11.1.1 A note on the source material currently available on "smart grids"

1. *Background material*

   There are currently numerous excellent textbooks and scholarly papers available on the subject of "smart grids," and much valuable research is currently being done into this subject. It is therefore not the intention of this author to attempt to compete with any of these sources of information, although some of them will naturally be referred to in this book.
2. *The effect of "smart grids" on the "system frequency"*

   In this book, we are chiefly interested in the potential effect (if any) of "smart grids" on the *stability and control* of the "system frequency." Before we can proceed to answer this question, it would be helpful for us to define exactly what we mean when we use the term "smart grid." There have been many definitions proposed, but we give a few in the following sections.

11.1.2 What is a "smart grid"?

1. *Definition 1*

   A "smart grid" may be defined simply as "the vision that we have of the power grids of the future" (please see Ref. [1], p. xxi for more on this definition).

   This is a rather broad definition of a "smart grid" and as such could refer to a wide range of possible developments in grid systems in the future. Although a valid definition as it stands, for our purposes we would prefer a more focused definition that better allows us to determine what, if any, is the potential effect of "smart grids" on the "system frequency."

Modern Aspects of Power System Frequency Stability and Control.
DOI: https://doi.org/10.1016/B978-0-12-816139-5.00011-4

In the next paragraph, we shall therefore consider a somewhat narrower definition of the "smart grid."

2. *Definition 2*

Alternatively, a "smart grid" may be defined in a more specific and formalized way as: "the smart grid involves the application of advanced communications and control technologies and practices to improve reliability, efficiency, and security which are key ingredients in the ongoing modernization of the electricity delivery infrastructure" (please see Ref. [4] for more details of this definition).

This definition is, because of its greater focus, a more suitable one to work with in working toward our goal of determining the potential effect, if any, of the "smart grid" on the "system frequency." We see from the earlier definition that the "smart grid" is envisioned as principally a communication and control infrastructure-based concept. As such, it raises the question of whether this technology essentially aids or conflicts with frequency control technology in its traditional sense, or whether the two can coexist independently.

In the next section, we consider a third possibility for defining a "smart grid," this time regarded from the point of view of the functionality of "smart grids," in other words, what can they do?

11.1.3 What can a "smart grid" do?

Another way of viewing "smart grids" is to ask the question: What can "smart grids" do?

As a possible answer to this question, let us look at a proposed definition for our example:

1. *Definition 3*

This definition also comes from the US Department of Energy (DOE), this time from the year 2009:

The smart grid:

i. *enables informed participation by customers so that consumers can become an integral part of the electric power system, modifying the way they use and purchase electricity, participating in load balancing and helping to ensure reliability;*
ii. *accommodates all generation and storage options, including all distributed energy resources (DER), even being diverse and widespread and in the form of renewables. DG [distributed generation] and energy storage;*
iii. *enables new products, services and markets, managing independent grid variables such as energy, capacity, location, time, rate of change, and quality;*

iv. *provides the power quality for the range of needs of different end users, proving varying grades of power quality with variable prices;*
v. *optimizes asset utilization and operating efficiency, increasing the efficiency of maintenance procedures, decreasing losses and controlling congestions;*
vi. *operates resiliently to disturbances, attacks and natural disasters, reacting to such events isolating the faulted elements and keeping normal operation in the rest of the system.*

(Refer to Ref. [5] for more details on the definition.)

Let us comment on each of the earlier suggested contributory paragraphs in turn.

a. This paragraph emphasizes the role of the customer in the "smart grid" concept. The customer or consumer is to take an active part in determining the functioning of the system for the supply of electricity.
b. This paragraph highlights the concept of including all available options in both the generation and storage of electricity. Nothing is to be left out.
c. This paragraph introduces the concept of allowing a wide range of novel market activities as well as established ones.
d. This paragraph introduces the idea of providing graded products to suit the different requirements of different end users.
e. This paragraph refers to concepts relating to the increasing of operational efficiencies.
f. Finally, this paragraph relates to the need for robustness of "smart grids" against being compromised by faults both unintentional and deliberate.

Taken all together, the earlier components supply quite a detailed description of the "smart grid" concept. Indeed, one may be tempted to say that his definition covers pretty much everything to do with the production and consumption of electricity.

2. *The "smart grid" objectives of the European Union (EU)*

In parallel with the developments toward developing "smart grids" in the United States and other countries, we shall mention here the work ongoing in the countries of the EU.

The work in the EU has been driven by several factors which include the following:

a. *The "Kyoto Protocol"*

The "Kyoto Protocol" was set up as an International Treaty in 2005 as a continuation of the 1992 United Nations Framework Convention on Climate Change (UNFCCC). The "Kyoto Protocol" requires signatories to commit to reducing "greenhouse gases" from the earth's atmosphere to combat "global warming."

b. *The "Paris Agreement (Accord de Paris)"*

The "Paris Agreement" or "Accord de Paris," in French, is a more recent agreement than the "Kyoto Protocol" and was adopted on the December 12, 2015. It is also a UNFCCC agreement and becomes active in the year 2020.

c. *A list of the "greenhouse gases"*

According to Appendix A of the "Kyoto Protocol," there are six so-called greenhouse gases, which are listed there as: carbon dioxide (CO_2), methane (CH_4), nitrous oxide (N_2O), hydrofluorocarbons (HFCs), perfluorocarbons (PFCs), and sulfur hexafluoride (SF_6). Commitment is required from participating countries to reduce all these substances that are deemed to be currently polluting the earth's atmosphere to combat "global warming."

d. *Year 2020 targets*

The EU has strategies and policies in place for the year 2020 which include measures for the low carbonization of the energy industry, the increase of energy efficiency and conservation, and the development of a "green" and sustainable economy.

e. *The role of "smart grids"*

In the EU, "smart grids" are regarded as an integral part of the overall approach to meeting the "Kyoto Protocol" and the EU's own targets for the year 2020 (please see Ref. [6] for more details on this). As such the role of "smart grids" is central to the fight to limit the effects of climate change.

f. *The next section*

In the next section, we examine in more detail the potential effects that "smart grids" could have on the power system frequency in the future.

11.2 The possible structure of "smart grids"

11.2.1 Introduction

1. *Voltage levels*

When we are talking about "smart grids," we are usually referring to technologies based on the "low voltage" (LV) or "distribution system" (DS) levels and possibly on the "medium voltage" (MV) levels (if any) between the "DS" levels and the "high voltage" (HV) or "supergrid" transmission level voltages as well (please see Ref. [1], Chapter 1 for more details on this).

2. *The daily load cycle*

Clearly, however, there needs to be coordination, too, at the interface between the *HV* "supergrid" level and the lower *MV* (if any) and *LV* levels. For example, we need to ensure that the requirements for inertia and other technologies mimicking the effects of inertia are being provided in adequate amounts during the, somewhat new-looking, "daily load cycle"; in which, for example, peak demands may have shifted from their traditional positions in the day, and overnight demands have increased; the general picture is one of the "leveling out" of the demand across a 24-hour period.

3. *The future of "smart grids"*

   At the very lowest domestic, industrial, and commercial voltage levels, "smart grids" are likely to be all-pervasive in the future. The level of involvement will in theory vary according to the wishes of the consumer and the policy decisions of local and higher governmental planning bodies.

11.2.2 Proposed architectures for "smart grids"

As noted earlier, much research work has been, and still is being, devoted to this area. Any proposed architecture will tend to depend to some extent on areas of the highest usage of, for example, electric vehicle (EV) charging stations and "controllable" refrigerators and other domestic appliances.

Progress has, however, already been made in countries, such as the United Kingdom and the United States. In California, for example, EVs are already being used in large numbers, whereas in other states of the United States they are less so. In the United Kingdom, EV charging stations are already being sponsored at the corporate level.

11.2.3 In the next section

In the next section, we shall be considering the optimization of "smart grids." This is a necessary requirement that goes with the creation of sometimes highly complex system architectures of "smart grids."

11.3 The optimization of "smart grids"

11.3.1 Introduction

If "smart grids" are to be a successful component of the power systems of the future, they will require careful optimization at many levels of the power system architecture. This is a very challenging objective to meet because many systems are being proposed to run in coordination with other systems, such as gas supplies, potentially being optimized as well as electrical supplies. In addition to this, there are potentially very many nodes that would require optimizing in the most detailed of the proposed "smart grid" architectures (please see Ref. [1] for more details on this).

11.3.2 Potentially useful optimization techniques

1. *Linear optimization techniques*

   Linear optimization techniques based usually on the "simplex method" are by far the most common, for several good reasons:

   a. They are *faster* in terms of total execution time than nonlinear methods.

   b. Solutions involving one source can be added to solutions involving another source to give the combined solution (the "principle of linear superposition").

   c. The solution is *unique*, that is, there is only one solution corresponding to a given set of input conditions.

   d. Many weak nonlinear systems can be approximated at each iteration by a linearized form of the governing equations that have the above-mentioned advantages. The solution is normally updated at the end of each iteration to stay in line with the original, nonlinear formulation.

2. *Nonlinear optimization techniques*

   These techniques do not make any attempt to linearize the problem being addressed. They are more difficult to program than linear or linearized techniques, and nonlinear techniques can suffer from the following disadvantages compared with linear techniques:

   a. Nonlinear mathematical equations usually have more than one solution.

   b. In nonlinear problems, there exists the phenomenon of branching, whereby at some point in the iteration process the current solution may divided into two or more branches or routes; each route must therefore be identified and compared with the others to determine which is the optimal route.

   c. More than one solution may exist for the same value of the "objective function." This is a well-known feature of nonlinear systems. However, in practice, there may be one "sensible" solution and a number of solutions that can be discarded on the grounds of not being "sensible."

   d. In nonlinear mathematical algorithms, there is no guarantee that a solution can be found at all, although the systems being optimized may be initialized with parameter values that are sufficiently close to the solution for this not to be an impediment.

   e. Nonlinear techniques are, however, available on the market and used in practice, the proviso here being that one must know in advance which are the "sensible" solutions and which are not "sensible" to be able to be confident of the outputs being generated.

3. *"Mixed integer" programming techniques*

   "Mixed integer" programming techniques are those for which there is a mixture of "real" and "integer" variables in the formulation of the problem being solved. In an example taken from the very popular "optimal power flow" techniques, we can have a variable such as the phase-to-phase voltage at a system bus that is a continuous quantity represented by a "real" number and at the same time have discrete quantities such as transformer tap positions that are discrete quantities represented by "integer" numbers, such as +3 or −5, for example. "Mixed integer" problems may be linear or nonlinear. In the linear case the objective function and constraints, except the integer constraints, are linear (please see Ref. [1], Chapter 2 for an example application of this method).

4. *Genetic algorithms*

   "Genetic algorithms" are "heuristic" solution methods that rely on a process of trial and error (learning) to find the best or optimal solution according to a process that mimics real-life genetics. A starting set of candidate solutions is gradually improved by processes, such as "mutation" and "crossover," until an acceptable solution is found that fulfills all the requirements that have been specified for the problem. Along the way the "fittest" candidate solutions are retained, while the other candidate solutions are discarded (see, for example, Ref. [2], Chapter 5). In the next two sections, we will look at two more "heuristic" solution methods.

5. *Artificial neural networks (ANNs)*

   The basic principle on which ANNs work is that they mimic the way that the human brain functions (see, for example, Ref. [2], p. 109). "ANNs" require "training," that is, they have first to be given a working example which they then learn. As such, they are tuned to a specific problem through a learning process.

6. *Expert systems (ESs)*

   ESs consist essentially of a configuration in which the user interface has two parts: a "knowledge base" and an "inference engine." Facts are fed into the "knowledge base" and processed by the "inference engine." The results are subsequently fed back to the user as "expertise" (see, for example, Ref. [2], p. 111 for more details on this).

7. *Other optimization algorithms proposed for "smart grids"*

   Many other optimization algorithms have been proposed for the possible application to the monitoring and control of "smart grids" (please see Refs. [1,2] for more examples of these).

8. *In the next section*

   In the next section, we consider in more detail the potential effect that the implementation of "smart grids" could have on the system frequency and its stability.

11.4 The potential effect of "smart grids" on the "system frequency"

11.4.1 Introduction

As noted earlier in this book, the system frequency depends essentially on the active power balance between generation and demand (or load, if we are neglecting active power losses in our discussion). Therefore in our study of the possible effect or effects of the introduction and operation of "smart grids" into the DS on the bulk system frequency, we need to examine what kinds of sources and sinks of power would be present under the "smart grid" environment. This we do in the following sections. We shall see that many of the constituent elements of "smart grids" are already present in, for example, the Great Britain power system. The development opportunities largely concern the subject of possible "smart grid" architectures, data measurements, control systems, and such like.

11.4.2 Combined heat and power schemes

1. *Introduction*

   Combined heat and power (CHP) schemes were among the first to appear in the Great Britain power system, and as such have already been in use for several decades on it. We give here a description of CHP schemes, also known as "cogeneration" [7]:

   > *Combined heat and power (CHP) is a highly efficient process that captures and utilizes the heat that is a by-product of the electricity generation process. By generating heat and power simultaneously, CHP can reduce carbon emissions by up to 30% compared to the separate means of conventional generation via a boiler and power station.*
   >
   > *The heat generated during this process is supplied to an appropriately matched heat demand that would otherwise be met by a conventional boiler. CHP systems are highly efficient, making use of the heat which would otherwise be wasted when generating electrical or mechanical power. This allows heat requirements to be met that would otherwise require additional fuel to be burnt.*
   >
   > *For many organizations, CHP is the measure that offers the most significant single opportunity to reduce energy costs and to improve environmental performance with existing users of CHP typically saving around 20% of their energy costs.*

   As we can see from the given definition, "CHP" schemes have several benefits. One of these is to improve by a substantial amount the efficiency of the generation process. A second benefit is to decrease pollution by reducing carbon emissions into the atmosphere substantially. Not surprisingly, "CHP" is a popular energy solution.

2. *The possible effects of CHP on the system frequency*

   Since CHP schemes are well-established technologies, they should not by themselves change the way they affect the overall power system when "smart grids" become more established unless the number of such schemes were to diminish rapidly, in which case some inertia may be lost from the system with the associated destabilizing effect on the power frequency. In other words, conventional "CHP" assists the stability of the power system because the generators involved have inertia.

11.4.3 Solar photovoltaic cell technology

1. *Introduction*

   Solar photovoltaic (solar PV) cell technology is to be found both in large-scale "PV farms" and in a much more modest domestic setting on the roofs of homes throughout the United Kingdom and other countries. In the domestic setting, "solar PV" panels are being used not only to meet the energy requirements of households but also to feed energy into the local grid system thus acting as an electrical generator for the local distribution grid and possibly spilling over into the wider system.
2. *The possible effects on the system frequency*

   The vast increase in the installation and usage of "solar PV" in some countries such as the United Kingdom that has an independent synchronous system has meant that there has been a correspondingly significant decrease in the system inertia. Therefore it will be necessary to compensate for the lost system inertia from synchronous plant that is no longer needed.

   Given the current environmental policies of the EU/United Kingdom, great research efforts are currently being made on the subject of "synthetic inertia" (as introduced in Chapter 2: What can provide frequency control?, of this book) as part of the drive to maintain adequate levels of inertia in the power system and consequently the stability of the power system under disturbance conditions. In the meantime the maintenance of stability on the system could possibly sometimes present challenges.

11.4.4 Wind turbines in distribution systems (domestic wind turbines)

1. *Introduction*

   Some domestic users, farm owners, and others have their own private small-scale wind turbines. Currently, this is more prevalent in the United States than in the United Kingdom. These wind turbines can be used to service local electricity requirements, as well as to provide electricity for their own premises. There is also the possibility of feeding power back to sell to the local distribution grid.
2. *The possible effects on the system frequency*
   a. *Large wind turbines*

      As mentioned in the last chapter, wind turbines that incorporate back-to-back alternating current (AC)–DC converters effectively have inertia that is not useable by the power system to which it is connected. Most, if not all, larger wind turbines are of the "back-to-back" convertor type, mainly of the "DFIG" design.

b. *Small wind turbines*

However, the small domestic-size wind turbines in the kilowatt (kW) range are generally the simple induction generator type whose inertia *does* contribute to the total inertia of the grid system. Therefore in this instance, wind turbines can have a stabilizing influence on the system frequency, although this effect is somewhat diminished because the rotor is not locked magnetically to the stator as is the case for synchronous machines.

3. *The intermittency of wind power*

Perhaps the most fundamental property of wind power is its "intermittency": the wind is constantly changing. This could itself be viewed as a destabilizing factor for the system frequency due to the ever-changing generation to load balance that results from having large numbers of wind turbines synchronized to the system.

11.4.5 Biomass plants

1. *Introduction*

Biomass plants essentially burn waste products of various kinds, including wood, byproducts of agricultural processes, and animal and human waste as the means of generating electricity. However, the burning (combustion) of these materials directly is only one of the options available for power generation. For example, biomass materials can be made to produce gases for subsequent combustion by a process known as gasification. There are also other kinds of processes available for creating gaseous products that can subsequently themselves be burned. Biomass can be viewed in some respects as being a source of energy like fossil fuels such as coal or oil. Emissions from the combustion of biomass products must be carefully monitored and controlled to comply with regulations.

For complete description on the use of biomass for electricity generation (please see Ref. [8]).

2. *The possible effects of biomass plants on the system frequency*

a. *A summary of biomass generation*

It can be ascertained from the earlier description that there are several possible ways to convert biomass into electricity, but basically the method involves combustion, either directly or indirectly. In this respect the process is that of a normal synchronous generator that has inertia and, therefore, stabilizes the system frequency.

b. *Possible effects on the environment*

As an aside, it is worth noting from the final paragraph of the above description that using biomass is not necessarily as environment friendly as it may sound because there is an obligation to control emissions from the biomass combustion process.

11.4.6 Controllable domestic loads

1. *Introduction*

   Domestic loads, such as refrigerators, water heaters, air conditioners, washers, and driers, could be wired into proposed "smart grid" control architectures in such a way that signals may be sent by the local utility (distribution company) to switch the appliances "on" or "off" according to the needs of the optimization routine that would be controlling the grid demand (load).

   Thus, for example, during times of heavy load for the "transmission operator" (TO) domestic appliances like these, plus commercial and industrial loads could be switched "off," and conversely, during times of low demand for the "TO," these appliances could be controlled to the "on" position. In such a way changes to the "daily load profile" as measured by the "TO" are being made both smoother and less extreme than they were formerly.

2. *The possible effects on the system frequency*

   The net result for the system frequency with the strategy outlined in the previous paragraph ought to be that it should, if anything, be easier to control the system frequency, and hence that the frequency should correspondingly be more stable as a result, even if unplanned outages of plant or equipment were to occur (in other words, the system would be more robust to change).

11.4.7 Controllable commercial loads

Contracts for controllable commercial loads could be entered into with the "DS operator" or with the "transmission system operator" so that the system demand is made smoother during the day and night, with the same benefits as in the previous section. However, this may entail having staggered working arrangements in office buildings, for example, if discernable benefits are to be made.

11.4.8 Industrial loads

In the specific case of industrial loads, there is often a requirement to run a power intensive process (such as iron smelting), continuously at full output for a certain fixed period. Outside this period, the high level of power consumption is not required. We note, however, that this requirement forces certain industrial contracts to be less flexible than their commercial or domestic counterparts and less likely therefore to be part of the "smart" optimization scheme.

On the other hand, other industrial processes are not so power intensive and do not require power to be available continuously for lengthy periods during the day and so should be more amenable to "smart" optimization and the benefits for frequency stability that this should bring.

11.4.9 Control architectures of "smart grids"

1. *Introduction*

   Without too much exaggeration, it could be said that "smart grid" architectures are pretty much based on the principle of "everything is connected to everything else" (see, for example, the US DOE smart grid design called "architecture 1" that is reproduced in Figure 1.5 on p. 12 of Ref. [1]).

   Here we see considerable interaction between, for example, "transmission automation," "renewables integration," "distribution automation," "demand participation signals and options," "energy efficiency," and "system operations," to name some of the elements of the proposed architecture. Thus we are talking in terms of the need for a very large optimization algorithm indeed.

2. *The possible effects on the "system frequency"*

   Assuming that by this stage we are still talking in terms of having a majority "AC" system, rather than a minority "AC" system or even a "DC" system, what could be the likely effect of all these proposed future controls on the "system frequency"?

   The principal advantage for the "system frequency" that "smart grids" would have over traditional architectures is that the system demand can, up to a point, be manipulated to smooth out the "daily load profile," as discussed earlier.

   However, to achieve this goal would require a reliable optimization of the "smart grid" components in the chosen architecture. In the next section, we discuss one possible approach that aims to do this—"multiagent systems" or "MASs."

11.4.10 Multiagent systems

1. *Supervisory control and data acquisition (SCADA)*

   Traditional control architectures for electric power systems rely heavily on the principles of SCADA. (For a full description of these see, for example, Ref. [3], Section 5.4.)

   In traditional "SCADA"-type monitoring and control architectures, first there is some kind of centralized control, usually at a control center of some kind. Feeding into this central control hub in general terms we have then a further two components.

   The first of these components concerns security assessment and control, a vital element of which is "network analysis." This element provides information about such things as power flows, bus voltages, substation running arrangements, line outages, and the like in real time or with a slight delay.

The second of these components involves the control a scheduling of generation. This component varies from system to system depending on whether, for example, there is a competitive market in generation and whether there is an element of "automatic generation control" in the power system under consideration.

2. *MASs for "smart grids"*

   In contrast with the "SCADA"-type systems the "MAS"-type systems, which are becoming increasingly popular today, can be defined as "computation systems in which several agents cooperate to achieve a desired task" (Ref. [2], Section 2.4).

   As such, these systems emphasize the cooperation of so-called agents such as "arbitrator agents," "system monitoring agents," "fault detection agents," "diagnosis agents," a "Judgment Index Agent," and a "scheduling agent."

   There is a sharing of information among the agents to determine the actions (if any) that should be taken. "MAS"-type architectures lend themselves naturally to distributed systems and are currently being considered one possible architectural solution suitable for "smart grids."

11.4.11 Microgrids

A "microgrid" in contrast with a "smart grid" can be defined as a localized grid system that can operate by itself in "stand-alone" mode (islanded) or operate in "interconnected mode" with a larger grid system. It can be powered by generation, such as gas turbines or renewable energy, and typically is connected via inverters by a "DC link" to the larger system. It does not have sophisticated controls or wide-area decision-making capabilities like the "smart grid" has (Ref. [2], Section 2.5).

11.4.12 In the next section

In the next section, we work through a model in MATLAB to examine the possible effects of a "smoothed out" daily demand cycle on the system frequency follow the loss of generation or demand from a "smart grid."

11.5 MATLAB example simulation 16: "smart grids" and the "system frequency"

11.5.1 Introduction

1. *"Smart grids" and the "daily load profile"*

   One of the possible effects of employing widespread "smart grids" may be an alteration to the shape of the "daily demand profile." For example, the traditional morning and early evening peak demand times may disappear along with the overnight minimum as electricity usage patterns are modified due to greater consumer control over demand and different kinds of usage, such as EVs.

2. *The steady demand profile*

   In this example, we shall consider a situation in which a 50 Hz transmission system traditionally with a daily peak demand of 45,000 MW and a daily minimum demand of 15,000 MW now has a demand that is pretty much steady at 28,000 MW. What differences do we expect to see in terms of the behavior of the "system frequency" following the loss of the largest generating set that feeds into the system?

11.5.2 A statement of the problem

1. *The system*

   Let us consider the system as described in the preceding section. We have a 50 Hz system with a demand of 28,000 MW and a maximum demand of 45,000 MW. Let us say that the largest generating set is in the form of an *importing interconnector* of size 1200 MW.

2. *System inertia and response availability*

   Let the system has an average prefault inertia of 4.5 seconds. If the importing interconnector described is suddenly lost, what will be the subsequent behavior of the "system frequency"? How low will the frequency go if (1) no response is available, or (2) 500 MW of battery storage provides response in the form of a ramp with a straight-line profile giving no output at 0 second and full output at 2 seconds, being constant at the full output thereafter?

3. *The tool*

   This time we shall use the "direct solution of the differential equation" method as described in the first part of Chapter 6, Analytical solutions, to solve the problem.

11.5.3 The general solution of the problem

1. *No frequency response is available*

   For this case, we need to use the "swing equation," for example, as given in Eqs. (10.1) and (10.2). We repeat these equations next for ease of reference:

$$\Sigma P_{\mathrm{GEN(NR)}} + \Sigma P_{\mathrm{GEN}\ n\mathrm{(R)}} - \Sigma P_{\mathrm{LOAD}\ n} - k_1 \cdot (f - f_n) - P_{\mathrm{TRIP}} \cdot u(t - t_{\mathrm{TRIP}}) + a_0 + a_1 t = 2 \cdot H \cdot f_0 \cdot \frac{\mathrm{d}f}{\mathrm{d}t} \tag{11.1}$$

Eq. (11.1) is the basic form of the "swing equation" that we shall solve. The parameter k_1 is given by

$$k_1 = \Sigma P_{\text{LOAD } n} \cdot \alpha + \Sigma P_{\text{GEN } n(\text{R})} \cdot k \tag{11.2}$$

where α is the "load-damping" constant which we shall assume to have the value 2% per Hz, and k is the constant associated with the "droop" of the speed governor of responsive synchronous plant. Since we shall not be considering response in this example, the second term on the right-hand side of Eq. (11.2) we shall take to be zero.

From Chapter 6, Analytical solutions, the solution of Eq. (11.1) with Eq. (11.2) is given by

$$f(t) = A_1 + B_1 \cdot t + C_1 \cdot \exp(D_1 \cdot t) \tag{11.3}$$

where

$$A_1 = f_n + \frac{\{\Sigma P_{\text{GEN(NR)}} + \Sigma P_{\text{GEN } n(\text{R})} - \Sigma P_{\text{LOAD } n} - P_{\text{TRIP}} \cdot u(t - t_{\text{TRIP}}) + a_0\}}{k_1} - \frac{2 \cdot H \cdot f_0 a_1}{k_1^2} \tag{11.4}$$

$$B_1 = \frac{a_1}{k_1} \tag{11.5}$$

$$C_1 = f_0 - f_n - \frac{(\Sigma P_{\text{GEN(NR)}} + \Sigma P_{\text{GEN } n(\text{R})} - \Sigma P_{\text{LOAD } n} - P_{\text{TRIP}} \cdot u(t - t_{\text{TRIP}}) + a_0)}{k_1} + \frac{2 \cdot H \cdot f_0 \cdot a_1}{k_1^2} \tag{11.6}$$

and

$$D_1 = \frac{-k_1}{2 \cdot H \cdot f_0} \tag{11.7}$$

are the expressions for the four coefficients that appear in the solution as given by Eq. (11.3).

For the "no response available" case, we need to put $a_0 = a_1 = \Sigma P_{\text{GEN } n(\text{R})} = 0$ in the abovementioned equations.

2. *Storage-based ramped response is available*

   In this case in the abovementioned equations, we should again take $\Sigma P_{\text{GEN } n(\text{R})} = 0$ and the storage-based ramped response may be modeled by choosing suitable values for the parameters a_0 and a_1.

   We recall that we have specified a storage-based ramped response that begins at 0 MW at $t = 0$ second and rises to 500 MW at $t = 2$ seconds and remains constant thereafter at 500 MW for the purposes of this example.

11.5.4 The input data

The input data has been pretty much covered in the earlier subsections.

11.5.5 The output data

For each of the two cases the output data consists of the function $f(t)$ given by Eq. (11.3) with the parameters A_1, A_2, A_3, and A_4 given by Eqs. (11.4)–(11.7), respectively.

11.5.6 The suggested *MATLAB* code

We suggest that the code given in MATLAB example simulation 16 be used here. We list this next.

```
%    "Smart Grids" and the "system frequency"
% Inputting the data for the problem in "per unit" :
    PNR = input('Please enter the non-responsive generation in per
          unit')
    PR = input('Please enter the responsive generation in per unit')
    PL = input('Please enter the system load in per unit')
    FN = input('Please enter the nominal frequency in per unit')
    PTRIP = input('Please enter the generation tripped in per unit')
    A_0 = input('Please enter the value of the constant a0 in per unit')
    A_1 = input('Please enter the value of the constant a1 in per unit')
    H = input('Please enter the value of the H-constant in seconds')
    F0 = input('Please enter the value of the initial frequency in per
          unit')
% Calculate the power balance at nominal frequency as a parameter:
    PBAL = PNR + PR - PL - PTRIP
% Calculate the value of the parameter k1:
    K1 = 0.02*FN*PL
% Calculate the values of the constants A1, B1, C1 and D1:
    A1 = FN + (A_0 + PBAL)/K1
    B1 = A_1/K1
    C1 = F0 - (A_0 + PBAL)/K1 + 2*H*F0*A_1/(K1^2)
    D1 = -K1/(2*H*F0)
% Calculate the solution for the frequency f at a sequence of values
of t from 0 to 10
% seconds at intervals of 0.1 second:
    for k = 1,100
        t = k/10
        f(k) = A1 + B1*t + C1*EXP(D1*t)
    end
% Note that the entries of the vector f are in "per unit" and so need to
be modified to
% be in Hz
```

11.5.7 Instructions for the user

Enter the appropriate input data for each case in turn when prompted by the *MATLAB* program. The program must be run twice. The program will output frequency data to allow the user to plot graphs of *frequency f* against *time t* for each of the two cases described earlier.

11.6 The next chapter: conclusion

In the next chapter, we shall summarize the contents of the whole book up to the end of this chapter and attempt to draw some conclusion on some important questions: (1) What is the best method for solving the "swing equation?", (2) Does a shortfall of inertia on the system matter?, and (3) Where do we go from here?

References

Books

[1] From smart grids to smart cities, in: M. La Scala (Ed.), Electrical Engineering Series, Advanced Smart Grids Set, ISTE Ltd and John Wiley & Sons, Inc., 2017. ISBN: 978-1-84821-749-2.

[2] J. Momoh, Smart grid: fundamentals of design and analysis, in: Wiley IEEE Press Series on Power Engineering, 2016 (reprint). ISBN: 978-81-265-5812-4.

[3] H. Bevrani, M. Watanabe, Y. Mitani, Power System Monitoring and Control, Wiley-IEEE Press, 2014. ISBN: 978-1-118-45069-7.

Reports

[4] U.S. Department of Energy, Smart Grid System Report. Available at: <http://energy.gov/sites/prod/files/2014/08/f18/SmartGrid-SystemReport>.

[5] U.S. Department of Energy, Smart Grid System Report. Available at: <http://energy.gov/oe/downloads/2009-smart-grid-system-report-july-2009>, 2009.

[6] European Technology Platform Smart Grids, Vision and Strategy for Europe's Electricity Networks of the Future. Available at: <http://ec.europa.eu/research/energy/pdf/smartgrids_en.pdf>, 2006.

[7] UK Department for Business, Energy & Industrial Strategy, Guidance: Combined Heat and Power, UK Department for Business, Energy & Industrial Strategy, 2018 (updated 20.02.18).

[8] U.S. Department of Energy Federal Energy Management Program (FEMP), Biomass for Electricity Generation, U.S. Department of Energy Federal Energy Management Program (FEMP), 2016 (updated 15.09.16).

Chapter 12

Conclusions

12.1 A summary of the first 11 chapters of this book

12.1.1 A summary of Chapter 1, The need for frequency control

In Chapter 1, The need for frequency control, we introduced the concepts of "frequency control" and "frequency response requirements." The need for frequency control was explained as an essential requirement of system operation. The possible effects of new developments on the transmission system were introduced.

The scope of the book was then defined, and definitions of "normal" and "abnormal" operation were introduced. Available sources of frequency response were mentioned briefly, and the subject of frequency control in the modern era and associated issues was introduced.

The subject of "system requirements" was introduced and their variation with the "daily load cycle." The effect on requirements of the degree of integrity of the power system was explained.

The system requirements for the "intact system," the power balance in the system, and a more in-depth treatment of the "daily load cycle" were discussed.

Next, the subject of the system requirements following the loss of a generating set, its possible causes and effects on the system were dealt with. The automatic governor action of synchronous generators was introduced. The frequency dependence of load (demand) was also discussed.

"Frequency response erosion" was defined. The concept of the de-loading of synchronous sets for "response" and "reserve" was introduced. "Manual actions" were described.

The loss of demand (load) and its effect on the system frequency was then covered. This led to a discussion of the "high-frequency event" and its effects. "Demand (load) shedding" was also described.

The "system requirements" following the loss of overhead lines, transformers, busbars, and other transmission equipment were then discussed.

The monitoring of the system frequency in real time and modern challenges in frequency control were then discussed. Finally, the modeling of the power system for the purposes of analyzing frequency behavior was introduced.

Modern Aspects of Power System Frequency Stability and Control.
DOI: https://doi.org/10.1016/B978-0-12-816139-5.00012-6

12.1.2 A summary of Chapter 2, What can provide frequency control?

In Chapter 2, What can provide frequency control? we concentrated on what are the available methods for providing "frequency control." First, we looked at traditional sources such as synchronous generators and their various modes of operation.

After the clarification of some terminology, we turned to the concept of "continuous response" and its effect on the "system frequency." We then looked at the concept of "step-change" response and its effect on the "system frequency."

Next in Chapter 2, What can provide frequency control? we considered what are the newer providers of "frequency control" such as wind turbine generators and solar photovoltaic (solar PV) panels. Next, we looked at the concept of "demand-side management," its objectives and applications. Then we introduced the idea of "smart grids," what they are and their possible link with frequency control.

In the next section the issue of "system inertia" and its important relationship with "frequency control" was discussed. The ideas of "synthetic inertia" and "fast frequency response" were introduced in relation to "frequency response" and the need for adequate "system damping."

12.1.3 A summary of Chapter 3, Per unit systems for frequency analysis

In Chapter 3, Per unit systems for frequency analysis, we covered aspects of the "per unit" system that is used in the remainder of the book. First, "per unit" systems in general are introduced, and their properties are described.

Then "per unit" systems as applied to individual "synchronous machines" are discussed. After this, the extension of "per unit" systems to "bulk systems," such as the whole power system, are described. The "MVA rating" and "moment of inertia" as applied to the whole power system are covered. This leads to a definition of the "*H*-constant" that describes the "moment of inertia" for the whole system.

We then went on to look briefly at some of the practical applications of the "per unit" systems considered later in the book.

In the next section of Chapter 3, Per unit systems for frequency analysis, we examined the fundamental relationship between "frequency stability" and "system inertia." The question of whether an "AC" power system could operate without "inertia" in the conventional sense of the word was posited.

12.1.4 A summary of Chapter 4, Initial analysis of the frequency control problem and a derivation of the swing equation

In Chapter 4, Initial analysis of the frequency control problem and a derivation of the swing equation, our main aim was to derive the "swing equation," which governs the change in frequency on a power system following an event that disturbs the balance in power between generation and demand.

We began by clarifying some of the terminology relating to the use of the term "intact network" and its relationship to the associated concepts of "faults" and "maintenance outages."

Next, we looked at the concept of the power balance on the wider system and of dynamic changes on it. We considered also the power balance within an individual synchronous machine.

The imbalance following the loss of generation or demand from the system was examined in relation to "rotational kinetic energy" and the power balance of a synchronous machine.

We then moved on to a step-by-step derivation of the "swing equation," first in terms of "physical units" and then converted to "per unit" format.

12.1.5 A summary of Chapter 5, Techniques available for calculating frequency response requirements

In Chapter 5, Techniques available for calculating frequency response requirements, we began by introducing the tools and techniques available for calculating "frequency response requirements." We noted that this involved constructing a useable standard form of the "swing equation" for subsequent analysis in the following chapters.

Next, we looked at the power system frequency and frequency requirements during "normal operation," including a consideration of the "quiet system" and of various "frequency events" that were likely to happen on it.

We then considered the time periods of frequency evolution following a system loss. We looked at a major loss of generation from the system and how plant, such as synchronous machines, reacted to such a loss. We also looked at the consequences of a loss of load (demand) from the system and the division into time periods of the time following a loss.

After that, we approached the question of how to solve the "swing equation" to help up with the critical issue of calculating the "frequency response requirements" for our system.

We then began to examine the structure of the "swing equation" and how to model the system. Different ways of solving the "swing equation," such as the "spreadsheet method", the "control diagram" approach, and so on were described.

We looked in more detail at the "swing equation" itself and how to model the different aspects of the dynamic changes within the power system with it. The system balance during the primary, secondary, and tertiary time periods were considered in detail, step by step.

The "variation of demand with frequency" effect was introduced and how to model it. The form of the variation of responsive power with frequency in the case of a synchronous machine was also analyzed.

The modeling of the tripping of demand blocks using the "unit step function" was introduced.

The output profile with time of a "general responsive device" was also introduced with a possible technique for modeling it.

We put together a standard form for the "swing equation" considering in detail each component. Finally, we listed the analytical and numerical methods that could potentially be used to solve it. These techniques would be analyzed in detail in the following chapters of the book.

12.1.6 A summary of Chapter 6, Analytical solutions

In Chapter 6, Analytical solutions, we turned our attention to the solution of the "swing equation" by analytical methods.

The first analytical method we considered was the "direct solution of the differential equation" method. The second analytical method we considered was the "solution by Laplace transforms."

In both methods, we started from a standardized form of the "swing equation" which contained the elements of "responsive generation," "nonresponsive generation," "demand (load)," "generator governor" representation, the representation of the "variation of demand with frequency," "tripped" generation or load, "general responsive profile," system *H*-constant, and system starting frequency.

In both cases, we were obliged to approximate the $f \times \mathrm{d}f/\mathrm{d}t$ term by $f_0 \times \mathrm{d}f/\mathrm{d}t$ to make the solution method workable.

In the "direct solution of the differential equation" method, we were able to solve the approximated "swing equation" by the method of applying an "integrating factor." We found that it was a straightforward, albeit tedious, process to relate the coefficients in the format of the solution back to the coefficients in the approximated "swing equation."

It was decided to restrict the polynomial in time t in the expression for the "general responsive profile" to the first two terms thus making it linear. The reasons behind this decision were that (1) the equation was much easier to solve; and (2) any arbitrary profile could be represented by a concatenation of linear segments.

It was found that although the "solution by Laplace transforms" looked analytically different from the "direct solution of the differential equation" method, the solution obtained was fundamentally the same. Hence, the closed solution provided by the latter method was if anything to be preferred although the former method dealt more easily with the step function.

Both the basic solution with the responsive synchronous machines initially at full output and the solution where the responsive synchronous machines were initially de-loaded were derived.

At the end of the chapter, there is a section dealing specifically with the advantages and disadvantages of the "direct solution of the differential equation" and "Laplace transform" methods.

12.1.7 A summary of Chapter 7, Numerical methods for solving the "swing equation"

In Chapter 7, Numerical methods for solving the "swing equation," we dealt with a range of *numerical methods* that can be used to solve the "swing equation" in the form of an "initial value problem (IVP)."

The numerical methods discussed in Chapter 7, Numerical methods for solving the "swing equation," were as follows: (1) the "Euler's" method, (2) the "modified Euler's" method, (3) the "Runge–Kutta" methods, (4) the "Milne–Hamming" method, and (5) the "Adams–Bashforth–Moulton" method. Chapter 7, Numerical methods for solving the "swing equation," begins by listing the numerical methods to be discussed. The "swing equation" is confirmed to be a member of the set of problems known as "IVPs" for short. The "swing equation" in its approximated form is also identified to be "linear" and of "first order."

The "Taylor series expansion" as used in numerical methods is introduced. "Forward difference" and "backward difference" representations are described. The concepts of "single-step" and "multistep" methods are introduced. The effect of the size chosen for the "time step" on the "stability" and "accuracy" of the numerical solution is discussed. The concept of the "truncation error" is introduced.

The "Euler's" method is the simplest method of the five methods considered in this book. It works by updating the function at the nth step by using the derivative of the function evaluated at that step. It is a "single-step" method that has a "local truncation error" of $O(h^2)$ and is therefore categorized as a "first-order method." Its accuracy can be increased by decreasing the "step size" h. Also, the "stability" of "Euler's" method may be improved by halving the "step size" at each iteration.

The "modified Euler's" method is an improved version of the "Euler's" method in which the time step is calculated in two stages: a "predictor" stage and a "corrector" stage. It is therefore a simple example of a "predictor–corrector" method in which the "corrector" stage uses an average of the derivative of the function at the nth and $(n+1)$th time steps. This method has a "local truncation error" of $O(h^3)$ and so is classified as a "second-order method."

The "Runge–Kutta" methods are single-step methods in which each step is divided into subintervals. Weighting factors are used to determine the function values at points within the step. "Runge–Kutta" methods come in different "orders." In this book, we have used the "classic fourth-order Runge–Kutta" algorithm to solve the "swing equation." This version of the "Runge–Kutta" algorithm has a "local truncation error" of $O(h^5)$ and is therefore a "fourth-order method."

The "Milne–Hamming" method is a "multistep" method. Function and derivative values are required at several steps to begin the algorithm. Another algorithm is used to accomplish this. In the example given in this book the "Runge–Kutta" algorithm was used. The "Milne–Hamming" method chosen in this book is a "predictor–corrector" method with a "local truncation error" of $O(h^5)$ and is therefore a "fourth-order method."

The "Adams–Bashforth–Moulton" method is also a "multistep" method and a "predictor–corrector" method. Like the "Milne–Hamming" method the version used in this book has a "local truncation error" of $O(h^5)$ and is therefore a "fourth-order method."

All the earlier numerical methods may be used for solving the "swing equation": each method has its own positive and negative features that were discussed at the end of the chapter. A MATLAB example was provided for each of the methods.

12.1.8 A summary of Chapter 8, The "control diagram" method

In Chapter 8, The "control diagram" method, we looked at an alternate way of approaching the solution of the "swing equation" by representing it in the form of a "control" or "block" diagram. This method uses a "control theory" approach to solving the equation.

First, we converted the "swing equation" into "state-space" format. Next, we converted the equation from "state-space" format into "transfer function" format. Following this, the equation was converted from "transfer function" format into a "control" or "block" diagram.

The "control diagram" format was then used to find the solution $f(t)$ for each time step in turn until the end of the chosen period of study to produce the complete evolution of frequency with time.

12.1.9 A summary of Chapter 9, Some important practical applications

In Chapter 9, Some important practical applications, we looked at some important practical applications of the "swing equation" and its solution: (1) the "rate of change of frequency" (RoCoF) calculation; (2) the "system response requirements" following a "low frequency excursion"; (3) the "system response requirements" following a "high-frequency excursion"; and (4) the "system response requirements" during "normal operation."

We saw that in the case of the "RoCoF calculation" it was not necessary to solve the "swing equation," it was only necessary to evaluate the system power balance immediately following the loss of an amount generation or demand from the system.

For the "system response requirements" following a "low frequency excursion" caused by the loss of an amount of generation from the system it was necessary to solve the "swing equation" using one of the methods described in the preceding three chapters to find the subsequent evolution of frequency with time and whether this complies with the relevant standards.

For the "system response requirements" following a "high-frequency excursion" the difference from the situation in the previous paragraph is that the loss is of demand rather than of generation. Otherwise the "swing equation" must still be solved to find the evolution of frequency with time and whether this complies with the relevant standards.

Finally, for the case of finding "system response requirements" during "normal operation" the foregoing two methods were combined to find both a positive and negative requirement relevant to cover relatively small variations in the system power balance from minute to minute in the absence of any significant loss.

12.1.10 A summary of Chapter 10, The challenges of operating systems with high penetrations of renewables ("low inertia systems")

In Chapter 10, The challenges of operating systems with high penetrations of renewables ("low inertia systems"), we dealt with "low inertia systems." These are systems in which there is high penetration of generators with either no inertia in the conventional sense [high-voltage direct current links (HVDC), solar PV installations] or the generator concerned does possess inertia in the conventional sense of the word, but it is inaccessible to the grid system [modern wind turbine arrays (wind farms)] and so cannot assist with frequency control.

We looked at four relatively new concepts: "synthetic inertia," "fast frequency response," "virtual synchronous machines (VSMs), and "energy storage devices," all of which can in theory contribute to the bolstering of the frequency stability of a system.

The effect of "HVDC links" on the "system inertia" and the provision of "frequency response" by "HVDC links" were mentioned. We saw that "HVDC links" can be either "importing" or "exporting," thus acting as either a generator or a load from the system point of view.

The modeling of wind farms and their effect on "system inertia" and consequently "system stability" were discussed. Different designs of wind turbine were looked at. "Converter-based" and "nonconverter based" wind farms and the provision of "frequency response" by wind farms themselves were discussed.

Finally, in Chapter 10, The challenges of operating systems with high penetrations of renewables ("low inertia systems") the modeling of "solar PV farms" was considered. Their ability to provide "frequency regulation" and "voltage regulation" (by means of a converter) was mentioned. Both the effect of tripping a "solar PV farm" and the provision of "frequency response" by a "solar PV farm" were covered.

12.1.11 A summary of Chapter 11, "Smart grids" and the "system frequency"

In Chapter 11, "Smart grids" and the "system frequency," we looked at the concept of "smart grids" and their possible implication for the "system frequency" and for providing "frequency control" themselves.

We began by defining what is meant by a "smart grid" and looked at some possible definitions. We mentioned two international agreements which have a bearing on "smart grid" development, the "Kyoto Protocol" and the "Paris Agreement (Accord de Paris)."

We then moved on to examine the possible structures of "smart grids" and their position in the electrical supply network.

Next, we looked at the possible methods available to optimize the operation of "smart grids," citing techniques such as "genetic algorithms" and "artificial neural networks."

We then looked at the potential effects of "smart grids" on the "system frequency." Elements of the power system such as "combined heat and power" schemes, "biomass plants," and "controllable domestic loads" were discussed.

The control architectures of "smart grids" were considered, including "multiagent systems" and their structural elements.

We finished off the chapter with an example in MATLAB of the possible effect of "Smart Grids" on the "daily load profile."

12.2 What is the best method for solving the "swing equation"?

12.2.1 A summary of the methods covered in this book

In this book, we have looked at two analytical approaches (the "direct solution of the differential equation" method and the "Laplace transform method"), five numerical methods ("Euler's method," the "modified Euler's method," the "Runge–Kutta method," the "Milne–Hamming method," and the "Adams–Bashforth–Moulton method"), and one further approach (the "control diagram method"). Which of these is the best? Before we can answer this question, we shall look at the characteristics of these eight different methods based on the following criteria: (1) "numerical accuracy," (2) "numerical stability," and (3) "ease of use."

12.2.2 The "direct solution of the differential equation" method

1. *Numerical accuracy*

   Except for the approximation of "$f \times df/dt$" as "$f_0 \times df/dt$," this method gives a "perfectly accurate" solution if, in addition, the step functions is treated as a boundary condition where two solutions join.
2. *Numerical stability*

   Numerical stability is not an issue with this method because it is an analytical method.
3. *Ease of use*

   This method is easy to use, except where there is a step change in the power balance other than that at $t = 0$, when extra work must be done to deal with the boundary conditions involved to ensure continuity of the "system frequency" there.

12.2.3 The "Laplace transform" method

Although the "Laplace transform" method is a different approach from the "direct solution of the differential equation" method, we found that the difference is only superficial because the same solution is reached in the end. Therefore it is not really an independent method.

12.2.4 The "Euler's" method

1. *Numerical accuracy*

   The "Euler's" method is not particularly accurate. It has a "local truncation error" of only $O(h^2)$. This problem can to some extent be overcome by decreasing the step-size h at the expense of increasing the number of iterations used.
2. *Numerical stability*

   "Euler's" method has been found in the past to have stability issues. The standard method used to overcome this problem was to halve the step-size h at each iteration.
3. *Ease of use*

   The "Euler's" method is the simplest of the numerical methods to be considered in this book and is the easiest of the numerical methods to use.

12.2.5 The "modified Euler's" method

1. *Numerical accuracy*

   The "modified Euler's" method is an order of magnitude more accurate than the simple "Euler's" method, having a "local truncation error" of $O(h^3)$.

2. *Numerical stability*

   The "modified Euler's" method is numerically more stable than the simple "Euler's" method.
3. *Ease of use*

   The "modified Euler's" method is more difficult to use than the "Euler's" method because at each iteration it requires function evaluations at more than one point.

12.2.6 The "Runge–Kutta" methods

1. *Numerical accuracy*

   The standard "Runge–Kutta" method used in this book has a "local truncation error" of $O(h^5)$ and so is two orders of magnitude more accurate than the "modified Euler's" method. More and less accurate "Runge–Kutta" methods are available.
2. *Numerical stability*

   According to Lapidus and Seinfeld (Chapter 7: Numerical methods for solving the "swing equation," Ref. [1]), all explicit "Runge–Kutta" methods are stable for small enough h. Also, as the order of the method increases, so does its "region of stability."
3. *Ease of use*

   Being a single-step method, the "Runge–Kutta" methods are relatively straightforward to use.

12.2.7 The "Milne–Hamming" methods

1. *Numerical accuracy*

   As with the "Runge–Kutta" methods, there are "Milne–Hamming" methods of different orders available. In this book, we have considered a "Milne–Hamming" method of fourth order, that is, whose "local truncation error" is of $O(h^5)$. This is comparable with the order of the "Runge–Kutta" method we chose and therefore of similar accuracy.
2. *Numerical stability*

   The fourth-order "Milne–Hamming" method has a "region of stability" about one quarter that of the fourth-order "Runge–Kutta" method.
3. *Ease of use*

   Being a "multistep" method, the "Milne–Hamming" method requires another method to initiate it, and function evaluations are required at multiple steps at each iteration. In these respects, it is more complicated to use than the "Runge–Kutta" method.

12.2.8 The "Adams–Bashforth–Moulton" methods

1. *Numerical accuracy*

   As with the "Milne–Hamming" and "Runge–Kutta" methods, there are "Adams–Bashforth–Moulton" methods of different orders available. In this book, we have considered an "Adams–Bashforth–Moulton" method of fourth order, that is, whose "local truncation error" is of $O(h^5)$. This is same as the order of the "Runge–Kutta" method and the "Milne–Hamming" method we chose and therefore of comparable accuracy.
2. *Numerical stability*

   The fourth-order "Adams–Bashforth–Moulton" method has a "region of stability" of about twice that of the fourth-order "Milne–Hamming" method, but only about half that of the fourth-order "Runge–Kutta" method.
3. *Ease of use*

   In terms of ease of use the "Adams–Bashforth–Moulton" methods are comparable with the "Milne–Hamming" methods.

12.2.9 The "control diagram" method

1. *Numerical accuracy*

   Along with the "direct solution of the differential equation" method and the "Laplace transform" method, the numerical accuracy of this method is limited only by the numerical accuracy of the arithmetical operations performed on the machine on which it is programed.
2. *Numerical stability*

   Again, numerical stability is not an issue with this method as it is an analytical method.
3. *Ease of use*

   Despite the attraction of its graphical representation in the form of a "control diagram," this method is not particularly easy to use because it requires quite a lot of arithmetical operations before the solution can be arrived at. As such, it is more difficult to use than, say, one of the numerical methods considered.

12.2.10 The verdict

Of the methods considered in this book the method with the best balance of (1) numerical accuracy, (2) numerical stability, and (3) ease of use is the *fourth-order Runge–Kutta method*, whose accuracy is comparable with the *fourth-order Milne–Hamming method* and the *fourth-order Adams–Bashforth–Moulton method* but having a "region of stability" significantly greater than that of both these two "predictor–corrector" methods. It also deals in a natural way with "step functions" introduced into the system power balance without having to worry about satisfying boundary conditions.

12.3 Does a shortfall of inertia on the system matter?

12.3.1 The original design structure of a purely "AC" system

In a traditional "AC" system the generation side is composed wholly of synchronous machines. The demand (loads) are composed of a mixture of synchronous and asynchronous (induction motor) elements. Inertia in this kind of system is provided naturally by the synchronous generators and the asynchronous loads. There is no shortage of inertia because the machines that generate power have inertia of their own provided by their solid, rotating parts. The asynchronous loads (induction motors) also have solid, rotating parts and, therefore, contribute to the total inertia of the system.

12.3.2 The mixed "AC" and "DC" system: the decline of inertia

When importing "DC links," "wind farms" employing AC–DC converter technology and "solar farm" are introduced into an "AC" system; the inertia of that system is reduced because these additional sources of generation do not have their own inertia. Since the average inertia is an important factor in maintaining the dynamic stability of the power system, some alternative means of inertia must be found to redress the balance.

12.3.3 Mitigation measures to counter a shortage of inertia on the system

If we assume that introducing more synchronous generators to the system is out of the question, there are basically two kinds of remedial methods to counteract the reduction of inertia on a power system: (1) add actual inertia contained in technology such as synchronous compensators to the system; or (2) use power electronics to simulate the behavior of synchronous machines. At present, this means employing "synthetic inertia" or building a *VSM* by means of power electronics. Work is currently underway on both of these concepts plus the concept of "fast frequency response."

12.4 Where do we go from here?

12.4.1 Basic trends

If you have worked through the *MATLAB* examples in this book, you will have noticed the following trends: (1) the lower the proportion of synchronous generation there is in the system, the lower is the inertia H and the more difficult it is to control the system frequency following the loss of generation or demand; (2) the lower the system demand is, the lower the overall inertia of the system is.

12.4.2 Manipulation of the system demand

The abovementioned scenario of too little inertia on the system may in part be avoided, if by "SMART" technology the system demand can be evened out during the day and incidences of what are currently the lowest demands on the system can be avoided by increasing the demand during these periods.

12.4.3 Adding inertia to the system

Inertia may be added to the power system in the form of rotating machines such as "synchronous compensators." However, apart from the extra purchase and running costs involved, this may not be a potential solution, as with synchronous machines, that fits well with current environmental policy. The addition of flywheels to the system has also been suggested.

12.4.4 The "Virtual Synchronous Machine"

A potential solution in the form of the *VSM*, mentioned earlier, is a possible solution provided that the important physical parameters of the "synchronous machine" such as inertia and voltage support can adequately be synthesized by electronic means.

12.4.5 "AC" or "DC"?

What began as a purely "AC" system has become a mixed "AC" and "DC" system in some countries due to the addition of "HVDC links" and "back-to-back converters," for example, as included in modern designs of wind turbines.

12.4.6 Solar technology

Solar PV technology is increasing at a rapid rate in developed countries. This is both reducing the amount of inertia available on systems and changing the daily "load profile" as mentioned earlier. These and other changes are making the job of the control engineer increasingly more challenging as the "system revolution" progresses.

Appendix A

The "exact solution" of the "swing equation"

The "exact solution" of the "swing equation" in the form as stated in Eq. (6.1), namely

$$\sum P_{\mathrm{GEN(NR)}} + \sum P_{\mathrm{GEN}n\mathrm{(R)}} - \sum P_{\mathrm{LOAD}n} - k_1 \cdot (f - f_n) - P_{\mathrm{TRIP}} \cdot u(t - t_{\mathrm{TRIP}}) + a_0 + a_1 t = 2 \cdot H \cdot f_0 \cdot \frac{\mathrm{d}f}{\mathrm{d}t} \tag{A.1}$$

is given as Eq. (6.27) of Chapter 6, Analytical solutions, in the form of

$$f(t) = A_1 + B_1 \cdot t + C_1 \cdot \exp(D_1 \cdot t) \tag{A.2}$$

with the parameters A_1, B_1, C_1, and D_1 given by Eqs. (6.34–6.37), respectively, which we reproduce here for convenience:

$$A_1 = \frac{f_n + \left(a_0 + \sum P_{\mathrm{GEN(NR)}} + \sum P_{\mathrm{GEN}n\mathrm{(R)}} - \sum P_{\mathrm{LOAD}n} - P_{\mathrm{TRIP}} \cdot u(t - t_{\mathrm{TRIP}})\right)}{k_1} - 2 \cdot H \cdot f_0 \cdot \frac{a_1}{k_1^2} \tag{A.3}$$

$$B_1 = \frac{a_1}{k_1} \tag{A.4}$$

$$C_1 = \frac{f_0 - f_n - \left(a_0 + \sum P_{\mathrm{GEN(NR)}} + \sum P_{\mathrm{GEN}n\mathrm{(R)}} - \sum P_{\mathrm{LOAD}n} - P_{\mathrm{TRIP}} \cdot u(t - t_{\mathrm{TRIP}})\right)}{k_1} + 2 \cdot H \cdot f_0 \cdot \frac{a_1}{k_1^2} \tag{A.5}$$

$$D_1 = -\frac{k_1}{2 \cdot H \cdot f_0} \tag{A.6}$$

where

$$k_1 = \alpha \cdot \sum P_{\text{LOAD}n} + k \cdot \sum P_{\text{GEN}n(\text{R})} \tag{A.7}$$

in which the constant "α" is defined by the equation:

$$\sum P_{\text{LOAD}} = \sum P_{\text{LOAD}n}\{1 + \alpha \cdot (f - f_n)\} \tag{A.8}$$

and the constant "k" is defined by the equation:

$$\sum P_{\text{GEN(R)}} = \sum P_{\text{GEN}n(\text{R})} \cdot \{1 - k \cdot (f - f_n)\} \tag{A.9}$$

Also, equivalently we have

$$\sum P_{\text{GEN(R)}} = \sum P_{\text{GEN}n(\text{R})} \cdot \left\{1 - 100 \cdot \frac{(f - f_n)}{D(\%)}\right\} \tag{A.10}$$

where

$$\frac{100}{D(\%)} = k \tag{A.11}$$

in which $D(\%)$ is the generator governor droop in "%."

Some typical values for these two parameters are

$$\alpha = 2\% \text{ per Hz and } D = 4\%. \tag{A.12}$$

Appendix B

The "exact solution" of the "Swing Equation" when the power set point is not the nominal power output

When the power output of a responsive machine at the nominal frequency f_n is not chosen to be the rated or nominal power output, the equations in Appendix A require slight modification. First, let the proportion of the nominal output power at nominal frequency be designated by "p." Then the "Swing Equation" in the form given by Eq. (6.1) of Chapter 6, Analytical solutions, becomes

$$\sum P_{\mathrm{GEN(NR)}} + p\sum P_{\mathrm{GEN}n\mathrm{(R)}} - \sum P_{\mathrm{LOAD}n} - k_1\cdot(f - f_n) \\ - P_{\mathrm{TRIP}}\cdot u(t - t_{\mathrm{TRIP}}) + a_0 + a_1 t = 2\cdot H\cdot f_0\cdot\frac{\mathrm{d}f}{\mathrm{d}t} \tag{B.1}$$

where "p" is a parameter in the range 0–1 but a typical value of which is 0.9.

The form of the solution of Eq. (B.1) remains as given in Eq. (6.27) of Chapter 6, Analytical solutions:

$$f(t) = A_1 + B_1\cdot t + C_1\cdot\exp(D_1\cdot t) \tag{B.2}$$

but the definitions of two of the four parameters in this expression must be modified. In this expression the parameters A_1, B_1, C_1, and D_1 originally given by Eqs. (6.34–6.37), respectively, now become

$$A_1 = \frac{f_n + \left(a_0 + \sum P_{\mathrm{GEN(NR)}} + p\sum P_{\mathrm{GEN}n\mathrm{(R)}} - \sum P_{\mathrm{LOAD}n} - P_{\mathrm{TRIP}}\cdot u(t - t_{\mathrm{TRIP}})\right)}{\mathrm{k}_1} \\ - 2\cdot H\cdot f_0\cdot\frac{a_1}{k_1^2} \tag{B.3}$$

$$B_1 = \frac{a_1}{k_1} \tag{B.4}$$

$$C_1 = \frac{f_0 - f_n - \left(a_0 + \sum P_{\mathrm{GEN(NR)}} + p\sum P_{\mathrm{GEN}n\mathrm{(R)}} - \sum P_{\mathrm{LOAD}n} - P_{\mathrm{TRIP}} \cdot u(t - t_{\mathrm{TRIP}})\right)}{k_1} + 2 \cdot H \cdot f_0 \cdot \frac{a_1}{k_1^2} \tag{B.5}$$

$$D_1 = -\frac{k_1}{2 \cdot H \cdot f_0} \tag{B.6}$$

where

$$k_1 = \alpha \cdot \sum P_{\mathrm{LOAD}n} + k \cdot \sum P_{\mathrm{GEN}n\mathrm{(R)}} \tag{B.7}$$

In Eq. B.7 the "load damping" coefficient α is still defined via the relationship

$$\sum P_{\mathrm{LOAD}} = \sum P_{\mathrm{LOAD}n}\{1 + \alpha \cdot (f - f_n)\} \tag{B.8}$$

but the constant of proportionality "k" is defined by the modified equation:

$$\sum P_{\mathrm{GEN(R)}} = p\sum P_{\mathrm{GEN}n\mathrm{(R)}} - \sum P_{\mathrm{GEN}n\mathrm{(R)}} \cdot k \cdot (f - f_n) \tag{B.9}$$

but as before

$$k = \frac{100}{D(\%)} \tag{B.10}$$

Appendix C

List of MATLAB simulation exercises

Simulation no.	Section	Description
1	6.3	Using the "direct solution" method to solve the Swing Equation
2	6.5	Using the "Laplace transform" method to solve the Swing Equation
3	7.5	Using "Euler's method" to solve the Swing Equation
4	7.7	Using the "modified Euler's method" to solve the Swing Equation
5	7.9	Using the Classic Explicit fourth-order Runge–Kutta method to solve the SE
6	7.11	Using the Milne–Hamming method to solve the swing equation
7	7.13	Using the Adams–Bashforth–Moulton method to solve the SE
8	8.7	Using the "control system" approach to solve the swing equation
9	9.3	The rate of change of frequency (RoCoF) calculation
10	9.5	Calculating the response requirements for a low-frequency excursion
11	9.7	Calculating the response requirements for a high-frequency excursion
12	9.9	Calculating the response requirements for normal operation
13	10.3	The effect of HVDC links on the system frequency
14	10.5	The effect of wind farms on the system frequency
15	10.7	The effect of solar farms on the system frequency
16	11.3	Smart grids and the system frequency

Index

Note: Page numbers followed by "*f*" refer to figures.

A

AC. *See* Alternating current (AC)
Acceleration, 49
Accord de Paris. *See* Paris Agreement
Active power outputs, 41
"Adams–Bashforth–Moulton" multistep predictor–corrector method, 179–183, 294, 299. *See also* "Milne–Hamming multistep predictor–corrector" method
  formulating, 181–183
  fourth-order, 179
    truncation errors, 180
  to solving swing equation, 183–188
    general solution of problem, 184
    input data, 184
    instructions for user, 188
    MATLAB codes, 185–188
    output data, 184
    statement of problem, 183–184
  stability, 180–181
Alternating current (AC), 1, 3, 33, 245, 279–280
  frequency stability of AC machines in power systems, 246
  system, 44–45, 300–301
  transmission system, 245
Analytical methods, 81–84, 135
  "control diagram" approach to, 206
  "direct solution of differential equation" method, 82–83, 85–86
  solution by Laplace transforms, 83–86
  for solving "swing equation", 85–86
    examples to providing in MATLAB, 86
Analytical solution methods, 87–88, 292–293
  advantages and disadvantages, 132–133
    direct solution of differential equation, 132
  direct solution of differential equation, 87–99
  MATLAB example simulation, 100–105
  numerical methods for solving "swing equation", 133
  objective in, 87
  recalling equation to solving, 88
  solution by Laplace transforms, 87, 132–133
ANNs. *See* Artificial neural networks (ANNs)
Arbitrator agents, 283
Artificial neural networks (ANNs), 277
Asynchronous "squirrel cage", 256
Automatic
  frequency correction devices, 62
  generation control, 282–283
  governor action, 10–12
  response, 17
  voltage regulation device, 4
  voltage regulators, 35
Average power factor, 43

B

"Back-to-back" convertor type, 279–280
Backward difference
  methods, 138
  in numerical method, 138
  operator, 139
Base values, 51–52
Biomass
  combustion process, 280
  plants, 280
Bipolar "HVDC links", 248
Bulk generator base quantity, 41
Busbars, loss of, 18–19

C

C++ language, 69
Carbon dioxide (CO_2), 274
Cascade tripping, 9

CHP schemes. *See* Combined heat and power schemes (CHP schemes)
Classic explicit fourth-order "Runge–Kutta" method formulation, 161–163
Classic fourth-order Runge–Kutta" method, 168
"Collatz Rule of Thumb", 160
Combined heat and power schemes (CHP schemes), 278–279
Commercial service, 24
Communication and control infrastructure-based concept, 272
Commutation failure, 36
Continuous response, 25
  effect, 25
  example, 25
  system frequency measurement, 25
Control architectures of smart grids, 282
"Control diagram" method, 84, 191–192, 206, 294
  to analytical methods, 206
  differential equation, 191
  initial condition in, 199
  multiple iterations of, 201
  numerical methods, 206
  overall objective of process, 199
  practical applications, 206–209
  ramped response, 201–206
  reacts to status of "input", 200
  representation, 69–70, 192
  "single iteration" using, 200
  "state-space" format to "transfer function" format, 196–197
  swing equation, 192–193
    representation in "state-space" format, 193–196
  "transfer function" format to, 197–199
  uniqueness, 206
Control engineer task, 17
Controllable commercial loads, 281
Controllable domestic loads, 281
Controlled tripping, 61
Convergence
  of calculation process, 201
  tolerances, 201
Converter-based wind farm, 258
Converter-based WTGs, 257
Convolution theorem, 112–113
"Corrector" for "Milne–Hamming" method, 169

D

Daily demand variations, 19–20
Daily load cycle, 7–8, 7*f*, 211, 274–275
  modern changes to shape, 8
  system requirements variation with, 5–6
Daily load profile, 281–284
DC. *See* Direct current (DC)
De-loaded plant, 16
Deceleration, 49
Degree of part-loading, 221
Demand
  participation signals and options, 282
  shedding, 17
  trips/tripping, 16, 24, 33, 61, 67
    causes, 16–17
    relays, 27
Demand blocks, 76–77
  loss of, 16–17
    causes of demand trips, 16–17
    demand shedding, 17
    demand trips, 16
    high-frequency event, 17
    loss size, 16
    task of control engineer, 17
"Demand variation with frequency" effect, 74
Demand-side management (DSM), 27, 30–31
  applications, 31
  examples, 31
  objectives, 30
Demand-side response. *See* Demand-side management (DSM)
"Dependency of load on frequency" effect, 213–214
"Dependent" variable, 137
DER. *See* Distributed energy resources (DER)
DFIG. *See* Doubly fed induction generator (DFIG)
Diagnosis agents, 283
Differential equation, 191
  with additional simplification, 90
  solving, 91–93
Direct current (DC), 1, 3, 300–301
"Direct solution of differential equation" method, 82–83, 85–87, 250, 292, 297. *See also* Euler's method
  with additional simplification, 90
  advantages and disadvantages, 132
  a_0 and a_1 parameter, 98
  alternative method for relating constants, 96–98

checking on accuracy of solution, 93
"comparison of Laplace transform" solution with solution from, 120–123
H parameter, 98
$k1$ parameter, 98
noting form of solution, 94
relating constants, 94–96
remarks on statements, 89–90
to solving "swing equation", 100–105
solving differential equation, 91–93
solving mathematical problem, 88–89
variation of demand with frequency, 99
Distributed energy resources (DER), 272
Distribution automation, 282
Distribution system (DS), 274
wind turbines in, 279–280
DOE. *See* US Department of Energy (DOE)
Domestic loads, 281
Domestic wind turbines, 279–280
Doubly fed induction generator (DFIG), 28, 29*f*, 247, 255*f*
Driving function, 91
Droop characteristic, 11, 11*f*, 13
Droop profile, 71
of synchronous machine, 74–75, 75*f*
DS. *See* Distribution system (DS)
DS operator, 281
DSM. *See* Demand-side management (DSM)

E

Economy of "HVDC links", 248
Electric vehicles (EV), 31, 275
Electrical power output, 49
Electronic control of frequency, 5
Energy
efficiency, 282
storage, 31
devices, 247
Equivalent system inertia, 42
Erosion of inertia from power systems, 246
ESs. *See* Expert systems (ESs)
Euler's method, 144–147, 293, 297. *See also* Direct solution of differential equation" method; Modified Euler's method
formulation for solving "swing equation", 145–147
MATLAB example simulation, 147–149
general solution of problem, 148
input data, 148
instructions for user, 149
output data, 148
recommended code, 148–149
statement of problem, 147–148
numerical stability of, 144–145
"truncation error" associated with, 144
EV. *See* Electric vehicles (EV)
Evolution of power system frequency, 64–65
Evolutionary time periods "postevent", 55
Exact solution of differential equation" method, 114, 191
Exact solution of "swing equation", 84, 228, 233, 236, 239–241, 303–305
Expert systems (ESs), 277
Exporting
case, 249
HVDC link, 249
interconnector, 252
link, 250
External "HVDC links", 248

F

Fast frequency response (FFR), 35, 247
drawbacks, 35
techniques, 44–45
Fault detection agents, 283
FFR. *See* Fast frequency response (FFR)
First-order differential equation, 91
First-order ordinary differential equations, numerical techniques for solving, 143
First-order Runge–Kutta method, 159
Forward difference
methods, 138
in numerical method, 138
operator, 138–139
Fossil fuels, 280
power stations, 246
Fourth-order "Adams–Bashforth–Moulton" method, 179
truncation errors, 180
Frequency. *See also* High-frequency
error, 4
frequency-dependent loads, 74
effect, 12–13
per unit systems for frequency analysis, 36–37
solar photovoltaic farms in frequency studies, 262–263
stability, 44–45
of AC machines in power systems, 246
initial analysis of frequency stability problem, 45

Frequency control, 1–5. *See also* Traditional providers of frequency control
continuous requirement to regulating system frequency, 4
deviations of system frequency, 3–4
initial analysis of frequency control problem, 291
methods of regulating system frequency, 4
modeling of power system for analyzing frequency behavior, 21–22
modern challenges in, 20–21
in modern era, 5
monitoring of system frequency in real time, 19–20
need for, 289
new providers, 27–32
DSM, 30–31
high-voltage-connected batteries, 29–30
HVDC links, 27
smart grids, 31–32
solar PV system, 28–29
system inertia, 32–36
wind turbines, 27–28
normal operation of power system, 3
operating condition of synchronous machines, 3
providers, 290
and stability, 1
system at risk, 5
system impedances, 4
system requirements, 5–6
intact system, 6–8
loss of demand block, 16–17
loss of generation set, 8–16
loss of lines, transformers, busbars, and transmission equipment, 18–19
traditional aspects, 1
Frequency response, 24–25
erosion, 14–15
frequency-responsive mode, 23–24, 71, 213–214, 238
frequency-responsive plant, 219
by HVDC links, 249
by solar PV farms, 263
by wind farms, 258
Frequency response requirements, 65. *See also* High-frequency; System response requirements
approaching calculation, 52
evolutionary time periods "postevent", 55
modeling details, 52–53
during normal operation, 55
power system frequency and, 56–58
solution techniques, 53
swing equation
approaching solution of, 63–81
techniques for solving, 56, 81–85
usable form of, 55–56
techniques for calculation, 55, 291–292
time periods of frequency evolution following system loss, 58–63
"Fully loaded" machine, 221

G

Gain, 197
Gasification, 280
GEN (NR). *See* Nonresponsive generation (GEN (NR))
GEN (R). *See* Responsive generation (GEN (R))
General responsive profile (GRP), 77–78, 106
approximating "GRP" term, 80
modeling of, 77–78
Generation, 21
rescheduling, 15–17
Generation set, loss of
automatic governor action, 10–12
effect of frequency-dependent loads, 12–13
and generation rescheduling, 15–16
initial fall in frequency, 10
partial loading of generating plant, 13–15
primary causes, 9
RoCoF relays, 10
selection of governing action, 12
system requirements, 8–16
Generator response characteristic, 214
Genetic algorithms, 277
Governor
action effect, 213
"droop", 247
characteristic, 213–214
response, 12, 23–24
Green energy sources, 20
Greenhouse gases, 273
GRP. *See* General responsive profile (GRP)

H

H-constant for whole system, 43–44
High voltage (HV), 274
tolerance, 239
High-frequency. *See also* Frequency response requirements

event, 17
excursion, 208, 233–238
calculation of loss of demand, 232
procedure for calculating loss of generation, 232–233
problem, 233–234
response requirement, 238
High-level computer language, program coding in, 69
High-voltage direct current links (HVDC links), 5, 27, 245, 295
effect
on system frequency, 249–253
on system inertia, 248–249
frequency response by, 249
in system frequency studies, 248–249
High-voltage power system (HV power system), 28
High-voltage-connected batteries, 29–30
Higher order "Runge–Kutta" methods, 159
HV. *See* High voltage (HV)
HV power system. *See* High-voltage power system (HV power system)
HVDC links. *See* High-voltage direct current links (HVDC links)
Hybrid systems, 20–21
Hydrofluorocarbons (HFCs), 274

I

Imbalance, 49–52
assumption, 51
internal power balance, 49
"per unit" form
step by step conversion, 52
of "swing equation", 51–52
residual rotational power, 50
rotational kinetic energy, 50
step by step derivation of "swing equation", 50
Importing
case, 249
HVDC link, 249
interconnector, 251
link, 250
Independent variable, 137
Individual machines, 40
kinetic energies, 42–43
power balance, 49, 49*f*
Induction
generator, 254
motors, 12–13, 214–215
Industrial loads, 281
Inertia, 220
adding to system, 301
constant, 39–40
erosion, 246–247
shortfall of, 300
mitigation measures, 300
mixed "AC" and "DC" system, 300
original design structure of purely "AC" system, 300
Inertia-less power, 28
Inertia-less systems, system frequency in, 45
Inertial response, 12, 35, 60, 259–260
Inference engine, 277
Information distribution, 194
Initial frequency, 215
Initial value problem (IVP), 136, 293
Instantaneous demand tripping, 77
Instantaneous trip, 195
Intact system, 248
balance of power, 6–7
daily load cycle, 7–8, 7*f*
intact power system, 47–48
elements of fundamental balance in, 48–49
modern changes to shape of daily load profile, 8
solar PV farm in, 262–263
system requirements for, 6–8
"Integer" variable, 276–277
Integrating factor, 92
Interconnected mode, 283
Intermittency
compensating for, 31
of wind power, 280
Internal power balance, 49
Internal HVDC links, 248
Inverse Laplace transforms, 101
Isochronous governors, 24
IVP. *See* Initial value problem (IVP)

J

Jettisoning of blocks of demand for response, 24
Judgment Index Agent, 283

K

Kinetic energies of individual machines, 42–43
Knowledge base, 277
Kyoto Protocol, 273

L

Laplace transforms, 109–110, 297
advantages and disadvantages, 132–133
analytical solution by, 87
approach, 83–86
approximation in method, 107–109
piecewise linear representation, 108*f*
comparison with solution from direct solution of differential equation, 120–123
of constant terms, 110
equation, 117
of individual terms, 109
method to solve “swing equation”, 123–132
of quadratic form, 111
solution of swing equation by, 105, 111
approximating f(t) df/dt term, 114–115
approximating GRP term, 115
comments on solution, 120
dealing with f · df/dt term, 112
final stages of calculation, 116–117
form of solution, 119
needs to making additional approximation, 114
processing the individual terms, 105–107
putting solution together, 118–119
trying to using “integration by parts” on f · df/dt term, 113
unit step term, 115
of terms linear
in time (t), 110
in unknown function (f), 110
of terms of higher order, 111
of terms proportional to first derivative, 111
trying to using convolution theorem for, 112
of unit step function, 110
Large batteries, 29–30
Linear optimization techniques, 276
Linearity of equation, 137
Lines, loss of, 18–19
“Load damping” effect, 88, 99, 213
Load shedding. *See* Demand shedding
Load trips. *See* Demand trips
Load-damping
coefficient, 223–224, 235
constant, 284–285
Local truncation error
of Euler’s method, 144
of Runge–Kutta methods, 160
Longer term timescales, 48
“Loss of grid” condition, 212
“Loss of mains or islanding”. *See* “Loss of grid” condition
Loss(es), 18
in bulk of generation and demand, 18
of busbars, 19
of demand, 61–62
block, 16–17
of equipment generators or loads, 18
of generation set, 8–16
of grid/loss of mains/islanding protection, 212
of lines, 18
of transformers, 19
Low frequency excursion, 207–208, 236–237
response requirements for, 222–231
system response requirements for, 219–222
Low inertia systems, 22, 295–296
Low voltage (LV), 274
Low-inertia systems, 243
alternating current transmission system, 245
closure of fossil-fuel power stations, 246
developing challenge of system inertia, 245
energy storage devices, 247
erosion of inertia from power systems, 246
FFR, 247
frequency stability of AC machines in power systems, 246
MATLAB example simulation, 249–253, 258–268
modeling
of HVDC links in system frequency studies, 248–249
of solar photovoltaic farms in frequency studies, 262–263
of wind farms in system frequency studies, 253–258
provision of frequency response by HVDC links, 249
synthetic inertia, 246–247
VSM, 247
LV. *See* Low voltage (LV)

M

Machine torques, 49
Magnitude of “truncation error”, 143
“Major frequency events” on system, 56–57
Manual actions, 15–16
MASs. *See* Multiagent systems (MASs)

MATLAB code, 56, 228–231
"Adams–Bashforth–Moulton" method, 185–188
for "classic fourth-order Runge–Kutta" algorithm, 165–166
in "direct solution of differential equation" method
input data, 102–103
output data, 104
recommended code, 104
for Euler's method, 148–149
input data to, 218
in Laplace transform method
input data, 127
output data, 127
recommended code, 128–131
Milne–Hamming method, 175–178
input data to, 175
output data from, 175
for modified Euler's method, 157–158
output datum from, 218
suggested, 218–219, 236–238, 242, 252–253, 261–262, 267–268, 286
MATLAB example simulation, 216–219, 222–231, 233–238, 240–243, 249–253, 258–268, 283–287, 307
comparison of Laplace transform" solution with solution from method 1, 120–123
direct solution of differential equation" method to solve "swing equation", 100–105
exporting
case, 249
link, 250
general instructions for user, 219, 243
general solution to problem, 217–218, 223–228, 240, 250–251, 259–260, 264–265, 284–285
high frequency excursion problem, 234–236
importing
case, 249
link, 250
input data, 228, 236, 241, 251–252, 260–261, 266–267, 285
generating capacity left over after selecting responsive sets, 266
limit on number of sets selectable for response, 266
selection of sets, 266
instructions for user, 231, 238, 253, 262, 268, 287
"Laplace transform" method to solve "swing equation", 123–132
output data, 228, 236, 242, 252, 261, 267, 286
statement
of high-frequency problem, 233–234
of problem, 217, 222–223, 240, 250, 258–259, 263–264, 284
Mechanical power input, 49
Medium voltage (MV), 274
Merit order, 220–221
Methane (CH_4), 274
Microgrids, 283
"Milne–Hamming multistep predictor–corrector" method, 167–172, 294, 298. *See also* "Adams–Bashforth–Moulton" multistep predictor–corrector method
formulating, 170–172
to solving swing equation, 172–178
general solution to problem, 174–175
input data to MATLAB code, 175
instructions for user, 178
MATLAB codes, 175–178
output data from MATLAB code, 175
statement of problem, 173–174
stability of, 170
truncation errors, 169–170
"Minor frequency events" on system, 57
"Mixed integer" programming techniques, 276–277
Modified Euler's method, 150, 293, 297–298. *See also* Euler's method
formulation, 150–151
for solving "swing equation", 152–154
MATLAB example simulation, 154
general solution of problem, 155–156
input data, 156
instructions for user, 158
output data, 157
recommended code, 157–158
statement of problem, 154–155
"truncation error" associated with, 151–152
Moment of inertia, 41–42
Monopolar HVDC links, 248
Multiagent systems (MASs), 282–283
Multimachine system modeling, 70–74
power balance in preevent system, 71–72
during secondary and tertiary response periods, 73–74
system power balance

Multimachine system modeling (*Continued*)
immediately postevent, 72–73
during primary response period, 73
"Multiple iterations" of "control diagram" method, 201
"Multistep predictor–corrector" methods, 167–168
"Multistep" integration method, 139–140
MV. *See* Medium voltage (MV)

N

National transmission system, 1, 2*f*
"Negative feedback" systems, 10–11
Network analysis, 282–283
tool, 65–68
Newton, Isaac, 142
backward difference formula, 142
finite interpolation formulas, 142
forward difference formula, 142
Nitrous oxide (N_2O), 274
Nominal frequency, 4, 213, 239
Nonconverter-based wind farm, 257
Nonconverter-based WTGs, 257
Nonfrequency responsive mode, 23–24
Nonisochronous governors, 24
Nonlinear optimization techniques, 276
Nonresponsive generation (GEN (NR)), 212
Nonresponsive generators, 71–72
Normal band of operation, 239, 241
Normal operation, 208–209
calculation, 239
frequency, 239
range, 219
response requirements during, 239
Numerical accuracy, 297–299
Numerical methods, 84–85
advantages, 189
advantages and disadvantages of, 85
"control diagram" approach to, 206
disadvantages, 189
nature of, 84
for solving "swing equation", 133, 135–139, 293–294
backward difference methods, 138
backward difference operator, 139
defining time step for numerical solution, 139
Euler's method, 144–147
forward difference methods, 138
forward difference operator, 138–139
length effects of simulation time, 141
modified Euler's method, 150–154
"Runge–Kutta" type methods, 158–163
"single-step" and "multistep" integration methods, 139–140
size effects of time step on accuracy and stability, 140–141
for solving first-order ordinary differential equations, 143
stability, 161
Taylor series expansion, 137–138
"truncation errors" associated with numerical integration schemes, 141–142
types of mathematical problem, 136–137
Numerical stability, 297–299
considerations, 39–40
of Euler's method, 144–145

O

Objective function, 276
Operating systems with high penetrations of renewable, 243
Operational timescales. *See* Shorter term timescales
Optimization of smart grids, 275–277
potentially useful optimization techniques, 276–277
Order of equation, 137
Ordinary resistive loads, 12–13
Over-frequency, 10, 211

P

Paris Agreement, 274
"Part-loaded" generator, 219–221
Partial loading of generating plant, 13–15
Per unit systems, 39–40
for frequency analysis, 36–37, 290
frequency stability and system inertia, 44–45
individual machines, 40
initial analysis of frequency stability problem and swing equation derivation, 45
needs, 39–40
"per unit" quantity
dimensions, 39
for frequency studies, 40
and power system at large, 41–44
active power outputs, 41
average power factor, 43
H-constant for whole system, 43–44
incorrectness of adding inertias, 42

kinetic energies of individual machines, 42–43
moment of inertia, 41–42
system mega-voltampere rating, 41
step by step conversion, 52
of swing equation, 51–52
Perfluorocarbons (PFCs), 274
Photovoltaic system (PV system), 1, 262
Physical units, removal of, 39
PID controller, 263
Planning timescales. *See* Longer term timescales
Pole slip, 3
Power balance, 6–7
at nominal frequency, 195
in preevent system, 71–72
across wider system, 48
Power demands, 41
Power intensive process, 281
Power ramp, 195
Power system
analysis tool, 65
control centers, 211
erosion of inertia from, 246
frequency stability of AC machines in, 246
modeling for analyzing frequency behavior, 21–22
Power system frequency, 2, 4
control, 2
evolution, 64–65
and frequency requirements
"major frequency events" on system, 56–57
"minor frequency events" on system, 57
quiet system, 56
requirements for control during "quiet periods", 57–58
justification for analyzing, 2
normal operation, 3
Practical applications
different uses of swing equation, 211
MATLAB example simulation, 216–219, 222–231, 233–238, 240–243
operating systems with high penetrations of renewable, 243
RoCoF, 211
of swing equation, 299
system response requirements
for high frequency excursion, 232–233
for low frequency excursion, 219–222
during normal operation, 239
system RoCoF assessment, 211–216
"Predictor" for "Milne–Hamming" method, 169
"Primary response" period, 62
nature of, 63
system power balance during, 73
typical actions during, 63
Prime mover, 49
Processing individual terms, 105–107
Program coding in high-level computer language, 69
Programming by building control diagram representative, 69–70
Protection maloperation, 18
"Purely "AC" system, original design structure of, 300
"Purely electronic" response, potential drawbacks of, 36
PV system. *See* Photovoltaic system (PV system)

Q

"Quadratic form", Laplace transforms of, 111
"Quiet" system, 56

R

R language, 69
Ramp parameters, 235–236
"Ramp rate", 247
Ramped response, 201–206, 223–224
general solution of problem, 202
input data, 202–203
instructions for user, 206
output data, 204
statement of problem, 202
suggested MATLAB code, 204–205
Rate of change of frequency (RoCoF), 26, 34–35, 206, 211
calculation, 211, 216–219
loss of generation or demand, 211–212
calculation, 294
loss of generation or demand, 207
procedure for calculating, 212–216
relays, 9–10, 211–212
Rated power factor, 41
for individual machine, 43
Reactive losses, 4
Real variable, 276–277
Reference frequency, 195
Relatively small variations, 208–209, 239

Renewables
integration, 282
operating systems with high penetrations of, 243
Reserve, 15–16
Residual rotational power, 50
Response, 14
requirements, 220
for low frequency excursion, 222–231
during normal operation, 239–243
times, 5
Responsive generation (GEN (R)), 212
Responsive generators, 71, 235–236
Responsive modes, 12, 224, 235
Responsive power variation with frequency
demand blocks, 76–77
modeling of general responsive device profile, 77–78
relationship between constant *k* and droop of machine, 75–76
in synchronous machines, 74–75
Revised "swing equation", 108
RoCoF. *See* Rate of change of frequency (RoCoF)
Rotational kinetic energy, 40, 50
Runge–Kutta type methods, 158–164, 293–294, 298
characteristics, 158
classic explicit fourth-order Runge–Kutta method formulation, 161–163
first-order, 159
higher order, 159
mathematical form of, 158–159
MATLAB example simulation, 164–167
data usage, 164
general solution of problem, 164
input data, 165
instructions for user, 167
MATLAB code, 165–166
output data, 165
statement of the problem, 164
order of, 158
stability of, 161
truncation errors associated with, 159–161

S

SCADA. *See* Supervisory control and data acquisition (SCADA)
Scheduling agent, 283
Secondary response time period, 63
system power balance during, 73–74
Sharp cutoff, 36
Shorter term timescales, 48
"Single busbar" approximation, 66
"Single iteration" using "control diagram" method, 200
Single-step predictor–corrector" methods, 167
"Single-step" integration method, 139–140
Smart appliances, 31
Smart grids, 31–32, 268, 271–274, 296
MATLAB example simulation, 283–287
optimization, 275–277
potential effect of smart grids on system frequency, 277–283
source material, 271
structure, 274–275
architectures, 275
Solar photovoltaic system (Solar PV system), 27–29, 29*f*
cell technologies, 245, 279
farms in frequency studies, 262–263
frequency response by, 263
secondary applications of solar PV farms, 262
status of solar PV farm in intact system, 262–263
effect of tripping, 263
Solar technology, 301
Speed governor, 25
system, 4
timescales for speed governor response activation, 60–61
Speed of response, 5
Speed/torque converter plus synchronous generator, 256
Spreadsheet method, 68–69
"Stability and control" problems, 196
Stability of "Adams–Bashforth–Moulton" method, 180–181
Stability of Milne–Hamming method, 170
"Stand-alone" mode, 283
"State-space" format, 193–196
interpretation, 194–196
"stability and control" problems, 196
swing equation
application, 193–194
representation in, 193–196
to "transfer function" format, 196–197
Steady demand profile, 284
"Step-change" response, 26
Storage-based ramped response, 285
Sulfur hexafluoride (SF_6), 274
"Supergrid" transmission level voltages, 274

Supervisory control and data acquisition (SCADA), 282–283
Swing equation, 22, 55, 63, 87–88, 136, 191–193, 211, 250–251
  Adams–Bashforth–Moulton method, 183–188
  analytical methods for solving, 85–86
  approaching solution of, 63–81
    ability to calculating frequency response requirements, 64
    ability to model system, 64
    reasons for solving, 63–64
  approximating "GRP" term, 80
  approximating frequency term, 79–80
  best method for solving, 296–300
  calculating system frequency response requirements
    using power system analysis tool, 65
    procedure for calculating frequency response requirements, 65–68
    program coding in high-level computer language, 69
    programming by building control diagram representative, 69–70
    spreadsheet method, 68–69
  classic explicit fourth-order Runge–Kutta method formulation, 161–163
  derivation, 45, 47, 291
    per unit form, 51–52
    step by step, 50
  "direct solution of differential equation" method to solving, 100
    example, 100–101
    general solution to problem, 102
    input data to MATLAB code, 102–103
    output data from MATLAB code, 104
    recommended MATLAB code, 104
    solution of "swing equation" by Laplace transforms, 105–123
    statement of problem, 101
    steps, 101
  evolution of power system frequency, 64–65
  forms of responsive power variation with frequency, 74–78
  formulating Euler's method for solving, 145–147
  formulating modified Euler's method for solving, 152–154
  frequency response requirements, 65
  general mathematical form of, 78–79
  Laplace transform method to solving, 123
    example, 123
    general solution to problem, 125–127
    input data to the MATLAB code, 127
    instructions for user, 131
    statement of problem, 124–125
    steps, 124
  method, 228
  Milne–Hamming method to solving, 172–178
  "multimachine" system modeling, 70–74
  numerical methods, 133, 293–294
  representation in "state-space" format, 193–196
  simplifying substitution, 192–193
  standard form, 192
  structure, 64
  techniques for solving, 56, 81–85
    analytical methods, 81–84
    "control diagram" method, 84
    numerical methods, 84–85
  usable form of, 55–56
  variation of demand with frequency, 74
Synchronous compensators, 301
Synchronous generators, 10–11, 23–24, 33
  governors, 27
Synchronous machine, 247
  operating condition of, 3
Synthetic inertia, 34–35, 246–247
System damping requirements, 36
System demand, 6
  manipulation, 301
  modification, 32
System frequency, 219, 239, 249, 258–262, 271, 283–287, 296
  continuous requirement to regulating, 4
  deviations, 3–4
  effect of HVDC links, 249–253
  in inertia-less systems, 45
  measurement, 25
  methods of regulating, 4
  monitoring in real time, 19–20
    daily demand variations, 19–20
  potential effect of smart grids, 277–283
    biomass plants, 280
    CHP schemes, 278–279
    control architectures of smart grids, 282
    controllable commercial loads, 281
    controllable domestic loads, 281
    industrial loads, 281
    MASs, 282–283
    microgrids, 283
    solar PV cell technology, 279
    wind turbines in distribution systems, 279–280
  solar photovoltaic farms, 263–268
  studies

System frequency (*Continued*)
HVDC links in, 248–249
wind farms modeling, 253–258
System generation (P_G), 213–214
System impedances, 4
System inertia, 32–36, 44–45, 245
current research activity, 36
drawbacks of fast frequency response, 35
effect on HVDC links, 248–249
factors for achieving effective "frequency control", 32–33
inertia factor, 33
modern technologies, 32
regulation of output power of generators, 32
system demand modification, 32
time factor, 33
fast frequency response, 35
generators with massive, rotating parts, 34
generators without massive, rotating parts, 34
insufficiency of power changes, 35–36
potential drawbacks of "purely electronic" response, 36
requirement for minimum system inertia, 34
synthetic inertia, 34–35
system damping requirements, 36
wind farms effect on, 257–258
System load (P_L), 213–214
System mega-voltampere rating, 41
System monitoring agents, 283
System operations, 282
System power balance
immediately postevent, 72–73
during primary response period, 73
during secondary and tertiary response periods, 73–74
System requirements, 5–6
degree of integrity of power system and, 6
variation with daily load cycle, 5–6
System response requirements, 233–238. *See also* Frequency response requirements
for high frequency excursion, 232–233
loss of demand, 208
loss of generation, 207–208
for low frequency excursion, 219–222
calculation, 219
procedure for calculating loss of generation, 219–222
during normal operation, 208–209, 239
System RoCoF assessment, 211–216
calculation of RoCoF loss of generation or demand, 211–212
procedure for calculating "RoCoF", 212–216

T

Taylor series expansion, 137–138, 293
Tertiary response period, system power balance during, 73–74
Time constant, 197
Time delay, 60–61
Time factor, 33
Time periods of frequency evolution
immediately postfault, 58–62
inertial response, 60
loss of demand, 61–62
major loss of generation, 59–60, 59*f*
timescales for activation of speed governor response, 60–61
response time period of more than minute postfault, 63
time period from 1 second up to minute postfault, 62–63
Time step, 140
size effects on accuracy, 141
on stability of solution, 141
TO. *See* Transmission operator (TO)
Total system load $P_{L(TOT)}$, 213–214
"Total voltampere rating" of system, 43
Traditional providers of frequency control, 23–24
governor response, 23–24
jettisoning of blocks of demand for response, 24
Traditional synchronous generator-fed system, 219–220
"Transfer function" format, 196
to "control diagram" format, 197–199
"gain" and "time constant", 197
"state-space" format of "swing equation" to, 196–197
Transformers, loss of, 18–19
Transmission
automation, 282
equipment, 18–19
system operator, 281
Transmission operator (TO), 281
Trigger frequencies, 24

Trip
in colloquial system operations usage, 48
frequencies, 32
machine(s), 71–72
Truncation errors
with Euler's method, 144
fourth-order Adams–Bashforth–Moulton method, 180
magnitude of, 143
Milne–Hamming method, 169–170
with modified Euler's method, 151–152
with numerical integration schemes, 141, 143
example, 143
Newton's finite interpolation formulas, 142
order of, 143
with Runge–Kutta type methods, 159–161

U

Under-frequency, 10, 211
Unit step function, Laplace transform of, 110
Unit step term ($u(t-t_{trip})$), 115
United Nations Framework Convention on Climate Change (UNFCCC), 273
Unplanned changes, 20
Unresponsive generation, 213
Unresponsive modes, 12
US Department of Energy (DOE), 272–273

V

Variable speed
with full power electronics conversion, 256
with partial power electronics conversion, 255–256
Variation of load with frequency, 214–215
Virtual inertia, 247
Virtual synchronous machine (VSM), 35, 247, 295, 301
Voltage levels, 274
Voltampere base, 40

W

Weighting factors, 293–294
Wind farms, 258–262
modeling in system frequency studies, 253–258, 254*f*
frequency response by, 258
effect of wind farms on system inertia, 257–258
WTGs, 253
Wind turbine generators (WTGs), 27, 253, 254*f*, 255*f*, 256*f*
Wind turbines, 27–28
in distribution systems, 279–280
domestic, 279–280
large, 279–280
small, 280

FSC
www.fsc.org
MIX
Papier | Fördert gute Waldnutzung
FSC® C083411

Kolibri 360 GmbH
Ferdinand-Jühlke-Straße 7
99095 Erfurt, Deutschland
produktsicherheit@kolibri360.de